ADVANCED MACHINING
TECHNOLOGY HANDBOOK

Other McGraw-Hill Books of Interest

ADVANCED MACHINING TECHNOLOGY HANDBOOK

James Brown

McGraw-Hill
New York San Francisco Washington, D.C. Auckland Bogotá
Caracas Lisbon London Madrid Mexico City Milan
Montreal New Delhi San Juan Singapore
Sydney Tokyo Toronto

Library of Congress Cataloging-in-Publication Data

Brown, James.
    Advanced machining technology handbook / James Brown.
      p.   cm.
    Includes index.
    ISBN 0-07-008243-X (alk. paper)
    1. Machining—Handbooks, manuals, etc. I. Title.
  TJ1185.B8234 1998
  671.3'5—dc21   97-39192
                                     CIP

McGraw-Hill

A Division of The McGraw·Hill Companies

1 2 3 4 5 6 7 8 9 0 DOC/DOC 9 0 3 2 1 0 9 8

ISBN 0-07-008243-X

The sponsoring editor for this book was Harold B. Crawford, the editing supervisor was Andrew Yoder, and the production supervisor was Tina Cameron. It was set in Times Roman by Lisa M. Mellott through the services of Barry E. Brown (Broker—Editing, Design and Production).

Printed and bound by R. R. Donnelley & Sons Company.

McGraw-Hill books are available at special quantity discounts to use as premiums and sales promotions, or for use in corporate training programs. For more information, please write to the Director of Special Sales, McGraw-Hill, 11 West 19th Street, New York, NY 10011. Or contact your local bookstore.

Product or brand names used in this book may be trade names or trademarks. Where we believe that there may be proprietary claims to such trade names or trademarks, the name has been used with an initial capital or it has been capitalized in the style used by the name claimant. Regardless of the capitalization used, all such names have been used in an editorial manner without any intent to convey endorsement of or other affiliation with the name claimant. Neither the author nor the publisher intends to express any judgment as to the validity or legal status of any such proprietary claims.

 This book is printed on recycled, acid-free paper containing a minimum of 50% total recycled fiber with 10% postconsumer de-inked fiber.

Information contained in this work has been obtained by The McGraw-Hill Companies, Inc. ("McGraw-Hill") from sources believed to be reliable. However, neither McGraw-Hill nor its authors guarantees the accuracy or completeness of any information published herein and neither McGraw-Hill nor its authors shall be responsible for any errors, omissions, or damages arising out of use of this information. This work is published with the understanding that McGraw-Hill and its authors are supplying information but are not attempting to render engineering or other professional services. If such services are required, the assistance of an appropriate professional should be sought.

CONTENTS

Part 6   Surface-Treatment Processes   397

Chapter 30.   Titanium-Nitride (TiN) Coating   399

Chapter 31.   Hardcoating   411

Chapter 32.   Electrolizing   418

Part 7 Electrochemistry Processes   471

Part 8 Machine Deburring Processes   509

Index 569

INTRODUCTION

I have spent a lifetime in various industries as a design and manufacturing engineer. In my experience, the most significant reason for the decline of some world-class companies has been the reluctance of their management to change their traditional manufacturing processes and keep pace with advancing technology.

Here are a few examples of changes in manufacturing processes that have impacted profitability:

Some thin aluminum parts common to the computer industry have a tendency to warp slightly when ground. A double-faced grinder was tried because that takes off an equal amount of stock simultaneously from both sides, and normally doing both sides together eliminates the problem of heat. But this procedure was unsatisfactory. Lapping them on a double-wheel machine eliminated the heat problem and, at the same, time speeded-up the production considerably.

Forgings, large and small, are being replaced by castings, which are much cheaper. Finishing them with the "Hot Isostatic Process" makes them as strong and dependable as forgings.

Ultrasonics are now being used, not only to drill accurate holes in glass, but the process is fine for machining advanced composites. Not only is the process good for machining parts for aircraft, but it is bringing composites down to earth, to golf clubs, tennis racquets, and bicycles.

Formerly, many CNC machining operations were stopped to have the workpiece inspected on a coordinate measuring machine because extremely tight tolerances were involved. Now, it's possible to perform that same inspection operation on the CNC machine. This can be done only if the company is using computer integrated manufacturing and can use the same database for machining as was used for design.

Countless workpieces require more time to manually deburr internal cross-holes than to drill them. Now this difficult type of deburring can be processed swiftly in three different machines.

Workpieces can now be formed with practically no waste of material by selecting the correct process. This list of samples could go on and fill a book, which is precisely why this volume was written.

There is a significant reason for professionals to read and study this book. By professionals we mean those who have anything to do with the cost of products.

We should all be concerned with product cost. Countless companies supplying the American market with appliances, personal and home products, automobiles, industrial equipment, and entertainment devices, are now being squeezed by competition. The military wants a bigger bang for the buck. Industry wants foreign markets. None of this is possible until we learn to manufacture products at lower cost.

Usually, the designer is the individual most involved with determining product cost. That is true because he draws the line that indicates the manufacturing process he is contemplating, the one that would be most reasonable. But, his superiors who sign off on his drawings are also responsible, as are the program manager, and even the procurement department, which places orders for part manufacture and component purchase.

If these involved personnel could take a second look at the design and the quantities required, perhaps small design modifications would allow manufacture through a different, less costly process.

It is this knowledge and understanding of the various means of production that is so necessary for the selection of effective manufacturing processes. The contents of this book should improve economic production. Although many of the processes described in the book are well established, many of the people responsible for product cost do not fully understand their value. Complete coverage of a process in one chapter is not possible. However, I have tried to provide sufficient coverage to guide those anxious to "find a better way." If some individuals obtain assistance from the material contained in the book, the effort of writing it will have been worthwhile.

WELCOME TO THE THIRD MILLENNIUM

Welcome to a world with materials of such strength at high temperatures that conventional machining processes are impractical.

A world where complex shapes and fine finishes are what the design demands, and where the conventional approach to meet these demands just won't work.

A world where to save weight and resources, component designs rely on consistent, controlled, manufacturing process of quality unsustainable with present methods.

A world that you've got to survive in and that you've got to prepare for now.

This is the message given by President Rhoades of the Extrude Hone Co. to customers seeking "a better way." He claimed that in the third millennium, companies relying on labor-intensive, uncontrollable hand deburring, edge contouring, and surface improvement will not survive.

CONTENTS ABBREVIATED

Part 1 Design and Administrative

Chapter 1 A Brief History of Metal Cutting. In addition to providing historical information on metal cutting through the years, this chapter is an exploration into the successive use of high-carbon steel, alloy steel, high-speed steel, cemented carbides, cermet, and ceramic cutting edges on indexable insert tooling for turning, milling, and drilling. Before the modern manufacturing methods are covered, it makes sense to place them into the proper perspective with the past.

Chapter 2 Computer-Integrated Manufacturing. CIM is a means of running a complete business; managing not only the design and manufacture of a product, but also its finance and marketing (which includes sales, service, bar code systems, and warehousing)

—everything that occurs between the birth and death of a product. Several software companies advertise that is what they do. Nevertheless, these claims must be substantiated. Sometimes they mean that their programs integrate with others that integrate with others.

Part 2 Sheetmetal and Press Processes

Chapter 3 Fineblanking. If parts require smooth square edges for their periphery as well as for their holes and slots, consider stamping them on a fineblanking machine. Stamped on a standard press, those parts would have a "breakaway" at the die side of each edge. A secondary operation, such as boring, drilling or milling, would be necessary. This is an excellent process choice for making cams, gears and similar hardware.

Chapter 4 Cold Forging: Coining. This is a process of precision cold forging to "near net shape". It means forming a metal part at room temperature to a desired shape by forcing a lubricated slug into a closed die under extreme pressure. In this process the metal is reformed in continuous, unbroken lines that follow the contour of the part. This old process is now being used in an unconventional manner. One innovative vendor designed some clever tooling that permitted a conical part to be made complete in one minute. Previously, the part had been machined in a CNC lathe and it took close to 60 minutes.

Chapter 5 Orbital Cold Forging. The lower portion of this vertical press holds a fixed die, while an orbiting upper die, with very little noise, forms the workpiece as it is hydraulically raised in the lower die. The process produces 3 to 10 parts per minute holding tolerances as close as 0.002". Just as fineblanking is different from stamping, orbital cold forging is different from forging.

Chapter 6 Hydroforming. This process is sometimes referred to as "fluid-forming" or "rubber diaphragm forming". The hydroform machine operates in the vertical, like a press, with upper and lower sections. But, where a standard draw-press involves a punch and die, this machine doesn't require any die. In its place the machine has a soft diaphragm, which under hydraulic pressure, hugs the workpiece and forces it to assume the contour of the punch which is raised through the lower half of the machine. This diaphragm takes the place of a die and saves cost.

Chapter 7 Roll Forming. This is a high-production process of forming metal from strip, coiled or sheet stock. The material is fed through successive pairs of rolls—each pair progressively forming until the desired cross-section is completed. During the run, the material is notched, mitered, embossed, and cut off. This process should be considered for sheet metal cross sections, which would require secondary operations if made the customary way.

Part 3 Machine-Shop Environment Processes

Chapter 8 Trepanning. The only metalworking operation in which most of the metal is removed unnecessarily in drilling through holes. Just think about that statement. Drilling a through hole creates a pile of chips; a cylinder has been cut into small pieces. If that cylinder could be removed as a solid unit by cutting only the circumference, the center core, about 75% of the volume, would fall out.

Chapter 9   Metal Spinning.   Spinning is done on a lathe-like machine with the operator holding a tool between the cross-slide and the sheetmetal spinning against a chucked mandrel. For small-quantity production, this process would save some tooling cost. Whereas standard draw-press tooling involves two tools (upper and lower), metal spinning requires only one inexpensive tool, a mandrel. Consequently, the process is ideal for development work.

Chapter 10   Shearforming.   Shearforming is a sophisticated form of metal spinning. Very close tolerances (0.002") can be consistently held. The machine has a controller, so the process can be automated. Its appearance is similar to a lathe with a very heavy frame. The motor-driven spindle holds a mandrel contoured to fay with the internal configuration of the part to be shearformed. A massive hydraulically activated cross-slide carries heavy rollers, which form the part.

Chapter 11   Electrical-Discharge Machining.   EDM is a precision metal-removal process using an accurately controlled electrical discharge (spark) to erode any electrically conductive metal, regardless of its hardness. Since its acceptance as a tool-making device, EDM has lowered toolmaking costs considerably.

Chapter 12   Abrasive-Waterjet Cutting (AWC).   The velocity of a waterjet is about three times that of a pistol bullet. Any soft material in its path, like cloth or wood, will simply be removed. If the object is to remove something hard (such as metals, ceramic, concrete, or glass), they can also be removed by adding an abrasive to the jet.

Chapter 13   Lapping, Polishing, and Honing.   Honing is the use of bonded abrasives in stock form (for hollows) or annular wheels for flat honing. In one way, this is like grinding because that process also uses bonded abrasive wheels. But flat honing is more like lapping because both processes have loose workpieces that are free to move. And unlike grinding, flat honing is a cool, gentle process.

Lapping is an abrasive machining process where rolling abrasives are suspended in a liquid vehicle and used between the lap surface and the workpieces.

Polishing also uses loose abrasives, which embed themselves on a soft, felt disc and improve the lapped surface finish. These distinctions are described because all three processes are used more and more in high technology and must be understood to be used to full advantage.

Chapter 14   Industrial Saws.   Every product or part of a product made by man starts with a cut-off job. The raw material has to be cut. In machine shops all over the world, the turbine, bar stock, sheetstock, pipe, or plate must be cut. It is always the first step in production. The four alternative methods for doing this: hacksawing, cold sawing, abrasive cutting, and band sawing.

Chapter 15   Industrial Parts Cleaning.   Health, safety, and environmental regulations are redefining the workplace. Toxic chemicals and hazardous waste must be eliminated. Of major concern are traditional vapor degreasing and cleaning solvents. The potential effects to our health and the ecological consequences of these toxic solvents are well documented. Safe alternatives have been produced and research is continuing to find other acceptable, safe, cleaning products. Industries are now on notice that they are responsible for their hazardous wastes including waste solvents, forever.

Chapter 16 Hot Isostatic Pressing (HIP). HIP is a relatively new manufacturing process, whose frequency of use is increasing rapidly. The process uses the simultaneous application of inert gas pressure and elevated temperatures to close shrinkage porosity and internal voids of castings, resulting in a material consistency expected of forgings. It is used similarly in powder metallurgy to densify products. This permits PM to be used heretofore sufficient strength was lacking.

Chapter 17 Lathe Turning. Much development has taken place in cutting tools for lathe turning. Intermittent cutting should be avoided if possible. If not, try to have tool engaged at all times to try one of the suggestions made in book. Guiding rules are given for improving stability. New tools (such as coated cemented carbides, cermets, and cubic boron nitride) are described. You'll find feeds and speeds, cutting depth, and the shape of the cutting tool recommended for every material. We have guidelines for turning with ceramics, for parting and grooving, and for threading.

Chapter 18 Drilling. Drilling is the most common machining operation performed, with most small holes made by helical drills. A large share of holes are made by drills fitted with inserts. When large holes are required, trepanning is the choice. We speak of the importance of machining feeding and disposal of chips. Hints are provided for drilling to size. Gun drilling is described and so is the Ejector system, which can be used by most machines.

Chapter 19 Milling Machines. Milling performs metal cutting by a coordinated movement between a cutting tool and the feed of a workpiece. We have dual cutters for high production and CNC machines for special contour cutting. They come in all sizes and types, such as universal, vertical, horizontal, gantry, CNC, or machining centers. The differences between down and up milling are described. Surface texture is described and so is the significance of the pitch of milling cutters. Facemilling, tilting the spindle, and the cause of vibration are described. The end mill is designed to machine axially, which makes it more susceptible to vibration. Sometimes it is necessary to install a flywheel on the arbor when doing side and face milling. Hints for good, stable machining are given.

Chapter 20 Machining Centers. These centers can perform milling, turning, drilling, boring, and reaming. Generally, the entire job can be completed at one site, saving part handling. All machining centers have some kind of modular, quick-change, tooling. Every possible aid to production is used. Some centers are semi-automatic and others fully automatic, with computer controls even changing tools. In the more advanced shops, the tools have an identification disc attached, which provides information to the computer, eliminating possibility of human error. A plan of administering the computerized equipment is provided. Advice is provided for selecting modular tools.

Part 4 Fastening and Joining Processes

Chapter 21 Industrial Adhesives. Currently, many companies are involved in the manufacture of reliable adhesives. Their products include cyanacrylates, urethanes, putties of various materials, stick adhesives, underwater adhesives, multi-purpose adhesives, and epoxies filled with a multitude of various metals and other fillers. There are adhesives filled with carbides, ceramics, quartz and zinc for special applications. There are room-temperature vulcanizing (RTV) silicones. There is an adhesive for any fabrication or repair application conceivable.

Chapter 22   Magneforming.   This is a high-energy-rate cold-forming technique used to form and assemble components. When an electric current generates a pulsed magnetic field near a metal conductor, a controllable pressure is created that shapes without contact. Pressures up to 5000 PSI move workpieces at velocities up to 300 meters per second.

Chapter 23   Ultrasonic Technology for Production.   In industry, ultrasonic machines are commonly used for two completely different purposes: machining hard materials like glass or steel, and welding thermoplastic materials together. For welding, the high-frequency vibratory motion creates very satisfactory welds. Machining involves the use of an abrasive slurry which flows between the workpiece and the tool face. The tool vibrates vertically with an amplitude of only a few thousandths of an inch. But cycles at a frequency of around 20,000 times per second.

Chapter 24   Introduction to Aluminum Brazing.   Strong, uniform, leakproof joints can be made inexpensively in aluminum. Joints that are inaccessible and parts that are not joinable by any other means often can be brazed. Complicated assemblies with thick and thin sections, odd shapes with different wrought and cast alloys, can be turned into one integral all aluminum part by a single trip through a brazing furnace or dip pot.

Chapter 25   Welding.   Welding contributes to the production processes of most manufactured products, and to the maintenance of many other businesses. In this chapter, references to a variety of welding methods are displayed as a sort of menu of welding processes available. Its purpose is to facilitate the selection of the best method for the job.

Chapter 26   Induction Heating.   For 70 years, induction heating companies have been supplying solutions to complex heating problems. These non-contact techniques have yielded superior results in production efficiency, energy saving, and product quality. Typical applications include brazing, soldering, many types of heat treatment, epitaxial growth, vapor deposition, shrink fitting, and a dozen less-known jobs.

Chapter 27   Lasers.   Shortly after the invention of the laser in 1960, researchers all over the world began investigating the use of laser radiation to weld, drill, and cut materials. The laser provides accuracy of measurement for straightness and alignment. The field of medicine has already discovered many successful laser applications. Surgeons and dermatologists are using the device more each month.

Part 5   Injection-Molding Processes

Chapter 28   Powder-Injection Molding.   It is now possible to produce intricate ferrous parts by injection molding a predetermined mixture of fine metal powder and plastic binder. It is a dramatically different process from conventional powder metallurgy. The restrictions of other, commonly used metal-working methods, are eliminated. Actually, it is a successful marriage of two well-known processes: plastic-injection molding and powder metallurgy.

Chapter 29   Injected-Metal Assembly.   Injected-metal assembly uses a special die-casting technique to assemble small parts. Components to be joined are held in position with a tool, while a small volume of molten metal, usually zinc, is injected to form a hub at the intersection of the parts. The alloy quickly solidifies, forming a strong, permanent lock. With few exceptions, the parts can be of almost any structural material and shape. As soon as the assembly is completed, it requires no finishing steps and can be used immediately.

Part 6   Surface-Treatment Processes

Chapter 30   Titanium-Nitride (TiN) Coating.   Cost savings and productivity improvements associated with these coatings are so significant that we can no longer ignore them. Cutting tools can increase quantity of pieces produced by up to 400%. Machine feeds and speeds can be increased considerably. This coating has an Rc 80 hardness compared to an Rc 65 for hardened, uncoated H.S.S. tools.

Chapter 31   Hardcoating.   Hardcoating provides aluminum parts with a surface hardness close to that of a hardened alloy steel while maintaining the light weight of aluminum. This feature saves considerable money in toolmaking because it allows free-machining aluminum to be used in many places where steel is normally the other alternative.

Chapter 32   Electrolizing.   This is a proprietary, high-chromium coating with unusual success stories. The alloy is applied in a cold environment with a hardness of Rc 70 to 72. The coating is preceded by a unique cleaning of the base metal. During this cleaning procedure, the adhesive characteristics and qualities of electrolizing are generated. The process results in unusual lubricity, increased wear resistance, and precision thickness.

Chapter 33   Poly-Ond.   This liquid-bath process chemically deposits nickel phosphorous impregnated with polymers on the surface of metal parts. It is a proprietary process, first marketed in 1976, which creates an infusion of polymers throughout the thickness of an Rc 70 coating.

Chapter 34   Magnaplate.   Magnaplate coatings are a series of multi-step, proprietary processes which become an integral part of the top layer of the base metal. Each series of coatings protects a specific metal or a group of metals. Some coatings are utilized to solve specific problems.

Chapter 35   Powder Coating.   This application of fused plastic particles to a metal surface provides a durable and cost-effective alternative to conventional electroplating or painting.

Chapter 36   Spray Coating.   This is another method of treating metal surfaces. Spray coating of various plastics has become significant because of the many product design uses discovered for it. Recent advances in this procedure (involving new materials, application methods, surface preparation techniques, and curing technologies) have revolutionized traditional coating-job shapes and the services they offer.

Chapter 37   Industrial Finishing Systems.   This book contains several chapters on deburring and surface finishing, but those subjects would be incomplete without a few words on industrial finishing systems, such as dryblast, impact treatment, and vibratory-finishing systems. The vibratory tubs and sand-blasting equipment of World War II have been superseded by technologically superior equipment.

Part 7   Electrochemistry Processes

Chapter 38   Electrochemical Machining.   This controlled, rapid metal-removal process has virtually no tool wear. It removes metal atom by atom. Like the deburring process of electrochemical deburring, this process is based on an anode-cathode relationship. The tool (cathode) is shaped to provide the form desired in the workpiece (anode). An electrolyte is pumped under high pressure between the tool and workpiece. In the meantime, a mechanically driven ram feeds the tool into the workpiece.

Chapter 39   Electrochemical Grinding.   The mechanism of electrochemical grinding is similar to that of electrochemical machining. It uses the deplating power of electrolytes driven under pressure and guided to the workpiece, to remove stock much faster than conventional grinding, and it doesn't create the stress or heat caused by conventional grinding.

Chapter 40   Electropolishing.   Electropolishing streamlines the microscopic surface of a metal object by removing metal from the objects surface through an electromechanical process similar to, but the reverse of, electroplating. In this procedure, the metal is removed ion by ion from the surface of the object being polished. Electrochemistry and the fundamental principles of electrolysis (Faraday's Law) replace traditional mechanical finishing techniques (such as grinding, milling, blasting, and buffing) as the final finish.

Chapter 41   Electroforming.   This process involves electroplating a pure metal or an ally onto a mandrel. The plating continues until the required thickness is reached, after which the mandrel is removed and the plated metal is retained as the desired part. A variety of complex parts have been made, using metals ranging from soft copper to high-nickel alloys.

Part 8   Machine Deburring Processes

Chapter 42   Electrochemical Deburring.   This deburring process uses electrical energy to remove burrs in a very localized area, as opposed to thermal energy deburring, which provides general deburring. The part to be deburred is charged positively. The electrode (working tool) is charged negatively and an electrolyte is directed, under pressure, to the gap between the electrode and the burr.

Chapter 43   Thermal Energy Deburring.   In this process, the manufactured parts are placed in a thick-walled chamber, which is sealed and pressurized with a mixture of oxygen and natural gas. The combustible mixture is ignited by a spark, which creates a 6000°F heat wave. All burrs are vaporized in milliseconds.

Chapter 44   Abrasive Flow Machining and Abrasive Flow Deburring.   In this process, the manufactured parts are placed in a thick-walled chamber, which is sealed and pressurized with a mixture of oxygen and natural gas. The combustible mixture is ignited by a spark, which creates a 6000°F heat wave. All burrs are vaporized in milliseconds.

Part 9 Materials

Chapter 45 Ceramics. Ceramic parts are being used extensively in high technology. Parts can vary in size from small miniatures to 5 feet long by 2 feet in diameter. They are formed by isostatic pressure, as well as near-net-shape dry pressing. The parts can be coated for brazing; they can be machined for close dimensions. These parts are tough in corrosive, heat, and wear environments. As hard as sapphire, they outperform tool steels in some situations. Their compressive strength is around 300,000 PSI. Modern technological products have high demands in chemical resistance, thermal conductivity, dielectric strength, and electrical resistivity, which can best be met by ceramics. Currently, automotive companies are experimenting with the use of ceramic pistons and engine blocks.

Chapter 46 Advanced Composites. A composite is created by combining a reinforcing material with a plastic matrix. This is done to provide special desired characteristics. Currently, industry uses the term *advanced composites* to describe a material that has a modulus of elasticity greater than 16,000,000 PSI and a fiber-to-resin ratio greater than 50% fiber. They are now being used by the military and others, wherever great strength, high modulus, and heat resistance is required.

ACKNOWLEDGMENTS

Acknowledgments and grateful thanks are due to the following people and companies for information supplied and for their kind permission to publish data, photographs and illustrations. Their cooperation is appreciated because the book could not have been written without it:

1. *History of Metal Cutting*   Sandvik Coromant Tech. Editorial Dept., Sweden
2. *Computer-Integrated Manufacturing*   Valisys Corp., Santa Clara, CA
   Computer Corp., Bedford, MA
   MAX, Inc., Foster City, CA
   SAP America, Inc., Wayne, PA
   DataWorks Corp., San Diego, CA
   Icamp Inc., Bolton, CT
3. *Fineblanking*   California Fineblanking Corp., Huntington Beach, CA
   Fairfield Tool Co., Bridgeport, CT
   Feintool Co., Switzerland
   Schmid Corp. of America, Goodrich, MI
4. *Cold Forging: Coining*   Schmid Corp. of America, Goodrich, MI
   Cousino Metal Products Inc., Toledo, OH
5. *Orbital Cold Forging*   Schmid Corp. of America, Goodrich, MI
6. *Hydroforming*   American Metal Stamping Assoc., Richmond Heights, OH
   Roland Teiner Co., Everett, MA
7. *Rollforming*   Rollform of Jamestown Inc., Jamestown, NY
   Johnson Bros. Metal Forming Co., Berkeley, IL
   Samson Rollformed Products Co., Skokie, IL
8. *Trepanning*   Hougen Mfg. Inc., Flint, MI

28. *Powder-Injection Molding*   Prof. R.M. German, Rensselaer Polytech Institute, NY
New Industrial Techniques, Inc., Coral Springs, FL
Parmatech Corp., San Raphael, CA
Remington Arms, Inc., Wilmington, DE
Engineered Sinterings & Plastics Co., Watertown, CT

29. *Injected-Metal Assembly*   Fishertech, Peterborough, Ontario, Canada

30. *Titanium-Nitride Coating*   Balzers Tool Coating, Inc., Agawam, MA

31. *Hardcoating*   Sanford Process Corp., Natick, MA

32. *Electrolizing*   Electrolizing, Inc., Providence, RI

33. *Poly-Ond*   Poly-Plating, Inc., Chicopee, MA

34. *Magnaplate*   General Magnaplate Corp., Linden, NJ

35. *Powder Coating*   Plastonics, Inc., Hartford, CT

36. *Spray Coating*   Precision Coating Co., Dedham, MA
American Durafilm Co., Holliston, MA
E.I. DuPont Co., Wilmington, DE

37. *Industrial Finishing Systems*   Guyson Corp., Saratoga Springs, NY
Vibrodyne, Inc., Dayton, OH
Almco Corp., Albert Lea, MN

38. *Electrochemical Machining*   Anocut Corp., Elk Grove Village, IL
Extrude Hone Corp., Irwin, PA

39. *Electrochemical Grinding*   Anocut Corp., Elk Grove Village, IL

40. *Electropolishing*   Electropolishing Systems, Inc., Plymouth, MA
Delstar Electropolishing, Inc., Princeton, NJ

41. *Electroforming*   Gar Electroforming Corp., Danbury, CT

42. *Electrochemical Deburring*   Electrogenics Div., Elk Grove Village, IL
Bosch Corp., Surftran Div., Madison Heights, MI

43. *Thermal Energy Deburring*   Bosch Corp., Surftran Div., Madison Heights, MI

44. *Abrasive Flow Deburring*   Extrude Hone Corp., Irwin, PA

45. *Ceramics*   Wesgo, Inc., Belmont, CA
Morgan Matroc, Inc., Milwaukee, WI

46. *Advanced Composites*   Textron Specialty Materials, Lowell, MA

P · A · R · T · 1

DESIGN AND ADMINISTRATION

CHAPTER 1

A BRIEF HISTORY OF METAL CUTTING

Although the history of metal cutting is fascinating, the main portion of it, as seen from our perspective, is not very old. There were some good mechanics and engineers, such as Leonardo di Vinci and others. After all, people from centuries ago built the pyramids and Stonehenge. Yet, the way we view progress, it seems as though it went hand in hand with the Industrial Revolution of the eighteenth and nineteenth century and then accelerated in the twentieth.

Although this outlook can be called fair from a productivity oriented point of view, it can seem grossly unfair to the progress of craftsmanship prior to the industrial revolution. A lot of personal know-how and skill in metalworking was developed long before industry as we know it today.

In his book *The Myth of the Machine*, Lewis Mumford says, "What is usually treated as the technological backwardness of the six centuries before the Industrial Revolution represents instead a curious backwardness in historical scholarship. Significantly, the great technical advances of the eighteenth century occurred in the earliest Biblical times. It was the large capital investments in the metallurgical (military) industries that spurred the pre-Industrial Revolution."

Metal cutting by machine tools is something relatively new, as are the tool materials and science that have shaped this century's development. The immense craving for manufacturing has spurred the quest for productivity. Metal cutting as such, then has played a significant role in the advance of civilization.

Into the eighteenth century, wood was the dominant workpiece material and the machining of metal was very limited and crude. Until the nineteenth century, it was a slow, blacksmith task until machine power became available from the steam engine, and later electricity (Figure 1.1).

Machine tools developed in response to newly found power transmitted throughout the workshops by means of axles, belts, and pulleys to start with. Then came the early milling and planing machines, as well as the lathes, which could cut threads. The introduction of the cross-slide on lathes was another step forward. That meant that tools did not have to be

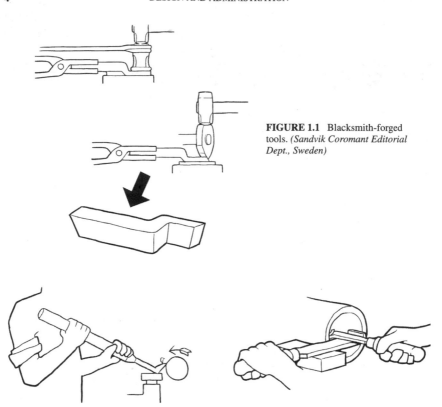

FIGURE 1.1 Blacksmith-forged tools. *(Sandvik Coromant Editorial Dept., Sweden)*

FIGURE 1.2 Early hand-held turning and boring, before lathes had tool posts (1850–1900). *(Sandvik Coromant Editorial Dept., Sweden)*

FIGURE 1.3 Pre-1900 Machine tools: lathe and shaper. *(Sandvik Coromant Editorial Dept., Sweden)*

held by hand any longer. They could now be secured in a tool post. See how the tools were hand-held in Figure 1.2.

Machining was very slow. Shaping was a widely used operation. This was a predecessor to face milling. It was performed with a turning tool mounted in a tool post, which made reciprocal movements across the surface as it was fed across the width of the face (Figure 1.3).

The cutting depth and the length of stroke were set and the machine was left to run. This method has almost completely been replaced by face milling.

In the nineteenth century, the military led the way, designing machines as needed, and producing interchangeable parts with standardized measurements. By mid-century, they introduced universal milling and grinding machines (Figure 1.4). The lathe turret was introduced as a method of quick changing tools. Then came the automatic screw machine. Now drills, lathes (both vertical and horizontal), and milling machines were being built

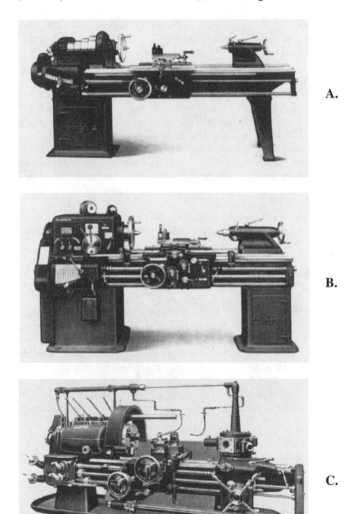

A.

B.

C.

FIGURE 1.4 Lathes: A. 1910, B. 1940, C. 1950. *(Sandvik Coromant Editorial Dept., Sweden)*

as required. The millers weren't producing fast enough, so twin milling machines were developed.

The emphasis was always on production. Make them faster, better, and cheaper. Now that computers have entered the arena, you'll find one man running several machines. We have advanced to the point where jobs are running day and night. Some factories are even running without any humans present.

One company was about to purchase a second machining center for about $100,000. A replacement they'd just hired for the third shift, produced a volume of work equal to the other two shifts together. This machinist had attended classes at the machining center's factory and knew the secret of using the right tool for the job. The chips were coming off his machine blue. Yet, when the machine was stopped for reloading, the anxious foreman was surprised to find that he could place his hand on the cutter. It wasn't hot, it was simply warm. The heat had been taken off in the blue chips.

We now know that trained manpower is important to production, and world-class factories are guided by the best-trained men available. Schools have sprung up all over. You can have your people take hands-on training in hydraulics, assembly techniques (such as the installation of bearings), and the use of tools and techniques (such as Electrical Discharge Machining, Abrasive Waterjet Cutting, Shearforming, and Electrochemical Machining). To return to the subject of cutting tools, such as milling cutters, we now know that it is important to use a differently designed cutter for each material. Aluminum, stainless steel, copper, and bronze ideally would have a special cutter. Some companies spend $1000 to $2000 for a milling cutter.

The following picture shows three nineteenth-century turning operations. Figure 1.5 shows turning iron and steel, turning soft metal, a surface-finishing operation. During the nineteenth century, various iron (then steel) processes were developed to produce tool steels. High-carbon and alloy carbon steel were the best tool steels available. Blacksmiths heat treated them to a high degree of hardness, but they became soft again with only a little work. The heat generated by cutting speeds of only 10 feet per minute were enough to reduce tool life to a very short duration.

Experiments with manganese led to air-hardening, and the find of tungsten in the steel led to a Mushet steel, which doubled production. This aroused much interest in metallurgy. The biggest event occurred at the turn of the century, when Frederick Tayor, at the Paris Exposition in 1900, performed unheard of feats of metal turning. He machined steel at such speeds and feeds that the chips came off blue. The cutting tool was red hot, but it stayed sharp.

Metallurgy soon became a popular subject in college. Many mechanical engineers were caught up in the enthusiasm and companies and schools participated in thousands of

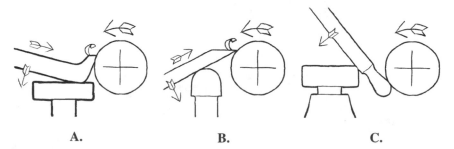

        A.   **B.**   **C.**

FIGURE 1.5 Turning steel, soft metal, and a finishing operation. *(Sandvik Coromant Editorial Dept., Sweden)*

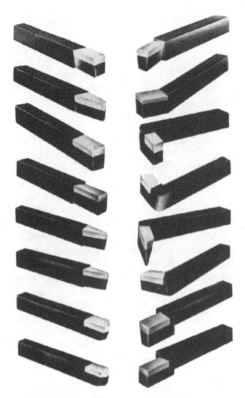

FIGURE 1.6 Lathe turning tools. *(Sandvik Coromant Editorial Dept., Sweden)*

tests. This led to the development of the high-speed steel, the cutting tool that took us into the twentieth century. Practically speaking, that meant that a turning operation which required 100 minutes with a high-carbon tool took only 26 minutes with a high-speed tool.

Cast alloys were introduced in 1915 and another step forward in the evolution of cutting tool materials was taken. *Cast alloys* is the collective name for some non-ferrous alloys based mainly on cobalt, chromium, tungsten, etc. (Figure 1.6). These castings contained around 50% of hard carbides. Some of the popular names were Stellite, Tungaloy, and Speedaloy. These had relatively high hot-hardness, 800 degrees C, and high resistance to abrasive wear, but were very brittle and difficult to make into tools.

Cast alloys are forerunners (and to some extent, related to) cemented carbides through their composition and, in some ways, their quality. Stellite tips can, in fact, be welded onto shanks of carbon steel. They are however, melted and cast, not a product of powder metallurgy as are cemented carbides.

Cast tips were brazed onto steel toolholders and milling cutters, as were cemented carbides. Although they offered considerable improvement over the high-speed steels at this time, they had only about half of high-speed steel's (HSS) toughness. Some operations that had taken HSS 26 minutes, took only 15 minutes with cast alloys.

Super HSS, which was HSS with cobalt added, appeared in 1930. This became an excellent tool for cutting aluminum and magnesium. At this time, turning and milling tools were made from HSS, spade drills were the forerunners of twist drills, and hollow drills made their first appearance.

The times put more pressure on engineers to speed up production, and they designed new machines. Power saws, large planers, shapers, radial drills, lathes, and boring machines, were developed and built. The nineteenth century became the twentieth. This was the era of the cemented carbides as a cutting material. What took HSS 26 minutes to machine, and cast alloys 15 minutes, took cemented carbides only 6. These powder metallurgy products contained more than 90% haed carbides in a binder. By 1934, 134 patents were issued for cemented carbides

These new machines acquired a significant role during World War II. They cost so much that only a few sources could afford them. Naval shipyards, such as Boston, had a large boring mill and planner that were rented to companies that needed them (Figure 1.7). I'm sure that other large facilities had similar arrangements for the benefit of the war effort.

Experiments with aluminum oxides started in the 1950s and continued up to the 1990s. Now we have a choice of several ceramics, and cutting tools with super-hard materials, such as cubic boron nitride and polycrystalline diamond, which have a more-limited application area, The big advance in tooling was the means of holding them. Instead of brazing cast alloys or carbide tips to a steel shank, they are now held on by mechanical means, which saves a lot of time in tool preparation. In fact, they now are indexable. Each tool has 2 to 4 cutting edges.

The man behind the drive for indexable cutting tools was Sven Wirfelt, one of the pioneers of the Sandvik Coromant Company since the early 1950s. His legendary role in the evolution of cutting tools was one of the major contributions towards making the company one of the world leaders in cutting tool technology. Much of the data in this chapter is from the book *Modern Metal Cutting*, published by the Sandvik Coromant Technical Department in Sweden. Anybody interested in machining of any type would find this book interesting. The book contains hints to machining aluminum, magnesium, copper, zinc, several types of cast iron, uranium, and composite materials.

When these inserts no longer required brazing or grinding, the new emphasis was placed on metal cutting. Now steps were taken to break the Achilles heel of cemented carbides, namely the compromise between wear resistance and toughness. By mixing alloys of tungsten carbide, cobalt, titanium, tantalum, and niobium, they found a grade of cutting tool that would resist wear and deformation at high speeds. This pretty well did it for cutting tools until the cermets were made for the latest machines of the twentieth century.

The only other facts that must not be omitted in a chapter about cutting tools of the 1990s is the coatings. Titanium nitride deposited on the surface of any tool, drill, reamer, milling cutter, or lathe tool, with a thickness of only 0.0001" has a hardness of about 80 Rockwell C and a coefficient of friction only ⅓ that of the matrix metal. This increases the life of the tool while expediting the work.

Uncontrolled swarf is important in lathe work and milling—especially in lathe work. If the chip is allowed to roll on and on, you soon have a situation that must be stopped. The chip must be broken.

The machine is halted while the chips are removed and the tool cutting geometry changed to break up the chips. Chip control is one of the key factors in turning and drilling. Figure 1.8 shows an indexable, triangular, chip-breaker tool, with the chips breaking as they come off the tool. Milling creates a natural chip length because of the limited length of the cutting-edge engagement. In drilling and boring, chip control is vital because of limited space inside holes. Chips have to be fairly exact so that they can be evacuated with little congestion.

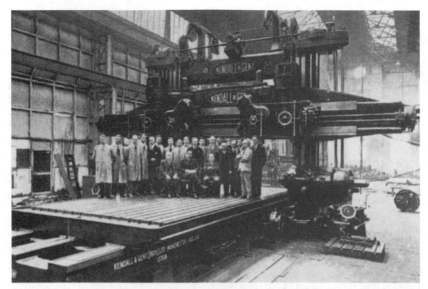

A.

B.

FIGURE 1.7 Large, expensive machine tools, like the two at the Boston Navy yard: A. 1930s plano/mill, B. Three trucks demonstrate the capacity of the boring mill. *(Sandvik Coromant Editorial Dept., Sweden)*

FIGURE 1.8   Lathe turning with an indexable cutting tool. *(Sandvik Coromant Editorial Dept., Sweden)*

Now that the machining processes have been perfected to a high degree, we are seeking other ways to save money by increased production. The purpose of this book is to demonstrate some unconventional manufacturing processes. In the following pages, you will find a few dozen described.

CHAPTER 2
COMPUTER-INTEGRATED MANUFACTURING

The concept of Computer Integrated Manufacturing (CIM) really started with Eli Whitney's revolutionary approach to manufacturing back in 1798. He introduced the idea of making interchangeable parts for ease of assembly. Then, about 170 years later, around 1970, we witnessed the introduction of Computer-Aided Drafting (CAD). This was a giant step forward.

CAD and its sister CAM (Computer-Aided Manufacturing), already proven technologies for design and drafting, tools, and products, now extended into numerical-control (NC) manufacturing, inspection, and quality control.

But even that is not really Computer-Integrated Manufacturing (CIM). Not yet, anyhow. CIM is not complete until the entire company, not just the manufacturing arm has access to the CAD data base and uses it to conduct the business of each department. The Accounting departments, General Accounting departments, Accounts Receivable departments, Cost Accounting departments, and Accounts Payable departments use it. The Production departments, Production Control departments, Procurement departments, Shipping and Receiving departments, and Quality Control departments use it. Business Planning and Engineering, Sales, and Field Service and Maintenance use it. Everybody gets to use it.

Marketing, Finance, and the total administration of company affairs should be able to make use of it This is a communal data base for their departmental business. Currently there is sufficient software available that makes it possible for anyone in a global organization to use that same data base. The software should be able to support the concurrent work of different groups and provide life-cycle management from concept through production, inspection, warehousing, sales, and complete record maintenance.

Two young toolmakers bought a CAD program and began making tools and dies, and made a success of their business, based on one strong advertising gimmick. They promised their customers to keep records of all the tools they made. A telephone call one year or five years later could get them started machining a replacement part or an entire punch and die.

No matter how small the shop, if you wanted to grow, you had to participate with advancing technology. A shop without computer assistance soon relegated itself to crumbs of work.

WHAT IT CAN DO

Hypro, Inc. of Waterford, WI wanted to achieve a full CIM environment, which meant their process planners needed a workable communication tool to automate the creation and display of work instructions and methods for manufacturing parts and assemblies. They wanted to accept engineering drawings directly from CAD/CAM systems, incorporating text and graphics at a reasonable cost. Hypro researched the Gerber System of South Windsor, CT. After talking to several Gerber customers, Hypro picked Cimplan to run on a SABRE CAD System. Gerber Systems guided Hypro through a 5-year planned expansion to a full CIM environment, which was their goal.

Some of these CAD programs require as much as five years to install, but once they are learned, they go along and there's no stopping them. Cimplan is a network licensed database software product employing UNIX-based platforms. Cimplan was conceived based on a simple observation that manufacturing companies required a vehicle to get easy-to-read instructions to the shop floor. Prior to getting Cimplan, the machine shop was lucky to have the latest print to read.

Peugeot Automotive of France is in the middle of their five-year plan to install the same type of computer integrated manufacturing plan and they appear to be successful. They are using a suite of nine Valisys software modules to integrate with CAD. These software modules create a 3-D CAD data base to perform inspections and quality-control tasks. At the time Peugeot investigated the market, it was their conclusion that Valisys led the market with complete CAD integration.

These are large companies with many employees and departments. Smaller companies achieve the CIM goal in much less time. Some software suppliers guarantee a complete switch to a computer-driven operation in six months, but it definitely takes time and money. The Thomas Register lists dozens of software suppliers, some of whom sell one hundred different programs, not all of them compatible.

If you want to be set up with a larger system, SAN /3 applications provide management information for mid-size companies, as well as for large multinationals. R/3 applications are installed across a broad spectrum that is not limited by company size or national borders. R/3 solutions are hard at work in various vertical industries. Automobile manufacturers use R/3 to build flow factories where just-in-time materials and assemblies flow from the vendor into production. Working with this program is no more difficult than dealing with desktop solutions.

Obviously, changing from a paper-run company to one computer-run is a significant decision for any company to make. Normally, this would be a positive choice, but it means that some hard decisions must be made. Maybe workstations will be required for your company instead of PCs. Maybe you want your own computer expert to guide your choice. You want to foresee what the future holds for you. Perhaps you need a platform powerful enough to take on some expansion in two years. Maybe automating inspection next year or supporting quality-control analysis would overload your present equipment. Perhaps your long-term goal of a paperless, peopleless, and defectless operation will obsolete the purchases you make today, if you're not guided properly.

DIFFERENCE IN COMPANY-USER SIZE

Several software companies recognize this dilemma and have prepared for it. If you call Computervision for one of their programs, you will find that they have an Optegra System with a Vault and just enough to get you started. And as you require additional units later, their's will be compatible.

If you deal with DataWorks Corp., you'll find that they have done pretty much the same thing. Their Vista package is prepared for small companies. Vista fully integrates 13 core business functions. The DesignWare feature permits users to define their own screens, add fields, change colors, change grid sizes, and drag choices from menus to the desktop.

They also have a Vantage package, which handles companies with annual sales of 5 to 20 million. And they have three more packages: Man-FactII, DataFlo, and Enterprise Server, which cover the needs of any size company.

The Optegra package enables designers, engineers, and anybody else with need to browse and interrogate 3-D models to gain a more-complete understanding of their surroundings. This reduces time-to-market and development costs. It eliminates the need for many physical prototypes. Engineers can digitally inspect and evaluate without the need for several physical models. Marketing and sales organizations can leverage the investment in no time at no added expense.

In most companies, the most significant single document is the engineering drawing. This contains the geometry of the part, the details are there, the dimensions, revisions, and approvals. The other matters such as parts list, bill of materials can be quickly accessed. So, the choice of a CAD program is important.

Then consider the other features you will require. DataWorks has its REP solutions for any manufacturer. SAP's R/3 program facilitates rapid response to changes by making you more flexible so that you can change to your advantage. SAP's R/3 System is for large companies, giving them the opportunity to optimize their size.

Computervision's Optegra provides development teams with the ability to take a flight, like virtual-reality to ensure the retrieval of any product information that they dig up. What benefits to be derived? Costly changes are reduced via fly through.

Optegra is one system that can build products electronically. Everything is put on a model. The wires, the cables, the piping, everything is designed in. This is called *Enterprise Data Management (EDM)* and its vision is to tie everything into one, unified, cohesive communications system. A truly major-domo, this person. The acronym EDM has many meanings and they should be understood.

COMPLETE COMPUTER-INTEGRATED MANUFACTURING

Presently, software programs will permit computer-integrated manufacturing. You might be able to acquire the complete CIM System from one supplier. But maybe his CAD is weak and you want that area strong. Well, you could buy a CAD System from Catia or Unigraphics and the Optegra or DataWorks could fill-in the remaining software to have a complete CIM system in operation. You could have all the other functions of business administration occurring simultaneously.

The most important decision you have to make is selecting software to fit your program. It must be compatible. You don't want to sift through all the software that floods the marketplace. It is quite possible that the most favorable system for you is an amalgamation of two or three programs. This is when a knowledgeable computer expert is needed.

If this computer expert were to ask you what an ideal CIM System would do for you, a good answer would be: "I could sit down at my PC, open the window, and have up-to-the-minute data from any point in the company at my fingertips. I could then select data and transfer it to an Excel spreadsheet or any other desktop application with the click of a mouse."

Now that dream is a reality. But, if you can't automate the entire enterprise, you're sunk before you start. You won't derive the full benefit that the ERP or MRP should deliver, like the maximum reduction of costs, the fastest possible time-to-market, and the highest possible product quality.

TRAINING

One of the reasons for the fast startup is the training your people receive. The student body is trained as though they are participating in the new software. Inevitable differences between the old methods and the new show up immediately and they are resolved as business changes. Because users work with authentic data in real-world settings, nearly all problems are solved before the cut-off date.

ERP and MRP are systems and they must be integrated with other systems. To fully understand ERP or MRP, you must be able to understand the following about the system.

1. How many users will be served by the system?
2. Where are the users located?
3. What type of network is in place?
4. What is the starting load and what will be the existing load on the network?
5. What is the network plan?
6. What type of computer system is in place?
7. Is network connectivity in place for all potential users?
8. What operating systems are in place?
9. What type of document files will be stored?
10. How will they need to be protected?
11. What informational attributes will be associated by document type?
12. How will the users find the documents?
13. How will the users store the documents?
14. Will remotes be required?
15. What are the legacy CAD Systems?
16. What are the interface issues?
17. What are the e mail connections?
18. What and where are the metrics needed to be implemented?

WINNERS AND LOSERS

The decade leading to the new century will be remembered as a period of fundamental change in the way new products are brought into the marketplace, and in this manner, the winners and losers in this race are defined. One determinant will certainly be who has shifted from paper-based new product definition to *Electronic Product Definition (EPD).* This is a significant distinction and heralds a process change at the very heart of new product development.

Entrepreneurs who exploit this change because they recognize its importance will leapfrog the competition. In the past, there have been too many occasions when design engineers, not in widely separated locations but in the very same building, have wasted large sums of money and months of time simply because they were "looking at the elephant while wearing different colored glasses." They simply perceived the problem a little differently.

The variability of definition that the tradition approach breeds must be stopped cold. The painstaking process of reconciling these varying definitions makes new product development an extended process that inherently contains a large amount of nonvalue activity. These multiple definitions must be replaced with one master definition and the virtual product must become the central focusing point, a central reference for the total effort.

THE VIRTUAL PRODUCT

Then, by making that dynamic virtual product available to every unit within the extended enterprise, it becomes a mechanism for managing the product development process and a "time-to-market machine" of unrivaled power. The prizes associated with Electronic Product Definition (EPD) are accessible to companies of any size and in any industry. Computervision Bedford, MA is pushing this approach in their worldwide advertising campaign and it might eventually make CIM a reality with fewer enabling programs required.

Computervision now offers Next Generation Solutions for enterprise management. They claim it optimizes management by sharing product and process knowledge throughout the plant. It increases productivity, improves automation, and build-to-order goals. Users can browse, locate stored information, and define configurations for customer proposals and product development, but alterations to the product can only be made by specified personnel.

Manufacturers who are accustomed to CAD/CAM techniques might at first glance perceive EPD as just sophisticated modeling. This would unfortunately, miss the revolutionary nature of the promise held out by EPD. CAD/CAM deals in geometry, but if you can believe the examples demonstrated by a few world class companies, there might be more here than meets the eye. EPD uses CAD-generated data to create a virtual version of the product. This virtual product development includes such elements as on-screen assembly of complex parts, wiring, piping, and test point connections. Structural integrity, operating features, all mechanical and electronic characteristics, and training and maintenance data all can be stored in the database.

The Anglo-French Aerospace suppliers, Hurel-Dubois, who make thrust reversers for jet aircraft, uses EPD to manage customer-generated changes. Through building the external components of its latest jet engine in virtual space, Rolls-Royce Aerospace eliminated two out of three full-size mockups. Applied Materials, a US-owned manufacturer of machines that make microchips, recently completed a pilot project before building it for real. Apparently, EPD cannot be neglected if you want your company to be world class.

Currently, a plethora of companies are producing software. Some of them have more than 100 programs for sale. Most programs are for a very specific use. I've just been handed one that changes pounds to inches. So, you've got to be careful that anything you buy is compatible with software already purchased.

In modern industrial competition, it is important to have software that follows the actual part design on CAD, then continues through manufacture and inspection, and finally tracks the part cost, and the trips through inventory control, marketing, and finance.

Computer integration of the different functions of an enterprise has been improving since its inception about 1960. The first concept has not changed; it is still the same. The thought has been to electronically integrate all of a company's activities including design, production, marketing, finance, and anything to do with total management. A number of different concepts have arisen. There are programs feeding off a central data base and there are islands of computerization linking administrative units of an enterprise.

This section has included information concerning the important management areas of finance, marketing, warehousing, and top-level administration. Now the chapter concentrates on the production area of manufacturing because, after all, your primary interest is in production.

The Valisys Corporation was formed in November 1987 after a few years of research. The company had been developing computer integrated manufacturing and although it does improve communication throughout the plant, its greatest impact was felt at the start of production. It works with several different types of CAD. At least a dozen other programs work as well, but when Valisys was introduced at Digital Equipment Corporation in 1988, I was working there and had the opportunity to see it in operation. In fact, I used it myself to solve a serious production problem. The entire scenario unfolded simply because drafting, manufacturing, and inspection could use the same data base, which, in this case, pinpointed the problem.

The Need for a Valisys-type Program

Before Valisys was ready, and while CAD/CAM and other factory automation technologies were available in abundance, a complex product could still require up to a year's time to advance from the design stage into routine production. Tracking the delay problem to its source revealed that engineering drawings were too often at fault. Situations causing delays included drawings incorrectly dimensioned, design intent miscommunicated, and tolerances based on feel, rather than on actual data. There was inappropriate placement of datums, inconsistent use of detailing terminology, omitted information, and the specific order of GD&T symbols was not followed. The realization that much of this wasted time could be saved instigated the formation of the Valisys Company.

Unlike the old bilateral tolerancing system, based on rectangular coordinates, GD&T specifications are based on the type and shape of features, rather than on distances from a point or edge. By adding symbolic callouts to each critical feature, a designer can provide all the tolerance information that the fixture designer and machinist require. There is an additional advantage. In the maximum material condition specifications, there is a bonus tolerance that stretches the allowable quality range without impairing the part's function. For example, if a hole is slightly oversized, the standard permits it to be slightly farther from its true position because it has enough slack to fit over the mating pins.

NEED FOR TOLERANCES

Tolerances are necessary to accommodate variations that occur when machining, molding, casting, or fabricating parts. A persistent dilemma is properly specifying tolerances in the design phase and communicating them to manufacturing and inspection. The consequences of this problem that costs manufacturers in this country millions of dollars each year are:

1. Designs that cannot be manufactured.
2. Manufacturing processes that are unnecessarily expensive.

3. Parts that cannot be easily and accurately passed or failed.

4. "Good" parts that defy assembly.

5. Arguments over what was meant by the drawing.

Valisys permits the use of CAD data base dimensions and tolerances throughout design, manufacture, and inspection. But first Valisys checks these specifications for compliance with the ANSI Y14.5 standards, eliminating errors of application and interpretation. Then these validated specifications are translated into programs for machining and inspection (Figure 2.1).

FIGURE 2.1   A cover from *Managing Automation* magazine, showing the Valisys System in action.

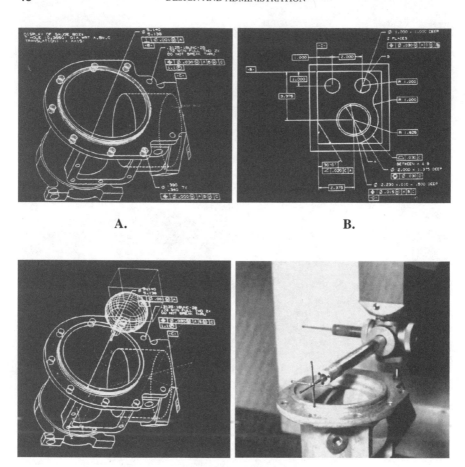

A.   **B.**

C.   **D.**

FIGURE 2.2 CMM in action: A. Softgauges (shown in grey lines), created for each part using GD&T call-outs, allow actual design specifications to be used in the manufacturing and inspection processes. B. Valisys allows a designer to determine whether the part description matches the design intent. C. Softgauges are used to automatically create inspection paths (shown in grey). Such paths guarantee that critical and major features are checked for assemblability and conformance to design intent. D. Valisys mathematically defines the probe cluster and uses the definition to ensure that probe tips will not collide with any part feature during the course of the inspection. *(Technomatix Technologies, Inc., San Jose, CA)*

For instance, in machining, Valisys orients tool paths with reference to datum surfaces (Figure 2.2). During inspection, Valisys generates proper inspection paths, executes inspections, and determines pass/fail figures. All of this is performed automatically, and these capabilities preclude many sources of both manufacturing and inspection errors.

Successful implementation of Geometric Dimensioning and Tolerancing (GD & T) results in a positive, universal language, a nomenclature for use throughout the manufacturing process, and quality assurance. That holds true in writing methods, material review, fixturing, machining, and inspection. This commonality and uniformity throughout the part description is a requisite for conveying accurate design intent downstream and for a successful integration of inspection with design and manufacture.

There are nine Valisys software modules that work in conjunction with CAD systems and interface with automated machine tools and inspection devices. The modules include V:Check, V-Gauge, V:Tolerance, V:Path, V:Interface, V:Inspect, V:Qualify, V:Track, and V:Control. Each module serves to check certain aspects of the design, manufacturing, and inspection process.

BACKGROUND

CAD systems are increasing output from design engineering by tenfold. NC machines are automating the machining process. Coordinate-measuring machines (CMM) have presented inspection with a flexibility and accuracy that was not possible just a few years ago (Figure 2.3). Automatic conveyorized systems aided by bar coding accelerate movement within the plant and permit the continuation of production with lower inventories. Nevertheless, these wonderful advances in technology have to be integrated in order to achieve the ultimate in production.

The first decision of the Valisys development team was that a universal design language had to be used to permit design information from various geographical locations to be exchanged. The task was simplified because of the universal acceptance of the ANSI standard Y14.S early in the 1980s. This standard describes geometric dimensioning and tolerancing (GD&T). The symbols used in GD&T provide labels for engineering drawings that not only provide dimensions, but also indicate functional relationships and other characteristics. Most important, this standard code prevents misinterpretations.

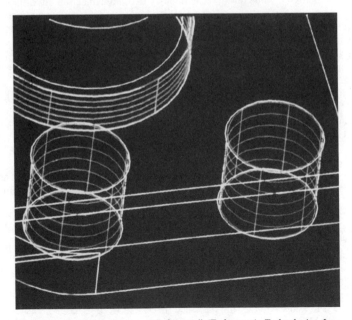

FIGURE 2.3 Valisys V: gauge "Softgauge." *(Technomatix Technologies, Inc., San Jose, CA)*

The general Valisys software strategy is to complement a CAD/CAM system by providing tools to verify the design process and to integrate functions important to quality control and inspection. In fact, the only step that Valisys doesn't take is the actual cutting of material on the NC machine. A separate tape is still required for that task, although Valisys will align the part in its X, Y, and Z axes, the same way a part is positioned in a coordinate-measuring machine (CMM).

CAD data bases contain complete part definition including design, dimensions and tolerancing data. If we could drive manufacturing and inspection directly from this CAD data base, the complete part manufacturing process could be more successfully automated.

As assemblies became more complex, industry needed a better method to describe the shape and dimensions of parts and how they fit together. This is not to belittle the very significant contribution of computer-aided manufacturing (CAM). That system accelerates drafting and research and development and is absolutely a modern miracle. And still, success towards acquiring a complete CIM system had been limited.

Without a doubt, automated tools have been a great advance for manufacturing. However, getting various types of equipment to work together has been a different story. We have needed some way to use CNC machine tools, CMMs, and other automated equipment in a related manner. Valisys provides a direct and simple linking system for all this equipment.

PROCESS DESCRIPTION

Complex fixturing is no longer required because machining and inspection paths are oriented by Valisys as the parts are positioned. This improves productivity considerably. The bottom line is increased value out of tremendously expensive machine investments.

When a part is placed in a machine tool, traditional manufacturing technique is to precisely fixture the part within X, Y, and Z planes. Any inaccuracy in positioning will adversely affect subsequent production. Valisys software simplifies these part setups by providing the intelligence for a machine tool probe to automatically determine the part's exact orientation to its own X, Y, and Z planes via datums referenced to part features, rather than the fixture itself.

There has to be a way for manufacturing engineers to communicate with design engineers and the shop floor too. The Valisys Corporation has designed software that fills this gap. This software automates the comparison between the product design database and the actual machined part. It even permits the part to be inspected before its removal from the machine.

And if the part is removed for some reason, it permits the inspection of the part on a CMM automatically, by comparing the part with the product design data base. The paperwork generated by Valisys improves quality control and production management (Figure 2.2).

In the average factory, approximately 10% of production is scrap and/or rework. To this waste, you can add the cost of machine downtime because of waiting for reinspection and the cost of decision making. Much of this costly waste is precluded by using Valisys or a similar system.

The story about Valisys is a story about quality. At a high-level engineering symposium recently, there was universal agreement that the greatest problem confronting American industry was quality.

If you want to know whether the parts just made will fit together in an assembly, they must be inspected. There are only three methods of verifying the fit.

1. *Hard gauges or gauging fixtures* Hard gauges are normally used in high-volume production. Sometimes two gauges are needed to check the maximum and minimum tolerances. This type of gauging is easy to use. For instance, the worker simply places the part in the gauge to see if it's within tolerance. As long as the part dimensions don't change, the gauge cost is not high.

   The disadvantage of these gauges is their inflexibility. Scrapping or reworking gauges can be expensive. Then, of course, the time to build them in the first place can be a month or more.

2. *Coordinate-measuring machines* CMMs are precise measuring instruments that can provide an electronic, digital readout that can be understood by a computer. They work slowly compared to hard gauges. They are also very expensive: Starting at $25,000 for a small manual machine, they can cost as much as $1,000,000 for large, fully automatic machines.

   Despite these disadvantages, CMM's are growing in popularity. In fact, since some models started to be produced with ceramic material used for the structural parts, these CMM's are being placed in open shops so that the machine operators can inspect alongside the production machines. CMM's are flexible and can be reprogrammed easily.

3. *Special inspection machines* Special inspection equipment has an important role in several large industries—especially the automotive. Pistons, camshafts, crankshafts, and gears are inspected automatically. Lasers and other vision systems are often used with the special equipment.

   Several manufacturers, including CMM producers, have been trying to link CAD systems to CMMs since the late 1980s. But so far, Valisys is currently leading the pack because of their very complete list of software packages. This situation is very volatile and probably will change soon, if it hasn't already, because there's too much business out there and advances in technology always invites competition.

Besides Valisys, several other companies are engaged in supplying similar software. So far, I have found none as complete or as universal in nature as Valisys. However, the business situation is capricious, and could change at any time. Accordingly, it would be advisable for anyone considering the acquisition of CIM-type software, to investigate several possible suppliers besides Valisys. At this time, the author has found a few possible competitors.

1. Cimline, 700 Nicholas Blvd., Elk Grove Village, IL 60007

2. Computervision, 100 Crosby Dr., Bedford, MA 01730

3. Cadam, Inc., 1935 N. Buena Vista St., Burbank, CA 91504

4. Automated Systems, Inc., Brookfield, WI 53005

5. Gerber Systems, Inc., Windsor, CT 06095

6. Icamp, Inc., Bolton, CT 06043

7. DataWorks Corp., 5910 Pacific Center Blvd., San Diego, CA 92121

The Thomas Register is an excellent source for names and addresses. As mentioned, the situation is volatile and there is heavy competition in the software field. More than one company claims it had more than 100 different software programs for sale. Recently, there have been many acquisitions and buy-outs in that industry. Some companies involved in software have modified their focus.

COORDINATE-MEASURING MACHINES

CMMs are now designed for use in the factory environment. As a consequence, manufacturers will soon demand the advantages of superior, real-time communication between shop and manufacturing engineering and between manufacturing engineers and design personnel.

It has been difficult to create software linking CAD to CMMs. Although machine tools and inspection programs appear to be similar, there are significant differences in the information required to generate both types of programs. NC machine operations are motion oriented: moving, turning, cutting, starting, stopping, and turning on the coolant. The action of a CMM-automated program requires other types of steps: select probes, determine inspection points, collect and analyze data, and report data.

To understand why the CMM-CAD link is so important, it is helpful to review how the CMM works. In all CMMs, a switch, called the *probe*, is mounted on a mechanism that allows the probe to move along three perpendicular axes. The mechanism contains electronic sensors that continually provide a digital readout, monitoring the position of the probe. When the probe touches a surface of the part being measured, the location of the probe is recorded.

Part of the problem of linking CAD and CMMs is that CMMs from various vendors differ in the way that they analyze inspection data, and thus they produce different analyses of the same part. Although each machine will inspect the same points, the internal processing algorithms that process and report the data, differ from each other.

CAD users are very interested in CMMs that move their probes by means of servo motors. The CMMs generally are controlled by small computers similar to those that operate numerically controlled machine tools. These CMMs can be programmed to follow a predetermined path and stop at selected points to measure those positions on the part being inspected.

The data gathered this way can be stored and analyzed by other computer programs. Programs can be prepared that will automatically accept or reject parts by comparing dimensions taken from CAD data with those acquired by the CMM. If you gather data from a quantity of CMM inspections, they could be used for statistical process control.

Because computers are now less expensive, CMM manufacturers are producing more sophisticated systems. CMMs are available that can store complex programs. Unfortunately, these programs are in proprietary language and cannot be transferred from one machine to another. The Valisys programs are unique in that they will run on several different CMM machines.

THE VALISYS PROGRAM

The main problem confronting computer-integrated manufacturing is the difficulty communicating engineering data from product designers to manufacturing engineering to the shop floor. Now, the Valisys Corporation's software, which improves the flow and accuracy of engineering data in a manufacturing environment, will be described.

The following is a description of nine Valisys software packages: V:Check, V:Gauge, V:Tolerance, V:Path, V:Interface, V:Inspect, V:Qualify, V:Track, and V:Control.

V:Check

V:Check facilitates and validates dimensioning and tolerancing in CAD. The results are greater productivity for designers, fewer design errors, and reduced design review periods.

V:Check makes sure that the correct graphic representation of GD&T symbols are picked, using standard CAD methods. and thereby reducing the designer's keyboard input. It also verifies that the design dimensions and tolerancing are valid, consistent, and in conformance with design standards. If an error is detected, V:Check calls attention to it, describes the error, and prompts for a correction (Figure 2.1).

Both plus/minus and GD&T are supported in conformance with ANSI standard Y14.5. Because of V:Check's error-correction support, it helps designers improve their skills in using GD&T. V:Check performs syntax checking. After the designer has finished dimensioning and assigning GD&T symbols to a CAD drawing, he uses the cross-hair cursor of the CAD system to designate each symbol that he wants to be checked for syntax. For example, its proper use, location, and sequence. The designer displays a graphic of the part as described by the GD&T symbols, which is then used to determine if the part, as described, meets functional needs. This checking step verifies whether or not the engineer has communicated design intent.

GD&T is an exact language and it must be used precisely. Reference datums are critical and symbols must be used in the prescribed sequence. Syntax errors make accurate interpretation of GD&T symbols difficult and unlikely, when a number of people have to view the drawings.

When GD&T is understood and used correctly, it reduces guesswork, lowers tool cost, reduces scrap, and provides a common engineering language. It is important for a facility to use the GD&T system because government contracts require its use. V:Check (or any similar system) ensures design integrity and reduces the number of engineering changes. In large military programs, the engineering changes can create an overrun because of as much as a 25% increase in charges. One large engineering firm estimates that each engineering change costs an average of $2000.

V:Gauge

Working from the dimensioning and tolerancing data generated in the design phase, V:Gauge generates a three-dimensional model of the worse case mating part. That means a software package that is the electronic equivalent of a hard tool is available to inspect parts. In most cases, this software package eliminates the need for hard tooling.

This model, known as a *Softgauge*, has more than one use. As a visual tool, it enables a designer to examine all features of his design. Then, it serves as the CAD design for a hard tool if and when one is desired. And finally, the Valisys program V:Path uses it to generate an inspection path for use by an NC machine tool probe during in-process inspection or by a CMM unit at final inspection. This softgauge graphically depicts the worst-case mating features of a part, again using the cross-hair cursor to identify the GD&T callouts, dimensions, and geometries (arcs, lines, etc. in the design). Valisys overlays the softgauge on the design display, using colors to distinguish between the two drawings (Figure 2.4).

V:Tolerance

This software is used to create worst-case software models of parts to make certain that the separate parts will assemble properly. Without this ability, parts that are within tolerance, as designed, might still fail to assemble properly. It is sometimes possible for parts that meet the extremes of individual tolerances not to assemble.

In addition, V:Tolerance generates the optimum sizes for both clearance and threaded holes. Given the fastener specified by the designer, V:Tolerance generates both the design

A.

B.

FIGURE 2.4 A coordinate-measuring machine at work: A. After using Valisys software to determine the part's true orientation, a process engineer operates a machine tool with confidence, knowing his part alignment is correct. B. An application engineer uses Valisys software on a coordinate-measuring machine to verify design specifications. *(Technomatix Technologies, Inc., San Jose, CA)*

geometry and drafting text describing those holes. The dimensions generated will permit maximum flexibility during manufacturing.

V:Path

V:Path creates an inspection path that can be run on a CMM. The program ensures that the inspection process incorporates checks of all critical features. It provides the same level of ensurance as a hard gauge because it works from the Softgauge, verifying fit and function, based on original design intent.

Inspection path definition with V:Path uses the graphics capabilities of the CAD system and is complete and automatic. The operator can preview the path on the screen and make any changes desired before running the inspection on a CMM. Dimensioning and tolerancing information in the part design, is transferred automatically into the inspection process, saving time and reducing chances of errors.

V:Interface

This program communicates Valisys functionality directly to the shop floor. It connects many devices and services, including terminals on the floor, machining, inspection equipment, and robotic part handlers. In other words V:Interface ties the shop floor directly into the CAD/CAM database.

This leads to a more accurate data acquisition on the shop floor because of direct monitoring of production processes. V:Interface brings the full sophisticated inspection capabilities to all inspection devices (not only CMMs, but CNC machine tool probes, in particular), allowing for true, in-process inspection. See Figure 2.5 for a view of material modifier errors and wrong callouts.

V:Inspect

Using inspection paths generated with V:Path and communicating through V:Interface, this program executes inspections by running CMMs, machine tool probes and other devices, such as laser systems.

V:Inspect accepts process-control information (lot number, serial number, operator, date, etc.), drives the probe to a specified point, takes a measurement and stores all the inspection data until a printout is requested.

V:Qualify

This program creates an as-built model from the measured data generated by V:Inspect and compares this model with the appropriate "Softgauge" model. If the "Softgauge" for each feature fits the as-built features, then V:Qualify signals that the part passes inspection.

If the part fails, V:Qualify decides whether or not the part can be readily reworked. If it can be, V:Qualify illustrates the rework graphically on the model of the as-built part and also provides numerical specifications. If the part cannot be reworked, V:Qualify illustrates the error and provides numerical data to support analysis by a material review board.

In addition, each set of inspection results are archived for later access. This is a benefit in government contract work, where traceability is required.

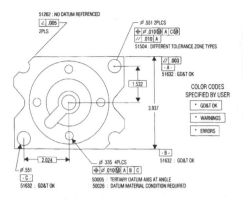

A.

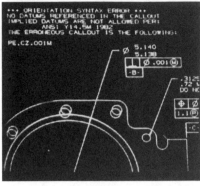

B.

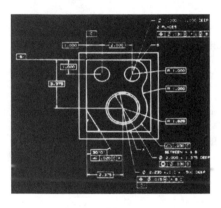

C.

FIGURE 2.5 Valisys corrections: A. Valisys/Design uses colors to effectively communicate information on tolerencing. B. Valisys software will review callouts and inform the operator of mistakes, such as missing datums, erroneous tolerence zones, or improper material modifiers. C. It is not unusual for material modifiers to be overlooked. *(Technomatix Technologies, Inc., San Jose, CA)*

V:Track

V:Track is used in the inspection phase to perform statistical process control. It accepts real-time data from the Valisys V:Qualify module to do this and also produces control charts from the same data source.

V:Control

V:Control is a programming language for automatically executing a series of Valisys functions on machines, accessible through V:Interface. The user defines a series of steps as a job by using V:Control language. These steps can be any Valisys function, a machine operation, an inspection, or any use of data previously acquired.

This job is stored as a named file. To run this job, the operator types the file name at the appropriate prompt. Without any further operator action, the steps contained in the job file are then performed automatically.

A single job in V:Control can handle up to 32 machines. The only limitation is that only one command can be executed at a time. An example of one V:Control job could be:

1. Load castings from a tray onto a machine tool.

2. Machine several features.

3. Inspect the part with a probe.

4. Load the good parts onto one tray and the bad parts onto another.

5. Print the inspection results.

6. Stop the machine if three bad parts in a row are made.

In process inspection is now a reality. You can put it to work today. Sophisticated inspections can be done while the part is still fastened to the machine tool. Formerly, industry had two alternatives when machining complex parts. Either perform all machining and then inspect, which is risky. Or, do a few machining steps, then take the part out of the machine to a CMM for inspection.

Then, back to the machine again and repeat the cycle.

Many hours can be wasted this way. Now it is possible to inspect the part as often as the engineer wishes without disturbing the setup at all. It's an automatic process that saves time and achieves quality results.

CASES

Learjet of Wichita, KS., recently found great satisfaction using Valisys. They had a large 60" J-shaped wing rib, which was a headache to inspect. Valisys was so useful there, it was used for inspection and verification of some master tooling. See Figure 2.6 for a picture of the J-shape and a vertical spar.

Consider the case of the European automobile manufacturer, Peugeot. In 1989, this company was determined to automate its many plants in Europe. Peugeot wanted to eliminate all manufacturing paper by the year 2000. The first step was to put all car parts on CAD/CAM. In August of 1989, the integration of Valisys into plant operation was started.

Valisys implemented much of Peugeot's intention of taking design data throughout the manufacturing process from CAD through engineering to manufacturing and final assembly. The software literally organizes the flow of the various processes, providing feedback from manufacturing to design and improving communication throughout. Valisys is only one step in the vast program pursued by Peugeot, but its contribution is significant.

I was employed as a consultant by Digital Equipment Corp. off and on for a four-year period. During this time, I had an opportunity to watch the implementation of the newly purchased Valisys System. It performed well doing all its assigned tasks. In fact, a serious problem in design was solved because I was able to coordinate output of a CMM inspection

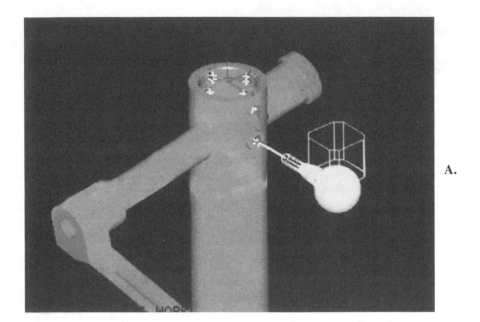

A.

B.

FIGURE 2.6 Valisys automates Learjet inspections: A. To meet rigorous production schedules and to improve quality, Learjet used Technomatix Valisys for inspection and verification of some of the master tooling, including the aircraft contour and the vertical spars. Spars are crucial machined parts that affix the horizontal stabilizer to the aircraft. B. J-shaped wing rib undergoes automated inspection. *(Learjet, Inc., Wichita, KS)*

of a troublesome package and compare it to its CAD drawing. The comparison, exploded to 10 times actual size, showed the places where the failures were occurring and the reason for them.

Valisys is a multi-faceted technology with a broad array of applications available for factory cell automation and statistical process control. In contrast to conventional inspection, which measures deviations in dimensions, Valisys measures ease of assembly. If a company wishes to consider "lights out, factory automation," it should possess a Valisys system or something just as good.

P · A · R · T · 2

SHEETMETAL AND PRESS PROCESSES

CHAPTER 3
FINEBLANKING

BACKGROUND

A German patent for "the fineblanking method" was issued to Fritz Schiess in 1923. For the next 15 years the process was used mainly in Switzerland, but was very slow to catch on elsewhere. Just before 1960, the Feintool Co. of Switzerland started building and selling machine tools for the process and has since provided the world with most of the fineblanking machines in existence. Because of the accuracy of the parts made by this process, at first it was used almost exclusively by the watch industry. Today, the process is used in many countries, in many industries.

Only two other companies build fineblanking machines—Schmid and Hytrel. Both are also Swiss companies and both have established assembly plants in this country. In order to popularize the process, both of these concerns do jobbing for any company that wants to try it. Fineblanking is a form of metal stamping, but it is also a form of extrusion because the part is extruded from the workpiece material as it enters the die opening under the "slow" pressure of the punch. The part does not undergo the fracturing that produces the rough, broken surface along the hole and edge peripheries that is characteristic of conventional stampings.

The reason for this is the main difference between fineblanking and conventional blanked workpieces. Fineblanking tools have less than 1% of the material thickness as clearance between punch and die in contrast to the 5–10%, which is customarily used in conventional stamping. Fineblanking cannot take the place of conventional stamping because it is a much more expensive machine and is much slower in operation. It simply cannot compete with most blanking or stamping jobs. However, for those applications where edge quality, flatness, and accurate dimensioning is required, fineblanking is the best method to use because it saves one or more secondary operations.

For a long time, the extra cost of fineblanking machines and tools just didn't seem worthwhile. If a comparison is made only of strokes per minute, a stamping press wins easily. In addition, this newer process requires careful attention to detail. Nevertheless, the main reason fineblanking was overlooked is that it was unknown or at least little understood.

For years since the fineblanking process was invented, it was regarded as a mysterious, little understood process, which produced parts by stamping, although those parts possessed the appearance and tolerances of machined components. Also, the only machine manufacturers were foreign and the replacement parts were difficult to obtain. The first toolmakers for the process were foreign, too. Time and the advance of the technology have now altered this concept and the process needs to be reintroduced.

The picture is beginning to change. Industry is becoming more receptive to anything that reduces cost of manufacturing. In the family of stamping processes, which include blanking, punching, coining, drawing, bending, etc., there is a variety of different machines to handle the different applications. The fineblanking process also requires its own special machine.

A trend that is boosting fineblanking is the growth of statistical quality-control programs. Fineblanking complements this movement because it is a technique oriented to quality. Quality is built into the machine and the tools. In addition, quality is easier to monitor when parts are produced on one high-quality machine, instead of on several machines using different processes. More and more companies are seeking suppliers who can deliver 100% good parts so that incoming inspection can be waived or at least minimized. They are willing to pay a premium for these products to save the cost of maintaining high inventories and a thorough quality-control procedure at incoming.

PROCESS

Fineblanking uses special presses and special tools to stamp metals so that the parts produced are sheared over the full material thickness in one stroke. In most cases only belt sanding or barrel tumbling is necessary to finish the part. In this process, part of the die, a V-ring follows the outer shape of the blanked workpiece. When the tool closes, this raised V-shaped ridge bites the stock material holding it tightly so that it cannot flow away from the punch during the blanking step. See Figure 3.1 for the sequence of action in the process.

Fineblanking involves three press motions, which distinguish fineblanking presses from conventional blanking presses. There is a closing force, a blanking force, and a counterforce; each separate and adjustable. One motion lifts the lower half of the tool, raising the material and clamping it against the upper half of the tool. At the same time, the V-ring bites into the material so that it cannot flow away from the punch when blanking occurs. In the next motion, the ram speed is reduced as the punch shears the workpiece into the die opening and against the counterpunch ejector. The ejector is exactly the same as the blanking punch, a mirror image, and is a close fit in the die opening. In the third motion, the ram reverses direction just when the workpiece is fully sheared. Then, the counterpunch pushes the finished part out of the die opening as the tool opens and the workpiece is blown onto a conveyor or a catch container. Heavy workpieces are swept out mechanically.

It is important to understand this process. Study Figure 3.1 again. The fineblanking action is repeated for emphasis. The operation begins with the tool open and the roll feeder pushing material into the die area. The ram then travels up at high speed until it reaches an adjustable preset position. At this place, the ram slows to another preset, adjustable speed. The tool touches the material and the tool safety of the machine determines if any debris or parts remain in the die area. If the area is clear, the ram continues upward pressing the V-ring into the material and preventing it from flowing away from the cutting surfaces during the shearing step. The blank punch contacts the workpiece and pushes (extrudes) the part into the die opening. While the shearing is proceeding, the counterpunch comes up to hold the part flat against the blank punch. After reaching "top dead center," the ram retracts, the scrap material is stripped from the punch, and the part is ejected from the die. The material is fed again and the cycle is repeated.

A typical fine-blanking tool is a single-station compound tool for producing a finished part in one press stroke. The only additional operation needed is the removal of a slight burr. The process requires a triple-action fine-blanking press. Closing force, counterpressure, and blanking pressure forces are individually and infinitely adjustable.

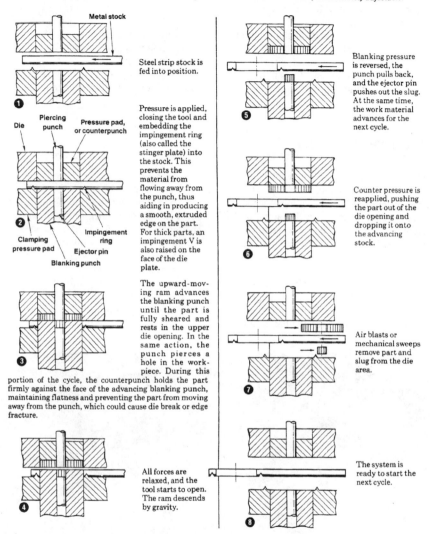

1 Steel strip stock is fed into position.

2 Pressure is applied, closing the tool and embedding the impingement ring (also called the stinger plate) into the stock. This prevents the material from flowing away from the punch, thus aiding in producing a smooth, extruded edge on the part. For thick parts, an impingement V is also raised on the face of the die plate.

3 The upward-moving ram advances the blanking punch until the part is fully sheared and rests in the upper die opening. In the same action, the punch pierces a hole in the workpiece. During this portion of the cycle, the counterpunch holds the part firmly against the face of the advancing blanking punch, maintaining flatness and preventing the part from moving away from the punch, which could cause die break or edge fracture.

4 All forces are relaxed, and the tool starts to open. The ram descends by gravity.

5 Blanking pressure is reversed, the punch pulls back, and the ejector pin pushes out the slug. At the same time, the work material advances for the next cycle.

6 Counter pressure is reapplied, pushing the part out of the die opening and dropping it onto the advancing stock.

7 Air blasts or mechanical sweeps remove part and slug from the die area.

8 The system is ready to start the next cycle.

Labels in figure: Metal stock, Die, Piercing punch, Pressure pad, or counterpunch, Clamping pressure pad, Impingement ring, Ejector pin, Blanking punch

FIGURE 3.1 How fineblanking works. *(Fairfield Tool Co., Bridgeport, CT)*

The forces are preset in this triple-action type of press, which is very rigid in construction and fully hydraulic. The parallelism of the upper bolster and the ram table is maintained very close because there is practically no clearance between the punch and die, and an out-of-parallel situation would quickly result in a shortened tool life.

Parts that are fineblanked are consistently stronger, flatter, and more precise than those produced on conventional presses. Material worked can be thicker. In fact, if the cross-section of the part has to be thickened to increase its strength, no modifications to the tooling are necessary. Only the stroke has to be readjusted. Features that are possible with fineblanking are thinner webs, smaller holes, narrower slots, and closer spacing between all shapes of holes. See Figure 3.2 for a picture of a part that would be difficult to make any other way.

A single station tool is capable of producing parts with offsets or bent-up sections up to 70 degrees, depending on the type of material and the length of the offset. The bending takes place as the die closes for V-ring impingement. The advantage of this timing is that all piercing is accomplished when the workpiece is in final form. Consequently, there will not be any variations from the accuracy built into the tool for hole spacings and inner and outer dimensions.

MACHINE FEATURES

Fineblanking machines have a tool safety to protect against scrap or parts not properly ejected. They have hydraulic tool clamping, which when combined with an adjustable, portable tool table, permits quick die changing to reduce downtime. There is a computer control to program closing speed, shearing speed, counter-pressure on the ejector punch, opening speed, top tool height, and V-ring force. It also controls the compressed air jets for workpiece and slug removal and it controls the operation of the stock feeding devices. At the same time the computer control keeps statistical information regarding the tool, the press, the production, and the operator. See Figure 3.3 for a view of two Feintool Corporation machines.

Two features of the fineblank press stand out. First is the adjustable solid stop for setting the top dead center or shut height of the machine. It is important to keep the punch from entering the die. If the punch is continuously extended into the die opening, the punch or die will eventually crack. The solid stop can be set to ensure the repeatability of

FIGURE 3.2 Designed with fineblanking in mind, these disk spacers for the computer industry take advantage of fineblanking's ability to produce smooth edges and coined shapes in one pass through the press. *(California Fineblanking Corp., Huntington Beach, CA)*

FIGURE 3.3 Two fineblanking machines marketed by the Swiss Feintool Corp.

the shut height. The second feature is the structural support that the machine gives the die. If the die is allowed to "breathe" under pressure, tool life will be shortened.

If the tool requires sharpening, the machine's shut height can be easily adjusted at the CNC console to compensate for material removed in grinding. The CNC includes diagnostics to alert the operator to faults in the machine. Under certain conditions, it will automatically shut down the machine to avoid damage to the machine or to the die. The slow speed of the machine, 30 to 160 strokes per minute, simplify the protection umbrella.

An important development was the introduction of coatings on tools to resist wear. The coating (read the chapter on titanium nitride coating) extends the useful life of the tool because it creates a surface hardness of about 70 to 80 Rc and adds lubricity to the surface. Lubricants have been developed especially for fineblanking. It is always wise to use lubricants that have been prepared for a specific task. That is true whether the process is turning, stamping, forging, or drilling.

The most common fineblanking machines range from 25 tons to 2500 tons, most of them are hydraulically operated. The fineblanking machine routinely maintains accuracies of 0.001" on all dimensions: hole sizes, center distances, form size, and all locations. Because the punch never enters the die, there is little tool wear. Generally, there is no difference between the first piece struck and the one millionth. Surfaces sheared will probably have finishes of 32 microinches or better, and certainly no worse than 63. Sometimes parts that ordinarily would be either machined from the solid or conventionally blanked and machined, can be fineblanked complete in one step. Samples of fineblanked components are shown in Figure 3.4.

A major benefit of this process is that a stroke of the press can deliver a finished part, and as previously stated, normally all that is ever required as the parts come off the press is a vibratory tumbling or high speed belt sanding. This eliminates the need for secondary operations such as flattening, milling, counterboring, countersinking, edge grinding and reaming.

There are other benefits to the process. Small holes, which are generally drilled after a stamping operation, can now be punched in during fineblanking. In conventional stamping, the smallest hole must be at least 1½ times the material thickness. The same minimum re-

FIGURE 3.4 Fineblanked parts. *(California Fineblanking Corp., Huntington Beach, CA)*

striction applies to spacing between holes and closeness to edges of cutouts. In fineblanking mild steel, the figure drops to 60% or 50% of material thickness. In copper or aluminum, the figure drops to 35%. The same holds true for narrow slots.

All presses have tool safety devices built in to prevent cycling to the next stroke until all slugs and parts have been cleared from the die. That means two things: Low tool breakage and less-expensive tools because the tools don't require separate safety devices. Figure 3.5 is a schematic of a Schmid fineblanking machine.

If greater strength is required for a part processed in a fineblank machine, all you have to do is increase the part thickness. The fineblank tool requires no change. This wouldn't be true if the part was being stamped in a conventional machine. Semi-piercing, which is a partial extrusion of holes, can be fineblanked along with the rest of the design. There are times when this is an important function. At the same time, you can countersink, counterbore, chamfer, and coin. And, of course, you never have to flatten the part.

DESIGN FEATURES

Fineblanking permits designers to incorporate many additional part features at no extra cost. In the same stroke which creates a part, the process can make inside diameter marks, coined sections, self rivets, contact points, bosses, spacers, female cam tracks, assembly pins and a variety of other functional shapes which would otherwise require machining. See Figure 3.6 for a comparison of a component made by machining versus fineblanking it. Semi-piercing, which is a partial extrusion of holes, can also be fineblanked along with the rest of the design, there are times when this is an important consideration.

Single-station fineblanking tools produce parts with forms, bends, and offsets. These parts are consistently within specifications. Because the part features are created as the tool closes, the dimensional accuracy is not affected and the features are not distorted.

Formerly, all fineblanking tools were single-station compound dies, now a growing percentage are progressive (Figure 3.7). That means some tools will have off-center loads; for this procedure, the press must be very rigid. When the workpiece design requires more than can be accomplished in a single-station tool, a progressive die should be considered. Progressive dies are considerably more expensive than compound dies, so it is imperative at this time to compare the cost of a progressive die to a compound die plus secondary operations that would be required.

In addition to the operations that a single-station die can perform: stamp, blank, pierce, semi-pierce, coin, chamfer, extrude, preform, make self rivets or contact points, and make

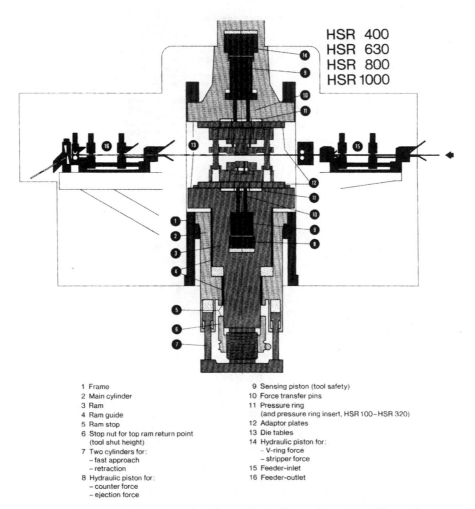

HSR 400
HSR 630
HSR 800
HSR 1000

1 Frame
2 Main cylinder
3 Ram
4 Ram guide
5 Ram stop
6 Stop nut for top ram return point
  (tool shut height)
7 Two cylinders for:
  – fast approach
  – retraction
8 Hydraulic piston for:
  – counter force
  – ejection force

9 Sensing piston (tool safety)
10 Force transfer pins
11 Pressure ring
  (and pressure ring insert, HSR 100–HSR 320)
12 Adaptor plates
13 Die tables
14 Hydraulic piston for:
  – V-ring force
  – stripper force
15 Feeder-inlet
16 Feeder-outlet

FIGURE 3.5 A schematic of the HSR series of Schmid fineblanking machines. *(Schmid Corp. of America, Goodrich, MI)*

MACHINE IT OR FINE-BLANK IT?

Trapped female cam for door actuators on a mainframe computer disk drive was initially to be machined from 5-in.-diameter, mild-steel barstock. The ⅜-in. hub would be turned, then the cam section parted off. The D-hole would be drilled, locking-screw hole drilled and tapped, and the cam track cut on an NC mill. Material and machining costs, based on a run of 50,000 pieces, was estimated at $15 each.

Cutaway shows part after removal of the pushed-down track.

Semipiercing raised the hub and pushed down the cam track.

Fine blanking, the alternative method, produced the parts from material stock 0.375 in. thick instead of 0.625 in. Tooling cost was $21,000.

In the first tool, the hub was raised by 0.300 in. In the second tool, the cam track was raised on the opposite side by semipiercing. At the same time, the D-hole was pierced. The locking-screw hole was then drilled and tapped, and the extruded cam track projection was trimmed off on a lathe. Cost, including the secondary operations was $3.50 per piece.

FIGURE 3.6 A comparison between machining a part and fineblanking it. *(Fairfield Tool Co., Bridgeport, CT)*

FIGURE 3.7 A progessive die built around 1984. Progressive tooling produced this valve plate, which was coined several times, then blanked out. Alternative production methods would have included several metal-cutting operations, casting, or powder metal—all at higher cost.

assembly pins, the progressive die can chamfer or counterbore both sides of the workpiece, make larger angle bends, and make cutouts and holes that are too numerous for a single station.

A wide variety of metals with good cold-forming characteristics lend themselves to the process. These include carbon and alloy steels, stainless steels, brass, aluminum, copper, bronze, and monel. Other materials can also be processed depending on their thickness, hardness, and configuration. European steel producers have introduced steels with grain structures that are more uniform in all directions. Fineblanking companies seek this quality because it improves cold-forming properties, so they have imported these steels until a domestic source complied with their needs. Some of our smaller mills are now making a superior grade of steel for them. See Figure 3.8 for a material suitability chart.

Components traditionally considered too heavy for conventional stamping (in excess of ½" thick) can be fineblanked at a definite cost saving over machining. Even the heaviest fineblanked parts have perpendicular, sheared edges, and straight-walled holes with superior finishes. When the part function requires full-bearing side walls, expensive milling, reaming, or grinding operations are not necessary.

On the subject of edge taper, there are a few things to consider. When working with tool steels, alloyed steels or exotic metals, all tolerance limits previously stated must change. Tolerances and minimum hole sizes increase slightly. Tests will indicate what to expect from a specific material. And sides are not perfectly square, as stated. They can have up to a one degree taper. That means a piece of metal ¼" thick could have as much as 0.004" taper. For most functional purposes, this small taper is not sufficient to require machining to eliminate it. The taper will be larger on outer surfaces than inner contours and holes. Compared to conventional stamping, which shears only about 30% deep and has a possible 30-degree taper from that point to the bottom, with a jagged edge on the remainder of the hole, the fineblanked surfaces are generally accepted as they come. Figure 3.9 shows a comparison of the two methods.

Die roll, shown on Figure 3.10, is also experienced. It is a slight rounding of edges, which occurs on all stamped parts. It happens on the side, which enters the die opening. So, if you must have a sharp edge, make it the punch side of the part. If the design cannot accept any die roll at all, simply start with a thicker workpiece material and grind or mill it flat. The harder the workpiece material, the smaller die roll you will have. And the thicker the workpiece material, the larger the die roll will be. A part having a die roll Y deep, will have a curvature extending toward the inside of the part for about a distance $2Y$.

As in all stamping operations, sharp corners should be avoided in tool design. This will prolong tool life. On a 90-degree corner, the radius should ideally be a minimum of 10 to 15% of material thickness. On corners with angles larger than 90 degrees, 5 to 10% of material thickness will suffice. On acute angle corners, 15 to 20% of material thickness is desirable. Tears might be caused by too small a radius. Increasing the radius won't cure all problems, but it will eliminate many, and in any case, it can't hurt. The solution might rest in changing the material to one that is more malleable. Too much punch clearance or a chipped punch can be a problem and so can a V-ring that isn't doing its job. When beginning a new job, examine the first pieces for tears on the edges and corners. This is the time to solve design problems, at the beginning of the production run.

Designers can enjoy opportunities beyond the scope of conventional stamping. For instance, they can fineblank holes as small as 40% of the material thickness, and interior openings can be located closer to each other or to edges without secondary operations. Designs with intricate or fragile cross-sections can be reproduced with maximum integrity and consistency.

The allowable variation in flatness is generally 0.001" per inch. Several factors affect this characteristic: the grain direction, the part configuration, and (of great significance) the flatness of the fed material. The machine setup must include a good roll straightener with the feed mechanism. Any fineblanker will present a list of suitable steels that will enhance the process.

NOT RECOM-MENDED	POOR	GOOD	EXCELLENT	MATERIAL	
			-O H	1100	ALUMINUM
			-O H2	3003	
	H 38 UP TO .187	H 36	H-32 H-34	5052	
		T-6 UP TO .125″	-O	6061	
	T-4 T-6	T 351 UP TO .100″	-O	2024	
			-O	5086	
	T-6		-O	7075	
LEADED		¾ UP TO .125″	¼ TO ½ HARD	**BRASS**	
SPRING RB 89-96		¾ HARD	¼ AND ½ HARD	**BRONZE**	
			¼ AND ½ HARD	**MONEL METAL**	
			¼ HARD RB 50-75	**NICKEL SILVER** **65% COPPER 18% Ni**	
			ANNEALED	**BERYLLIUM COPPER**	
			UP TO ½″	**SILVER**	
ABOVE RB 90	#1 TEMPER	RB 70-85	RB 50-75 #3, 4, 5, TEMPER	C-1010-1020	CARBON STEEL
		FULLY ANNEALED		C-1025-1035	
		SPHEROIDIZED ANNEALED		C-1035-1095	
		UP TO .375″	SPHEROIDIZED ANNEALED	8615-8620	ALLOY STEELS
	ABOVE .250″	SPHEROIDIZED ANNEALED		4130, 4140, 4620	
	¼-½ HARD		FULLY ANNEALED UP TO .090″	200, 300 SERIES	STAINLESS STEELS
			FULLY ANNEALED	410	
		FULLY ANNEALED UP TO .150″		416-420	
		UP TO 2½% SILICON FULLY ANNEALED		**SILICON IRON**	
		A-2 UP TO .032″		**TOOL STEEL**	

SUITABILITY OF MATERIALS FOR FINEBLANKING*

FIGURE 3.8 Material suitability chart. (*California Fineblanking Corp., Huntington Beach, CA*)

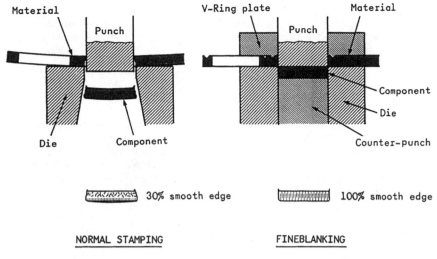

FIGURE 3.9 Stamping vs. fineblanking.

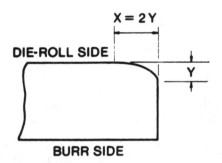

FIGURE 3.10 A die roll.

Fineblanking work hardens the workpieces. If the hardness is tested along the sheared surface of a hole, an increase in Rc will be noted as the test proceeds from the die-rolled edge toward the middle of the hole. Then, it decreases slightly.

Gears are commonly made by this process. For best results, the largest possible radius should be provided at the tooth peak and root. At their pitch diameter, the tooth width should be a minimum of ⅔ material thickness.

Offsets and bends must be in the proper relation to the part contour and feed direction. If there is any doubt what this means relative to your component, consult your fineblanking vendor for advice; most offsets and bends can be satisfactorily positioned in more than one place or in more than one direction.

When countersinking with a single-station tool, a bulge will be raised adjacent to the countersink. However, countersinking with a progressive tool will not create a bulge. Countersinks and chamfers, too, are not limited to round holes. Progressive tooling must be used if outside contour chamfering is desired.

Semi-piercing is another benefit of the fineblanking process. This step uses a punch that does not break through the part's material. The results of semi-piercing can be weld projections, self rivets, feet, contact points, pins for stacking, or female cam tracks.

If parts made by another process are not performing satisfactorily or are too expensive, consider a change to fineblanking. Remember though, it takes time to learn about a new process. Rather than take a chance of overlooking possible benefits, it would be a good idea to call on the experts for advice.

QUALITY

In fineblanking, the tolerances depend mainly on the accuracy of the tool. Other factors that have an effect are: material thickness and hardness, part configuration, and the tensile strength of the material. Tolerances can be held closer on inside contours than on outer surfaces. Some steels are better for fineblanking than others. Tolerances on cutouts, holes, and their spacing can always be held tighter on the superior steels, which have uniform structure in all directions and good malleability.

COST

The average minimum volume of parts for the fineblanking process should be around 5000 per month. And yet there are situations where because of part complexity with long machining cycles, as few as 500 parts per month can sometimes be economic to run. CNC operation coupled with equipment for quick tool changing makes this possible. If parts made by another process are not performing well or are too expensive, consider a change to fineblanking.

The cost consideration is significant. Not only are these machines more expensive than conventional presses, but the tooling is also more expensive. Remember that these machines should not be considered in the context of stamping. You should take account of the capabilities of the process. Compare what it would cost to process the parts through other equipment and consider the final costs. That would include the price of the other capital equipment, the floor space required, the operators, the energy use, and the cost of both scrap and extra inspection.

This is a good place to mention a job that I witnessed. A fineblanking press was running at a speed of 30 strokes per minute and consistently producing good sheetmetal parts. When I picked up a few finished pieces to check the surface condition, I was amazed to find that a hole in the $\frac{3}{32}$"-thick steel parts was tapped. At the bottom of each stroke, while the ram was stopped to change direction, a microswitch engaged a small tapping machine bolted to the ram and produced threads in that hole. This is an example of eliminating secondary operations.

PROCESS ADVANTAGES

The conventional stamped holes and cutouts must have a minimum dimension between holes and to outside edges, of material thickness when using material of 30,000 psi strength and as much as twice material thickness as the material strength approaches 90,000 psi.

The main benefits of fineblanking are smooth edges, flatness, tighter dimensions, and a finished part at each stroke without requiring secondary operations. When parts made by other processes are not consistently identical, performing properly, or simply costing too much to make, consider a change to fineblanking. Remember that this is a very different process and has to be studied carefully. The best step you can take is to confer with the experts—either the machine manufacturer or one of those companies doing jobbing. These sources of information can guide your final part design to best accommodate the process.

CHAPTER 4
COLD FORGING: COINING

PROCESS

Cold forging is big business. In 1980, Precision Metal Magazine stated that at that time, closed die forgings were going to top $2.5 billion for the year. I'll try to make you aware of what cold forging can do for you by making parts near net shape. This chapter will make you ask yourself what parts you use that are near net shape.

You make yourself more valuable to your company by having an understanding of the fundamental process of forging so that you can know if any parts fall into that category. If you can identify parts whose dimensions fall within the ability of the cold-forging process, that's all you have to know.

You might not know how to design the die, what kind of equipment to use, or how metal moves under pressure. But if you know something about the process, just ask yourself if it could make your part near net shape. You don't necessarily have to take the responsibility on your shoulders. If you think it's possible, let the cold-forging people look at the parts and tell you what they think of the possibilities.

Cold forging is a process which forms metal at room temperature to a desired shape by forcing a lubricated slug into a closed die under very high pressure. The volume of the slug is very important. Automatic sorters are sometimes used to separate the blanks by their gram weight into over, under, and correct gram weight. Usually a slight adjustment of the press can be made to compensate for the slight change in gram weight, as long as the parts in any one run are the same.

The force could be generated either by hydraulic or mechanical presses. The process deforms the metal plastically. In this process, the metal is reformed in continuous, unbroken lines that follow the contour of the part, increasing resistance to shear and breaking. Notice the lines in the deformation ratios of Figure 4.1. The lines are forced together, making the part much stronger. Even an untrained eye can see the deformation ratio increases as the job gets done. The bottom photo is just about twice the stiffness of the top one.

In cold forging, the slug (blank) contained in the die is shaped by compressive force of a punch. The metal can be forced back along the punch as a backward extrusion, it can be forced forward through the die as a forward extrusion, or it could be a combination of both.

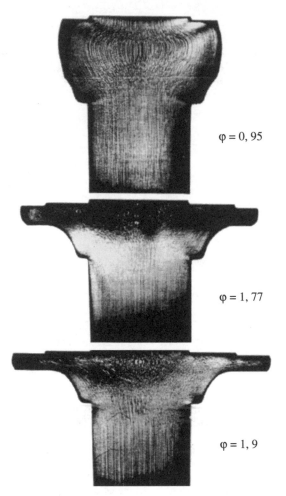

$\varphi = 0, 95$

$\varphi = 1, 77$

$\varphi = 1, 9$

FIGURE 4.1 Deformation ratios of a part under construction.
(Schmid Corp. of America, Goodrich, MI)

Frequently, the blank must be annealed before the first press operation or after it. Shearing the blank may harden the material. This might be fine for the finished part, but it might make the part too hard for further forming in the die. Therefore, it might require annealing before another press operation on it.

Cold extrusion is a high-production operation that can be automated to produce complete or near-complete parts without generating waste material. These cold extruded parts begin as raw material in the shape of rods, bars, coils, or billets. The individual pieces, called *slugs*, are produced by shearing, sawing, blanking, or any other convenient method.

Because blanks are placed into closed dies, the blank volume is important. Automatic sorters can be used to separate blanks by their gram weight into over, under, and correct weight. A slight adjustment of the press can compensate for slight changes in weight, as long as all slugs in each run are the same weight.

The part might, in some cases, be used just as it comes from the press, or further secondary processing (such as machining or annealing) might be required.

Frequently, annealing must be done during this process. Sometimes the blanking step might harden the piece. It will then require annealing before further forming can occur. Sometimes two forming steps are necessary with an intermediate annealing step.

LUBRICATION

When pressure is applied to a slug in the die, there is a considerable force that causes heat and resistance to movement. That is why the slug must be lubricated. The inside surface of the die must be protected. The protection is provided by dipping the slugs in a solution of zinc phosphate. It is important to give the slug a definite amount of lubricant. Too much would take room away from the slug and it wouldn't flow properly to fill the die. Too little would prevent a free flow, and damage to the die might result.

As a lubricant, zinc phosphate is far superior to oil for this process because it is dry and nonstaining. It doesn't soil hands or equipment by picking up dirt, as oil would, and it also prevents rust. Sometimes a part requires another press operation and it often gets a second zinc phosphate dip.

MATERIAL

Formerly, only low-carbon steels were cold forged. Now, almost any ferrous material can be processed, whether it is stainless, an alloy, or a high-carbon steel. Most nonferrous metals can also be cold forged. A cold-forged part in low-alloy steel can be stronger than a part machined from a high alloy. The ability to substitute low-alloy steel for high can save considerable material costs.

For example, AISI 4130 is easier to forge than AISI 4140, and as forged, will provide the same end specifications. General Electric originally specified stainless steel for a refrigerator door cam, but found it difficult to make. They changed the material to 1021 steel, cold forged it, and hardened the surface to make it wear resistant. The part has worked well for several million refrigerator doors. Don't allow material specifications to stop you from thinking about cold forging. See Figures 4.2 and 4.3 for samples of the process.

FIGURE 4.2 Cold-forged parts. (*Cousino Metal Products, Inc., Toledo, OH*)

FIGURE 4.3   Cold-forged parts. *(Cousino Metal Products, Inc., Toledo, OH)*

Automobile companies used to make parts from materials they were accustomed to. Now they investigate any possible material change because they use so much of each material. Many times, changing the manufacturing process to cold forging eliminates the need for heat treating.

Copper is a good metal for cold forging because it can be shaped into many configurations with much less tonnage than the ferrous metals. Part of the high cost of copper material can be saved when parts are cold forged from copper wire or bars, instead of machining the parts from solid stock and wasting the chips. The most common types of copper being cold forged are electrolytic copper and oxygen-free, high-conductivity copper. Copper parts are often cold forged to finished dimensions without requiring secondary operations.

Some aluminums are too brittle and they break before forming. Those successfully processed are series 1100, 2014, 3003, 6061, 6351, and 7075. Some huge aluminum forgings are being cold forged for airplanes and thousands of small ones are also being made.

GENERAL INFORMATION

Cold-forged parts pressed in moderate-sized machines can weigh as little as a few ounces to as much as 60 pounds. A few gigantic presses can handle larger pieces. Parts can be round, flat, long, short, or almost any shape. Various wall thicknesses can be accommodated. However, vendors prefer that the bottom of a cylinder be thicker than the walls, as is the situation with projectile cases.

Grooves and undercuts are normally machined in after the forging is pressed. Concentric parts are commonly pressed, but eccentric parts can be made as well. Critical dimensions can be held in this process. The accuracy is machined into the die that can hold the accuracy for hundreds of thousands of parts.

Depending on part complexity, lead time can vary from 4 to 24 weeks. Normal lead time is about 16 weeks. This, of course, is the time required to make dies. Because a die to make 100 parts could cost 50 to 100% as much as a die to make 1,000,000 parts, this process should not be considered for short runs. The initial tool cost is the customer's. After that, maintenance is generally the vendor's responsibility.

TOLERANCES

The weight of slugs can be held to ±0.5 grams. Flatness can be held to 0.005" across a 5" diameter surface. Tolerances of ±0.003" can be held on diameters and wall thicknesses of parts up to 2 inches in diameter. Center distances between the holes can be held consistently close.

The wall thickness of parts formed by extruding the material between the punch and the cavity of the die can be held closely along its full length. The inside and outside diameters will also be closely held. Tolerances of ±0.003" can be consistently held for the diameters and wall thickness.

Repeatability of cold-forged parts can be maintained for thousands of cycles; whereas in machining, a dulling tool will vary dimensions. The surface finish of cold-forged parts can be held to 30 microinches. With copper, it can be held to 10 microinches.

Sometimes dies for cold forging have to be developed by testing and changing. But when the desired configuration is developed, the parts will conform closely to the die and maintain dimensions for many parts (Table 4.1).

TABLE 4.1 Physical Properties of Steels Before & After Cold Forging.

	Before forging			After forging		
AISI	Rock B	Tensile	Yield	Rock B	Tensile	Yield
1010	58	52500	35950	88	91250	85300
1020	57	58000	40250	92	104800	101166
1040	81	74700	49750	94	110150	97875

COLD FORGING VS. COINING

Just about everything said about cold forging is true about coining. In addition, coining can be the forming of metal surfaces by a compression force sinking or raising the product material to the relief features of a tool and die, as in making coins (money).

Recently, an innovative coining vendor, Cousino of Toledo, Ohio, deliberated over the possibility of making a copper part shaped like a nose cone. Four other vendors had rejected the job, declaring it impossible to hold the tolerances. The inside and outside walls had to be concentric within 0.002" and the tapered diameters required were mathematical equations (polynomials) whose dimensions had a total tolerance of 0.002" each.

Cousino was able to complete this job successfully by using two sets of tools for two separate operations. The first set of dies moved about 85% of the material and the second set of dies moved the rest and "coined" the finish. Even with the cost of two sets of tools, this process was cost effective when compared with hydroforming plus machining, or orbital forging plus machining. Very often, the bottom line (cost) is a reflection of the vendor's ingenuity.

The work table of a coining press is smaller than that of a forging press. This permits less deflection of the ram support, sometimes called *breathing*. With less "breathing," smaller dies can be used and greater accuracy can be expected. This is one reason why you are urged to consult with vendors. If they can figure a way to make your part by coining or cold forging, there generally is a cost savings in there for you.

ADVANTAGES OF COLD FORGING

The major advantages are improved quality and lower production costs. The quality improvement is derived from the higher physical properties of the process: and the` unbroken grain flow, which improves toughness, hardness, strength, and fatigue resistance. The costs are reduced because secondary operations, such as heat treatment or trimming, generally are not necessary and improved metallurgical characteristics frequently permit the use of less-costly materials. Smoother finishes, closer tolerances, and greater uniformity of dimensions from part to part are obtainable.

Cold extrusion is virtually a chipless process with very little material waste. Less material is consumed per finished part than in machining or hot forging.

APPLICATIONS FOR COLD FORGING

Cold-forged parts are used in most industries.

Automotive:

- Bearing cups
- Differential pinions
- Piston pins
- Disc brake pistons
- Truck wheel nuts
- Disc brake caliper keys
- Valve spring retainers
- Tilt steering joints
- Transmission gears
- Spark plug shells
- Front wheel spindles

Construction:

- shock absorber tubes
- Bolts and nuts

Controls:

- Magnetic pole pieces for servo-mechanisms

Energy:

- Roof supports in coal mines
- Sucker rods for oil-well pumps

Farm equipment:

- Similar to automotive for many parts (such as gears)
- Ball joints for tractor hitches

Household appliances:

- Refrigerator hinges
- Compressor cylinder heads

Motors-electrical

- Magnetic pole pieces

Ordnance:

- Shell cases
- Fuse components

Sporting goods

- Golf club heads

GLOSSARY

Backward extrusion A process in which the metal is forced to flow in a direction opposite to the applied force, usually around a punch in a closed die.

Blanking Producing a preform or slug by a stamping press, where the part produced is to be used in extrusion or coining. The part is called a *blank*.

Closed die forging The forming of metal within a closed cavity formed by two halves of a die.

Coining The forming of metal surfaces by a compression force sinking or raising the product material to the relief features of a tool or die, as in (money) coins. The coining machine is often used to stamp parts to avoid encountering the deflection of a (larger bed) stamping press.

Cold forging The plastic deformation of metals at room or ambient temperature into precision parts, by extrusion or coining.

Combination extrusion A combination of more than one flow direction taking place simultaneously or progressively in a single-press cycle.

Drawing The process of reducing the cross-sectional area of, or producing a shape from, a rod, bar, tube, or wire, by forcing or pulling it through a die.

Extrusion The process whereby metal is forced by compression to flow in or through a die.

Forward extrusion A method by which the metal is forced to flow in the same direction as the applied force.

Heading An upsetting operation in which the enlargement of the cross-sectional area is confined to the end of the stock.

Hot and warm forging A forging operation in which the material is heated above room temperature before being placed in the forging die.

Impact extrusion An extrusion operation carried out under impact conditions. Generally, the term is used for the extrusion of nonferrous components under mass-production conditions in a fast-stroking press.

Indenting The process of forcing a punch into a workpiece to produce a blind hole of the required shape.

Ironing The process of smoothing (and possibly also reducing the wall thickness of) hollow components by forcing the material through a die with a punch.

Lateral extrusion A plastic flow condition in which the product material is made to flow at an angle to the applied force.

Nosing The process of using dies to close or reduce the opening in the end of a tubular component or cylinder.

Piercing The process of producing a hole through a thickness of metal by forcing a punch through the metal to form a hole the shape of the punch.

Preform or slug A piece of metal cut from coils or bar stock by sawing, impact cutoff, cropping, or machining, which is placed in a closed die for cold forging.

Sizing Usually, a final operation in forming a component to limited-size tolerance not obtainable in preceding operations.

Upsetting The compressive deformation of metal by a blow or steady pressure in the direction of the axis of the stock in order to enlarge the cross-sectional area of the entire length or just part of the length.

CHAPTER 5

ORBITAL COLD FORGING

PROCESS DESCRIPTION

Orbital cold forging represents a considerable increase in application possibilities for solid blank forming. The goal is to obtain a finished workpiece with the highest possible degree of accuracy in a single forging operation. During the process, the upper die makes contact with a rolling motion over the total workpiece surface. A characteristic of this process is an extremely high degree of deformation with relatively little force.

Both a fixed and a movable die are used. The fixed die is in the lower portion of the vertical press and it contains the slug (workpiece). An orbiting upper die, held at a slight angle (1 to 2 degrees), almost noiselessly forms the slug into a workpiece as the slug is hydraulically raised in the lower die. The quiet operation results because the material is displaced by constant pressure, rather than by a sudden hammer blow. The upper die can be made to orbit in any of four patterns. The operator simply dials in the pattern that he or she feels will be best for the job: Planetary, orbital, spiral, or straight line.

This process forms a cold-metal slug between two dies, with the upper die moving in an orbital motion. By means of the orbital motion, the forming force is concentrated in a small area of the upper die. This area is constantly shifted in the selected pattern across the entire piece part surface. This production method is so promising because accurate piece parts can be formed with a small amount of energy and little waste of material (Figures 5.1 and 5.2).

The dies are simple in design and low in cost. For this reason, it is profitable to forge parts even in small quantities—sometimes fewer than one thousand parts. The flash is usually removed by blanking. Depending on form and application, these parts can be finished by chip-making secondary operations.

The lower die presses the blank against the orbiting upper die. The orbital angle is normally about 2 degrees. This motion results in a progressive forging to final shape usually in 10 to 20 cycles and 5 to 10 seconds.

As mentioned previously, the workpiece is forged between the upper and lower dies. The lower die is mounted securely in the ram table, which travels in an upward motion. The

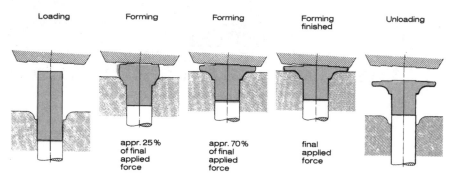

FIGURE 5.1   Sequence of operation. *(Schmid Corp. of America, Goodrich, MI)*

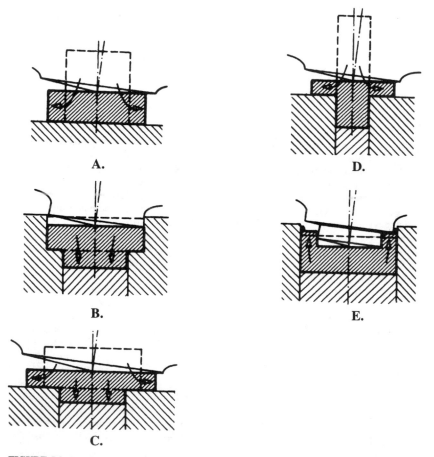

FIGURE 5.2   Possible types of forging: A. Upsetting. B. Forward extrusion. C. Combined upsetting and forward extrusion. D. Upsetting for flange. E. Backward extrusion. *(Schmid Corp. of America, Goodrich, MI)*

upper die is mounted in the orbital bell in the top half of the press. During the forming process, the upper die orbits in the preset pattern. In the sequence of operations, the slug is placed in the lower die. The ram then proceeds upward under high speed until it reaches an adjustable preset "forging stroke length."

At this point, the ram slows to the adjustable preset forming speed. The upper orbiting die comes in contact with the material and the forming starts. When the press reaches its closed (top dead center) position, the ram delays for an adjustable dwell time. This permits some finishing orbits of the upper die to finish "set" the workpiece. The ram then retracts, the ejector is raised to remove the part from the lower die, and the finished part is ejected. Sample orbital forged parts are shown in Figure 5.3.

In conventional forging, the pressure is applied simultaneously over the entire surface of the part being formed. This creates undesired sliding friction, which restricts radial flow of the material. Stress is therefore highest at the center of the part and decreases toward the edges. This sliding friction can generate a maximum stress, which is several times higher than the yield point of the material being formed.

In contrast, with an orbiting die in rolling contact against the part surface, there is little frictional force. The metal is allowed to flow radially. This keeps the surface pressure only slightly higher than the material's yield strength.

The orbital angle substantially determines the required force. If the angle is increased and the forming force remains the same, the time required for forming becomes shorter because there will be fewer orbits. Reducing the orbital angle from 2 degrees to 1 degree will increase forming time by 50%. Consequently, the maximum angle should be selected to increase production rates.

FIGURE 5.3 Varied products of orbital forging. *(Schmid Corp. of America, Goodrich, MI)*

ADVANTAGES OF ORBITAL FORGING

The orbiting die type of cold forging offers such advantages as short production times, reduced material requirements, increased material strength, smooth finishes, consistent accuracy, and close tolerances. And if large parts are to be formed, the die investment cost would be considerably less than with conventional cold forging.

A 4" diameter steel piece can be forged with an orbital press of 200 tons capacity, whereas a conventional press would require a 2000-ton capacity. Sometimes thin or complex-shaped parts can be successfully forged with an orbital press, but to do so with a conventional press would be difficult, if not impossible. That is because the required compressive load could exceed the strength of the tool.

The finish and accuracy of this process is superior to any other type of forging. By controlling the weight of the slugs, an operator can produce forgings almost without any flash. There is less die wear and unusually thin sections can be formed.

DIMENSIONS AND TOLERANCES

The surface quality of parts is governed mainly by the surface condition of the die. Using polished dies, this process can produce finishes down to 8 microinches. In conventional forging, repeatability and basic accuracy itself is often influenced to a degree by deflection in the die. Consequently, the ability to cold form a part with $\frac{1}{10}$ of the normal force is a significant and worthwhile improvement.

In 1985, a large defense contract was lost because the deflection in a huge conventional press could not maintain tolerances consistently. A competitor was successful by substituting the orbital forging process for the huge press. The technique of applying an orbiting force caused very little deflection. Reproducibility of dimensions in orbital forging ranges between 0.002 and 0.004".

The grain flow lines in the workpiece are favorably influenced by the cold-forming operation, which increases surface strength by at least 50%. It is therefore, possible to select less-expensive, low-alloy steels, which, after cold forming, display tensile strength values of high-alloy steels.

MATERIALS USED

Materials suitable for orbital forging encompass a wide range of those that are both ferrous and nonferrous. Any type of steel, either carbon or alloyed, can be formed. They must however, be ductile and have a spheroidized structure. This type of annealing makes the microstructure more conducive to severe forming. The ductility of steels depends on the content of carbon and alloy elements. The higher these contents, the lower is the ductility. Nonferrous materials, such as aluminum and copper, and high-nickel alloys are suitable for forming, but they must be in a low-temper condition. Also here, as in the case of ferrous materials, the ductility decreases with increasing alloy content.

The following is a list of commonly worked materials:

Steels:

1. 1008-1024. Soft. ¼ to ½ and full hard.
2. 1030-1095. Only fully annealed, spheroidized condition.
3. 4130-4140-8620-8630-8640. Same as 2.

Stainless steel:

1. 302-304-305 and 405. All rated A-very good in fully annealed condition.
2. 301-303-309-310-316-347. All rated B-good to bad. Use only fully annealed.
3. 403-410-442. Same as 2.

Aluminum:

1. Nonheat-treatable alloys (such as 1100-3003 and 5052 series) with excellent results.
2. All heat-treatable alloys (such as 2014-2024, 6061, and 7075) only in 0 temper with exception of 2024.

Brass:

1. Only brass without lead.
2. CA-230-260-270.
3. CA-235-265-268. Very good in soft, ¼ or ½ hard.
4. CA-237-263-274. Hard condition.

SLUG PREPARATION

The slugs for orbital forging can start from many forms. These include round or profiled rods, tubing, stamped blanks, forged blanks, or preformed slugs (Figures 5.4 and 5.5). Each slug must be lubricated. A dry lubricant (consisting of a phosphate coating, sometimes followed by molybdenum disulfide applied to the slugs before forming) has been successful. This not only maintains the press area cleaner, but also eliminates the possibility of a "hydraulic lock-up" in the tool, which can develop when using lubricating oils. Slugs can be fed manually, by gravity slide, and by robot.

THE ORBITAL PRESS

Orbital forging represents a significant technological breakthrough in the cold-forming process. It produces net-shape and near-net-shape components. These presses can be fully automatic and can function within a manufacturing cell directly tied to other machine tools. With the addition of available specially designed in-feed and out-feed material-handling robots, this system is very flexible and can handle long- and short-

FIGURE 5.4   The slug and its forging. *(Schmid Corp. of America, Goodrich, MI)*

FIGURE 5.5   Products of orbital forging: A. Eccentric screw. B. Cam disc. *(Schmid Corp. of America, Goodrich, MI)*

run jobs. With the added benefit of hydraulic tool-clamping equipment, an operator can change from one production run to another with the assistance of one helper, generally in less than 30 minutes.

There are currently three sizes of orbital presses in production. They all look like Figure 5.6. There is a 200-, 400-, and a 630-ton machine. They are fully hydraulic, and the shut height of each machine is adjusted by a mechanical stop. This ensures repeatability for dimensional control of the parts being forged.

The ram table and the upper half of the press are guided by four large cylindrical posts. This ensures concentricity of the upper and lower dies during forging. The orbital unit consists of the orbital head and gear drive. Within the head rests, the orbital bell holds the upper die (Figure 5.7). The orbital bell rides in a special spherical bearing that is designed to absorb the high forces created during forging. The gear drive rests on top of the orbital head. All four orbital motions are controlled from within this unit. The different movements (patterns) are generated by the combination of two eccentric rings, which are driven by a gear box.

Production rates vary from 5 to 15 parts per minute. This particular phase of cold forming produces near net shape more so than any other type of cold forming. Whenever dies are subjected to high pressures, they should be lubricated. In this process, a minimum of lubrication should be used. A Bonderite or Bonderlube would be fine.

All machine functions are controlled from a panel that is mounted on the right side of the press frame. This contains buttons, dials, and displays, and is divided, for simplicity, into two fields. The panel has a transparent, plastic cover that can be securely locked.

Little maintenance is required. It is basically limited to the periodic changing of oil filters, hydraulic and lubricating oil. A diagnostic system displays error codes and messages, thus facilitating troubleshooting. All danger areas are provided with covers and are electrically monitored. It is possible to add light curtains on all presses.

Neither impacts nor vibrations are produced by the orbital forging process. Noise produced by the drive and control devices is reduced to a minimum by a sound absorbing

T200

Orbital Cold Forging Press
force capacity 2000 kN
production rate 4–15 parts/min
max. workpiece diameter:
approx. 100 mm

T630

Orbital Cold Forging Press
force capacity 6300 kN
production rate 4–10 parts/min
max. workpiece diameter:
approx. 170 mm

FIGURE 5.6 The smallest and largest orbital forging presses. *(Schmid Corp. of America, Goodrich, MI)*

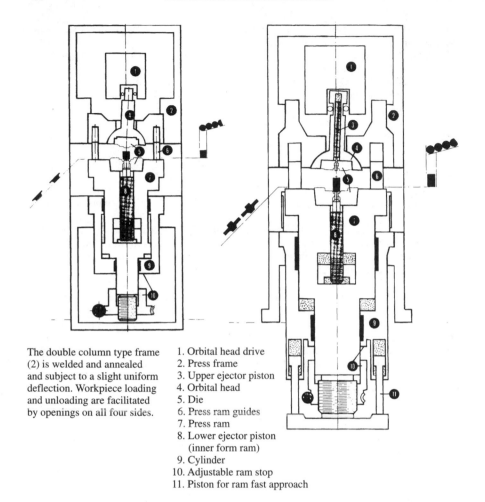

The double column type frame (2) is welded and annealed and subject to a slight uniform deflection. Workpiece loading and unloading are facilitated by openings on all four sides.

1. Orbital head drive
2. Press frame
3. Upper ejector piston
4. Orbital head
5. Die
6. Press ram guides
7. Press ram
8. Lower ejector piston (inner form ram)
9. Cylinder
10. Adjustable ram stop
11. Piston for ram fast approach

FIGURE 5.7   Press construction. *(Schmid Corp. of America, Goodrich, MI)*

enclosure. Leaking oil is collected by a sump provided with a float switch. Pressure-sensitive transducers initiate immediate retraction of the ram if anything hazardous to the dies occurs. For instance, if an incorrectly sized slug is present in the die.

It appears that orbital cold forging is still in its infancy with regards to industry awareness, but is destined to gain wider acceptance in the near future.

CHAPTER 6

HYDROFORMING

PROCESS DESCRIPTION

This process is sometimes referred to as *fluid forming* or *rubber diaphragm forming* because of the rubber, fluid pressure-forming chamber, which acts as the upper die member. The process was developed just after World War II. It is a lower-cost method of making small quantities of deep-drawn parts at less expense than the conventional method of deep drawing in a hydraulic or mechanical press.

Hydroforming uses a unique forming process that offers many advantages. The fluid-filled forming chamber serves as the upper blankholder and female die element—a universal die that accommodates any shape. In the lower portion of the press is a punch attached to a hydraulic piston and also a blankholder ring (draw ring), through which the punch moves. Normally, a thick neoprene pad is cemented to the rubber diaphragm to protect it.

At the start of the machine cycle, the top of the punch is level with the top of the draw ring (Figure 6.1A). The blank to be formed is on the blankholder ring and the punch just touches it. The forming chamber is lowered until the wear pad hits the blank at which time a small hydraulic force is applied to the blank. This force can be as high as 5000 psi (see Figure 6.1B).

The punch moves upward under increasing pressure (up to 15,000 psi). It pushes the blank into the flexible die member, forcing the blank to assume the shape of the punch. The diaphragm actually wraps the blank around the punch. All during the cycle, the blank is held securely by the two opposing forces (Figure 6.1C).

The workpiece metal was not stretched like a stamped part; it actually flowed. To complete the cycle, all pressure is released. The forming chamber rises and the punch is retracted from the completed part (Figure 6.1D). A Dayton Rogers or similar trim tool can be used to remove unwanted material, such as the flange. Normally, the radius adjacent to the flange is formed conveniently large to facilitate this cut-off (Figure 6.1D).

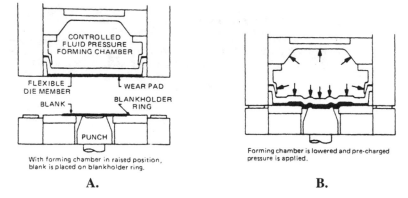

A.

Forming chamber is lowered and pre-charged pressure is applied.

B.

As punch moves upward, blank flows and wraps around punch.

C.

Pressure is released, forming chamber is raised, and punch is retracted from drawn part.

D.

FIGURE 6.1 Hydroform press operating cycle sequence. *(American Metal Stamping Assoc., Richmond Heights, OH)*

COMPARISON WITH CONVENTIONAL FORMING

In a conventional press, the blank is positioned on top of the lower die, then the draw ring is lowered to contact the blank. The upper die is then lowered to contact the blank and the part is drawn. Throughout most of the cycle, the blank is not touching one die. Sometimes this leads to tears or wrinkles because the blank is "out of control." At the same time, the part is stretching and work hardening. Figure 6.2 shows a popular hydroforming machine.

Often, more than one draw is required in conventional deep drawing, with annealing steps in between. In hydroforming, most asymmetrical parts can be made in one stroke without annealing. Because only a draw ring and a male punch are required for the hydroform, 40 to 50% of the usual deep draw die cost is saved. The hydroform setups are simple and fast because there is no die set. The tooling is self-centering and self-aligning. Practically all sheet metals capable of being cold formed can be hydroformed. This includes carbon and stainless steels, aluminum, copper, brass, precious metals, and high-strength alloys.

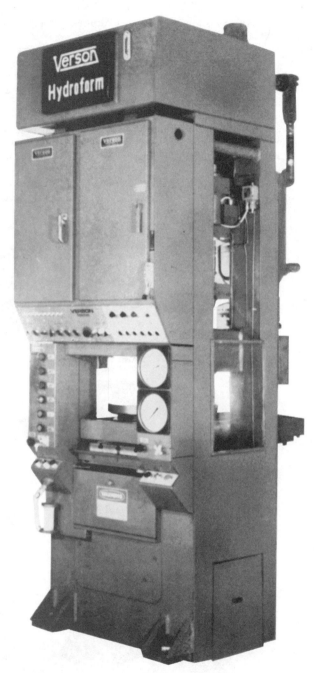

FIGURE 6.2 A popular hydroform machine. *(Verson Allsteel Press Co., Chicago, IL)*

When using a hydroform, hydraulic pressure in the forming chamber keeps the diaphragm in contact with one side of the blank while the punch is in contact with the other side of the blank. This keeps the blank in constant control and allows a smooth flow without scuffing, making stretch lines, or work hardening. After an R&D program in which a part is perfected during a spinning process, the spinning chuck can be used as the hydroform punch and make parts by this faster process. Because many draw rings are stored for ready use, you generally don't have to pay for one of them.

Irregularly contoured parts are easily formed using the hydroform process because matching dies are not required. Because hydroforming flows the metal, rather than stretching it, there is minimal material thinout. Sometimes this results in a worthwhile cost savings in material—especially when expensive alloys are used. There is a considerable savings in material for tools. Instead of expensive tool steels, this process uses cast iron or plain machinery steel—even for long runs. As a matter of fact, kirksite or cast plastics can be used for very short runs. Finally, if the thickness of the material is changed, no tooling modifications are needed for the finished part. See Figure 6.3. for typical parts.

FIGURE 6.3 Typical parts produced by hydroforming. *(Roland Teiner Co., Inc., Everett, MA)*

PROCESS EVALUATION

Although hydroforming enjoys distinct advantages over conventional deep drawing, certain basic factors should be evaluated before selecting it. The part should be analyzed to determine which factors create the most cost. Normally, for an order of 2000 parts or less, tooling will be the largest cost component. For an order of 20,000 parts, tooling and material costs will be about equal, with direct labor now creating a significant portion of the cost.

As quantities increase, the labor cost becomes more important. Accordingly, total manufacturing costs should be investigated before the process selection is made. That means you should evaluate the following expenses: material, tooling, die design, manufacture, development and setup, all labor (direct and indirect), and any finishing costs required.

A spinning might take 15 to 30 minutes to produce, whereas the same part probably would require less than one minute to hydroform. Generally, hydroforming can fashion a part without annealing it, but spinning the same part might require 1 to 3 anneals. Many odd-shaped parts can be hydroformed, but only symmetrical parts can be spun.

For price-comparison purposes only, the following generalizations are offered. On the average, spinning tools cost about $\frac{1}{10}$ of draw press tools. Hydroforming tools could cost about two or three times more than spinning tools. Sometimes there might be a small additional cost (perhaps as much as $300) for a draw ring. But generally a draw ring is available from the dozens kept in stock.

EQUIPMENT TERMINOLOGY

Hydroform presses are referred to by the size of the flat circular blank to be formed, the maximum forming pressure and the depth of draw. If a press can handle up to 8" in blank size, it is designated as an 8" machine. A 24" machine can handle up to a 24" blank. Draw depths can go up to 12". Forming pressures vary from 5000 to 15,000 psi.

If a machine can handle a 12" diameter blank, use a forming pressure of 15,000 psi and draw a blank 7" deep, it is referred to as a *12-15-7 hydroform press.*

HYDROFORMING DESIGN SUGGESTIONS

Try to design to the inside of the part because the part wraps itself around the punch. If you disregard metal springhack, the inside of the part should have the same dimensions as the punch. If the metal springs back, the spring back amount of stock can be removed from the punch in steps until the required dimensions are obtained consistently.

Remember that the open (flanged) end of the part will remain as thick as the starting blank. This can be important to the next step, whether it is welding, assembly, or whatever.

Zero draft can be used, if necessary. However, it is prudent to use a draft of 1 or 2 degrees to ensure longer tool life (less scratching and galling) and allow faster machine cycles.

In hydroforming rectangular pieces, the print might call for tighter corners than you get. At all times the minimum radius should be 2 to 3 times stock thickness. However, this radius can be decreased with a little extra work. Just lower the punch about $\frac{1}{16}$" and push the diaphragm down swiftly. This will produce a sharper corner.

An unusual advantage of hydroforming is in the simplicity with which certain drill jigs can be made. If a part requires drilled holes, it can be hydroformed first. A heavier piece of material is then hydroformed directly over the finished part. The finished part is then re-moved and the larger part has the holes laid out and drilled. Thereafter, the larger part is used as a jig to locate holes in the product parts.

Tight tolerances cost money whatever the manufacturing process used. So, common sense dictates loose tolerances, where possible. Draw depth tolerances are typically ±0.020", but ±0.010 can be held. Inside diameters of ±0.005" or even ±0.002" can be ob-tained. Often, it is the material that determines what tolerance to request.

Hydroform vendors should discuss the tools they are designing for your job. Although they must be responsible that those tools deliver the parts, as required by the drawing, you can acquire much knowledge by virtue of these discussions.

CHAPTER 7
ROLL FORMING

WHAT IT CAN DO

Most press or brake work, which can be done by roll forming, will save substantial money by that process. Roll forming has the additional advantage of being able to feed two different materials, one on top of the other, to form a single product. In this way, desirable properties, such as ductility, corrosion protection, tensile strength, and superior finish can be obtained.

Most metals can be roll formed, including anodized, prepainted, and plated materials. Thicknesses ranging from 0.005 to 0.188" can be fabricated into desired configurations. The roll forming process produces high-quality products. It produces consistent tolerances on both light- and heavy-gauge material, holding shapes and dimensions consistently also. Finished parts have an excellent appearance with no die marks—even on precoated or anodized materials. Figure 7.1 shows the type of rolls used to make the shapes, the products, in the photograph.

Applications are increasing steadily because designers are learning about the opportunities that roll forming presents. At this time, nearly every industry in the country makes use of the process. The "custom roll former" has had the experience necessary to help you change from other manufacturing processes to roll forming or to help you design new parts that are suitable for the process.

PROCESS DESCRIPTION

Roll forming is a high-production process of forming metal parts from coiled, strip, or sheet stock. This is done by feeding the materials longitudinally through successive pairs of rolls, each pair progressively doing its particular forming assignment until the desired cross-section is produced. On its way through the rolls, the metal has holes punched in it, it is mitered, notched, embossed, swedged, and edge rolled (Figure 7.2).

FIGURE 7.1 Forming rolls and samples of the process. *(Rollform of Jamestown, Inc., Jamestown, NY)*

Curving

Roll-formed shapes have uniform cross sections. This enables them to be easily bent. Shapes can be curved to uniform radii without wrinkles and without disturbing a prefinished surface. Designers should know that curves can be rolled into a continuous helix, then cut-up into sections. They should also know that material elongation is a characteristic that must be considered to prevent wrinkles. A material with a large coefficient of elongation is going to bend easier.

FIGURE 7.2 Roll form machining. *(Rollform of Jamestown, Inc., Jamestown, NY)*

Prepiercing

Prepiercing means punching a pattern of holes while the metal is in the flat strip before forming. It is a continuous part of the same operation of roll forming, but because it is a prelude to that roll forming and it eliminates additional and separate handling, it saves money.

Postpiercing

Postpiercing is punching a pattern of holes after the forming is completed. But again, it is part of a sequence of operations all of which are performed without handling. Sometimes this is preferred over prepiercing because the accuracy from the edge of the part would be greater.

Forming in Line

Tabs, stops, raised areas, and welding projections can be formed while the parts are being rolled. This process not only improves accuracy, but also saves money. Certain operations, which are normally secondary, can be eliminated by incorporating them into the roll-forming process. The material should be in as ductile a condition as possible because that permits crisp design, sharper corners, and easier bending. Bend radii specified by the mill should be followed when high-strength alloy steels, heat-resistant steels, titanium, and others are used.

GUIDES FOR ECONOMY

There are a number of design hints for those considering roll forming:

1. The depth of bend (channel) should not be too deep for its width.
2. If stiffness has to be increased, it can be done by design. Pressing ribs in flat sections or folding material over to double its thickness will increase stiffness.
3. If flat, wide areas are needed at edges, they can be formed by using stiffening ribs.
4. The leg of a channel or angle should always be longer than three times the metal thickness. This also applies when hemming or bending metal back on itself.
5. When using a piercing pattern that must be located a specific distance from the end, try to make this distance between ½" and 4".

TOLERANCES

These are only guidelines. If tighter tolerances are required, discuss them with your roll former. Dimensions in certain areas can be held closer than in others.

1. Cross-sections can be held to ±. 010" and angles to ±1 degree.
2. Straightness tolerances (bow or camber): 015" maximum deviation per foot of length.

3. Twist tolerance is ½" maximum deviation per foot of length.
4. Length tolerance for parts 0.015 to 0.025" thickness.
   ±0.020" on parts up to 36" long.
   ±0.047" on parts 36" to 96" long.
   ±0.093" on parts 96" to 144" long.
5. Length tolerance for parts .026" thickness and greater.
   ±0.015" on parts up to 36" long.
   ±0.030" on parts 36" to 96".
   ±0.060" on parts 96" to 144".

DESIGN HINTS

These suggestions might be helpful.

1. Use the largest bend radius permissible whenever using this process. An inside bend radius of less than the metal thickness will use more power and decrease the roll life.
2. Design the part to be as symmetrical as possible to preclude twist in the completed part.
3. Do not position holes, slots, and notches too close to a bendline to avoid distortion.
4. To maximize reduction of tool and part cost, do not request tolerances that are tighter than required.
5. The use of a CAD/CAM system for the design of roller dies will save time from the start to the completion of the schedule.
6. Figure 7.3 shows a mere handful of inexpensive, quality hardware made by this process.

The roll-forming process seems quite simple. The machinery and tools certainly appear that way. Nevertheless, there are many guidelines and technical solutions that designers should know to perform effectively. Figure 7.4 contains a list of unsophisticated rules that should guide the uninitiated.

EXPENSE

Normally, the roll former wants a minimum order of 10,000 feet. However, he'd probably run a 5000-foot prototype order. Rolls are inexpensive tools to make. Very often, the roll former can find the rolls you need in his tool room of stock dies. Dies are often available for all sizes. They can fabricate round or rectangular tubing, angles, or channels. The tubing wall thickness runs between 0.009 and 090". Angles run between 0.009" and 0.188" in wall thickness. The wall thickness of channels runs between 0.009 and 0.125".

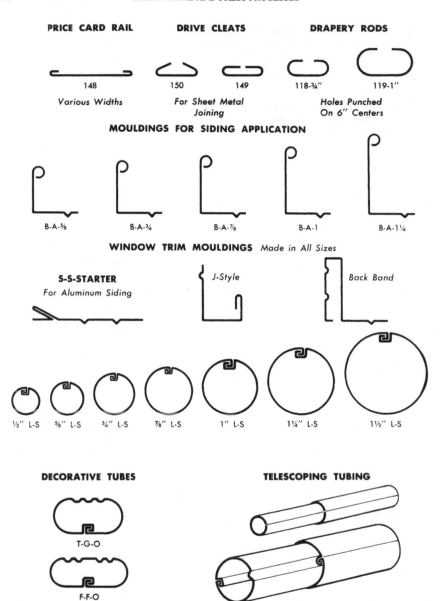

FIGURE 7.3 Simple hardware that is rollformed. *(Johnson Brothers Metal Forming Co., Berkeley, IL)*

Instead of this . . .   **Consider this**

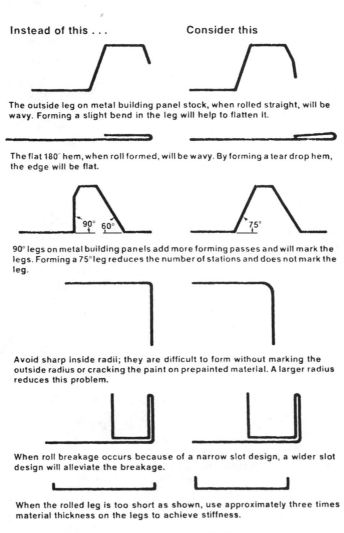

The outside leg on metal building panel stock, when rolled straight, will be wavy. Forming a slight bend in the leg will help to flatten it.

The flat 180° hem, when roll formed, will be wavy. By forming a tear drop hem, the edge will be flat.

90° legs on metal building panels add more forming passes and will mark the legs. Forming a 75° leg reduces the number of stations and does not mark the leg.

Avoid sharp inside radii; they are difficult to form without marking the outside radius or cracking the paint on prepainted material. A larger radius reduces this problem.

When roll breakage occurs because of a narrow slot design, a wider slot design will alleviate the breakage.

When the rolled leg is too short as shown, use approximately three times material thickness on the legs to achieve stiffness.

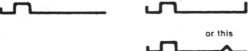

or this

When a section has a roll form on one end and a wide flat on the other end, put a leg on the end or a groove near the end to maintain straightness.

If a wide sweeping radius is impossible to control, put a bend in each end to keep it straight.

FIGURE 7.4 Design hints for roll forming. *(Samson Roll Formed Products Co., Skokie, IL)*

MACHINE-SHOP ENVIRONMENT PROCESSES

CHAPTER 8
TREPANNING

HISTORY

We've come a long way from the flat, one-piece spade drills of the early 1900s. After them came the carbon-steel helical drills, which prevailed until the high-speed steel helical drills that became popular around 1940. The carbon steel drills are still available for drilling wood and for those home mechanics who seldom use them. But in the last 50 years, machine shops have seen a large variety of changes to the common helical drills.

First, carbide tips were brazed to the two cutting lips. That permitted those drills to cut faster and through harder steel alloys than before. Then, the entire drill, shank, and cutting edges were made from a single solid piece of carbide. These, of course, had a special use where vibration was a problem. Replaceable tips (not brazed) were also in demand. Finally, drill manufacturers popularized titanium nitride coating (see Chapter 30) of the cutting tips.

The Hougen trepanning drills, described shortly, can also be titanium-nitride coated (TiN). However, those that are coated must be run faster than normal. Otherwise, the low-friction effect will allow the material to close-in (as occurs when drilling plastics). All these improvements permitted faster cutting and more effective drilling. These are fine accomplishments and we are still savoring them.

Trepanning, this new approach to drilling is here and doing very well. It will accelerate holemaking for any machine in the shop. Actually, this is not new. Trepanning has been used to put holes in sheetmetal since before World War II. In fact, if my sources are correct, German munitions manufacturers made cannons that way. They supposedly trepanned large-bore artillery pieces, then made smaller-bore cannons from the (slug) material cut out from the large gun barrels.

Trepanning is the standard method currently used by the sheetmetal trades for making large holes. When a carpenter installs a doorknob assembly, he trepans a solid piece of wood out of the door. Figure 8.1 shows what a small trepanning tool looks like.

FIGURE 8.1   Hole cutter for sheet metal. *(Hougen Mfg., Inc., Flint, MI)*

PROCESS DESCRIPTION

The only metalworking operation in which most of the metal is removed unnecessarily is drilling-through holes. Just think about it. Drilling a hole creates a pile of chips; a cylinder has been cut into small pieces. If that cylinder of chips could be removed as a solid unit by cutting only the circumference, the center core, about 75% of the volume, would simply fall away. Most holes, from ½" diameter to 6", can be cut out in this manner. It is called *annular tooling* or *trepanning*.

The word *annular* in this concept means relating to or forming a ring. These tools cut a ringed groove, the width of the tool, around a center slug and through the workpiece. Because the slug remains uncut, a minimum amount of metal needs to be removed to make a through hole. Drilling with annular tools is called *trepanning*. But the Hougen Company of Flint, MI, have improved the usual trepanning tool and their design is called *Rotabroach drills*. They are like a hollowed-out end mill. Figure 8.5 is a set of Rotabroaches. Compare these to Figure 8.1, which is a simple trepanning tool.

Although the technique has been known for many years, numerous shops are viewing it again with renewed interest in special cases to replace conventional drilling. The impetus of this activity comes from several directions. First, cutting 75% less metal to make the same size hole requires much less power. Eliminating the thrust caused by the center of a conventional drill also saves power. These two features permit smaller machines to cut larger holes in less time. A conventional helical drill has near-zero speed at the tip of the tool, as shown in Figure 8.2. The tool tip is actually being pushed through the workpiece, which is why thrust ratings are higher for solid helical drills.

Drilling large-diameter holes with annular tools is not a high-speed operation. Often, these tools are rotating at less than 100 RPM. Nevertheless, cycle time for these tools is less than for conventional drills because the multiple teeth on the annular tool permit much higher-feed rates.

ADVANTAGES OF AN
ANNULAR BROACHING DRILL

- One advantage is the saving of worthwhile scrap metal. Slugs are worth more than chips as scrap; often 2 to 3 times as much. In some cases, the shop can use the slug to make something else. In other cases, where an expensive metal is being machined, the slug is truly worth saving.
- Faster cutting rates.
- Lower power requirements.
- The ability to drill holes on an angle without a pilot hole or a starting flat and often without a guide bushing. A standard drill would walk on the workpiece and be deflected off-line—even by surface irregularities. An annular tool enters a cut like a milling cutter. It's able to take a chip immediately on contact with the workpiece. Where you have variances in surfaces, as in forgings and castings, the tendency for this tool to bite in immediately leads to more consistent accuracy (Figure 8.3).
- The multiple-tooth design of a broaching drill does well on interrupted cuts. This happens to be one of the rare occasions where a spade drill (current type has a replaceable, very hard tip) works better than a helical. But neither is as efficient as the broaching drill. And the usual finishing tool, the boring bar, is very bad for interrupted cuts. Accordingly, in these situations, the hole can actually be finished smoother with the broaching drill than the boring bar.
- This type of broaching drill is also used for machining annular grooves (Figure 8.4).

Trepanning is a roughing operation and will probably drill 0.003 to 0.005" oversize. Consequently, reaming or boring after trepanning is necessary to hold tighter tolerances or get a better finish. In that respect, it is no worse than most other types of drilling.

Several types of tools are used for trepanning. The simplest is a single- or double-point cutter (tool bits). You could use ¼" square tool bits (such as lathe cutting tools) or larger for

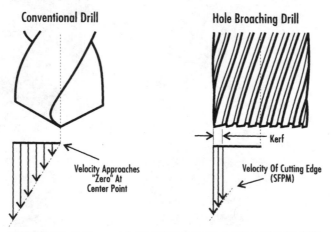

Conventional Drill   **Hole Broaching Drill**

Velocity Approaches "Zero" At Center Point

Kerf

Velocity Of Cutting Edge (SFPM)

FIGURE 8.2 Cutting speed comparison between annular and twist drills. Surface speed increases with distance from the cutter. Because of the "dead spot" at the center of the drill, this section of the drill is being pushed through the hole—requiring higher machine thrust. *(Hougen Mfg., Inc., Flint, MI)*

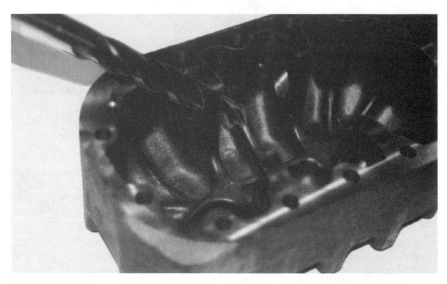

FIGURE 8.3 Drilling at an angle (simplified). Because of the hole-broach tool's tooth geometry, it can enter a workpiece without a pilot hole and often without a guide bushing at up to 45-degree angles. Workpieces with irregular surfaces can be drilled through in one operation. Without the need for a pilot hole, semi-circular holes can be drilled rather than milled. *(Hougen Mfg., Inc., Flint, MI)*

FIGURE 8.4 A new way to groove. Before a trepanning operation makes a hole, it makes a groove. This workpiece used a hole-broaching tool to cut this groove in 16 seconds with one Z-axis move. Previously, two ⅛-inch end mills took two hours to do the same job. *(Hougen Mfg., Inc., Flint, MI)*

either the single- or double-point trepanning tool. It orbits the spindle center line cutting the periphery of the hole. Usually, a pilot drill centers the tool and drives the orbiting cutter like a compass inscribing a circle on paper. These single- and double-point trepanning tools are usually adjustable within their work diameter. They are efficient and versatile, but begin to have rigidity problems when cutting large-diameter or deep holes.

A hole saw is another type of trepanning tool. Figure 8.1 is considered a hole saw. Hole saws have more teeth and, therefore, cut faster and with greater rigidity than single- and double-point tools. These tools pick up chips between teeth and carry them in the cut. Unless you peck drill, these chips can score the workpiece ID.

Hole-broaching annular tools are hybrid trepanners. They combine spiral flutes, like a helical drill, with a broach-like progressive tooth geometry that splits the chip so that it exits the cut along the flutes. With this design, the larger number of cutting edges and chip evacuation combine to reduce the chip load per tooth, so this drill can cut at much higher feed rates than the previously mentioned trepanning tools and hole saws. Like the hole saw, a hole-broaching tool has a fixed diameter; one size fits one hole (Figure 8.5).

THE ROTABROACH BY HOUGEN

The Hougen Company of Flint, MI, manufacturers a complete line of Rotabroach drills. These unique, patented hole cutters can be used on conventional machine tools and the equipment Hougen sells along with the cutters. They manufacture portable magnetic drill presses and Rotabroach cutter sharpeners. Their trepanning cutters can also be used in hand-held power drills. These cutters are for through-hole drilling in fractional sizes from ¾ to 4" diameters and 3, 4, and 6" in depth (Figure 8.5). They come with straight shanks and #2 and #3 Morse-tapered holders.

This annular hole broach is the ultimate in hole-cutting tools. The preceding information illustrates some of the versatility of annular tooling. Currently, all trepanning tools, whether simple one or two pointers, hole saws, or hole-broaching tools, make up only a small part of the arsenal of tools available to manufacturers. There is definitely a place for this kind of tool and this kind of process, just like there is a place for any method that helps improve a manufacturer's competitive edge. It is good to periodically remind ourselves that there are other ways to do things (Figure 8.6).

"53,000-SERIES" — 3" Depth Of Cut

"54,000-SERIES" — 4" Depth Of Cut

"56,000-SERIES" — 6" Depth Of Cut

FIGURE 8.5 Standard rotabroaches. *(Hougen Mfg., Inc., Flint, MI)*

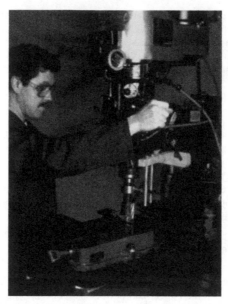

FIGURE 8.6 The 12000-series rotabroach cutter with R-8 tool holder. *(Hougen Mfg., Inc., Flint, MI)*

The Rotabroach found an effective place in building construction equipment, heat exchangers, ordinance projectiles, pumps, military aircraft, recreation vehicles, food-processing equipment, oil-field equipment, and many places in the automotive industry. In spite of the fact that they are more expensive than standard drills, the Rotabroach idea is often a money saver in the long run.

CHAPTER 9
METAL SPINNING

HISTORY

History records the fact that the Egyptians practiced an archaic form of metal spinning, making it the oldest-known method of producing hollow, circular metal components. It was introduced into our country in the middle of the nineteenth century. At first, the process was used solely to produce parts made from soft, easily worked nonferrous metals. It wasn't until shortly after World War II that harder, tougher, ferrous metals were worked. Now, because of its versatility, metal spinning is used for small- or medium-sized production lots.

PROCESS DESCRIPTION

Spinning is a chipless production method of forming symmetrical metal shapes. It is a point deformation process by which a metal disc or preform is plastically deformed by contact with a rotating chuck (mandrel) by axial or rotary motions of a hand-held bar or a roller mounted at the end of the bar. The shapes produced include cones, hemispheres, tubes, cylinders, and other symmetrical, hollow parts in a wide variety of sizes and contours.

Spinning is an economical, efficient, and versatile production method when the only alternative would require expensive stamping or deep-drawing dies. Manual spinning requires several passes to completely bend the metal blank into contact with the mandrel. During this process, there is only a slight reduction in blank thickness.

Metal spinning is one of the oldest-known methods of metal forming. It is done on a lathe-like machine using the strength and skill of an operator to form parts. A rather large mechanical advantage can be obtained for spinning thicker blanks especially of ferrous materials. Normally, manual pressure is applied to a lever (bar), which is forced through a fulcrum on the machine (lathe) compound to progressively flare the blank over the mandrel. See Figure 9.1 for a complete picture of metal spinning.

In the study of Figure 9.1, replacing the straight bar with a tool of scissors design achieves some mechanical advantage. The latest spinning machines have an option of using

A.

A wood or metal form is turned to the exact shape and
dimensions of final spinning

B.

A flat metal disc, the blank, is clamped into position in front of
the form. Both revolve at a high rate of speed.

C.

The disc edge is bent to insure rigidity and prevent the metal
from buckling. On heavy gauges, the edge is spun over on a
preliminary form or block.

D.

The disc edge is bent to insure rigidity and prevent the metal
from buckling. On heavy gauges, the edge is spun over on a
preliminary form or block.

E.

The craftsman, called a spinner, exerts pressure on the
disc with a long, blunt tool.

F.

This tool is deftly controlled with arm, hands and body.
It forces the metal to flow snugly over the from.

FIGURE 9.1 The mechanics of metal spinning. *(American Metal Stamping Assoc., Richmond Heights, OH)*

Trimming

Basically, all metal spinnings are produced by the method shown here; but special development and new techniques adapt the principles to a variety of setups, some of which are semi-automatic.

H.

I.

Spinning sheet metal is fascinating. It looks like liquid as it flows under the tool from the center out to the edge.

G.

Beading

J.

Very close tolerances are maintained.

K.

Completed spinning, a propeller cap, shown from blank to final form.

L.

FIGURE 9.1 The mechanics of metal spinning. (*American Metal Stamping Assoc., Richmond Heights, OH*) (*Continued*)

tool-holding carriages (compounds), which have air or hydraulic power provided to increase force.

In this process, a metal disc is spun at controlled speeds on a lathe (generally, an older type). The disc is held between a mandrel, secured to a chuck, and a follower attached to the tailstock. The mandrel corresponds to the inside contour of the part to be produced. Power is used to revolve the mandrel, disc, and tailstock follower.

Spinning tools are then forced against the rotating disc (traditionally, by hand). The operator forces the blank to assume the shape of the mandrel by means of a series of strokes. The metal flows in a manner similar to clay on a potter's wheel (Figure 9.2).

The size of the metal blank is determined by the surface area of the finished part. Actually, this size can be accurately calculated by formula. Pressure on the spinning tool is generally exerted by hand, although, as already stated, some machines have air power or hydraulics to perform this task. As a matter of fact, some automatic machines have template-controlled devices that speed up the operation somewhat.

Normally, parts for spinning are symmetrical and circular in shape because the machine rotates. The process is so simple and basic that pieces as large as 26 feet in diameter and thicknesses up to 2" have been worked. But most of the work processed is considerably smaller. You can see a sample of both parts in Figure 9.3.

The lathe-like machine turns both mandrel and disc, while a bar-like tool is levered through a pivot point on the machine compound to flow the metal over the mandrel. The pivot points can be moved on the compound to permit repositioning of the operator's bar.

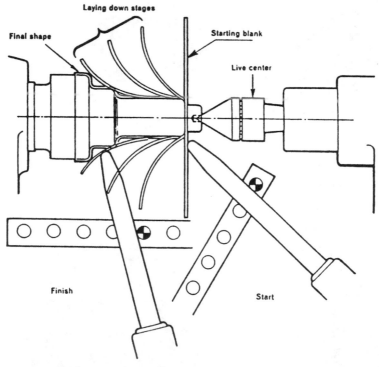

FIGURE 9.2 Setup for freehand spinning. Notice laying down stages to gradually make starting blank conform to shape of mandrel.

FIGURE 9.3 Two spinning machines that are very different in size: A. The automatic lathe is tracer-equipped for consistency in volume spinning. B. Large-diameter blanks can be spun on this huge hydraulic lathe. *(American Metal Stamping Assoc., Richmond Heights, OH)*

The operator must possess a reasonable skill to produce uniform parts. He exerts pressure on the disc, controlling a long, blunt tool with his hands, arms, and body. He also uses special tools to "finish" the surface, trim off excess metal, apply beads and flanges, and make parts with fairly close tolerances. Considering that this is all hand work (Figure 9.4), this requires moderate skill.

A sophisticated variation of this spinning process is called *shearforming*, which is covered in Chapter 10. This process is different from spinning in that it achieves a deliberate and controlled reduction in disc thickness and can make shapes that the conventional process cannot. It also speeds up the spinning process ten-fold.

TOOLING

The first step in spinning is to make the mandrel. Much of the accuracy of the spinning job depends on the mandrels. Even if metal mandrels are required, they are considered low cost, compared to the usual tools for a draw press. Nowadays, compressed wood or laminated wood is often chosen. Carefully selected woods, properly kiln dried, are cross laminated and glued together, or blocks of hard maple are machined to shape. Sometimes wood chucks are fitted with steel reinforcing rings at corners and areas where high-forming pressures are used. Up to 500 pieces could be spun on a wooden mandrel.

When quantities exceed 500, steel or cast iron should be considered for the mandrel. The mandrel is secured to the chuck; in fact, it is sometimes called the *chuck*. The blank, which is a flat metal disc, is centered and clamped tightly between the mandrel and the tailstock or an extension of the tailstock. If the disc is thin and large in diameter, it might first have its edge curled to establish a higher degree of rigidity (Figure 9.5).

Lubricants are generally used to counter the friction of the process and to prevent marring of the surface of the blank. Common lubricants for manual spinning include yellow naptha soap, hard cup grease, and combinations of petroleum jelly, paraffin, beeswax, and oils.

FIGURE 9.4 Typical parts produced by metal spinning. *(Roland Teiner Co., Inc., Everett, MA)*

ADVANTAGES OF METAL SPINNING

Because of the mechanical working of the disc in spinning, the grain structure is refined, thus providing metallurgical benefits. The heavy forces required to plastic flow the metal during spinning orient the grains parallel to the principal axis. This phenomenon is similar to the grain orientation created by forging.

According to many tests, cold working the metal during spinning increases tensile strength appreciably. Sometimes this increase actually doubles the strength. The spinning process allows the contour to be formed with little or no added machining. A cast part would probably have to be thicker to allow for a machine finish because you cannot cast thin sections. See Figure 9.6 for a list of materials that have already been used in the spinning process.

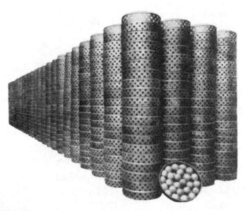

A. Fruit processing trays. Spun and perforated stainless steel trays used to hold fruit during processing replaced flimsy wire trays. A bead spun on the top edge adds strength, and a bottom lip facilitates stacking.

B. High voltage generator component. Doughnut-shaped device, three feet in diameter, prevents sparking in a 2.2 million volt generator. A combination of metal spinning, welding, and high-luster polishing was used to produce the unit.

C. Helicopter component. Progressive stages depicted here show how drawing, spinning, fabricating, finish polishing, and grinding were combined to make this 6061-T6 aluminum part. Critical finished wall thickness and weight tolerance necessitated grinding the final part.

FIGURE 9.5 Typical parts produced by metal spinning. *(American Metal Stamping Assoc., Richmond Heights, OH)*

Aluminum and Alloys

Non Heat-Treatable Types

1100	5086
3003	5454
5052	5456
5083	

Heat-Treatable Types

2014	6061
2024	7039
2219	7075

Also Clad Materials

High Strength — Low Alloy Steels

Cor-ten
Ex-ten-60
GLX-60W
High strength #2
Hi-steel
Hy-80
Hy-100
Hy-130/140
INX-60
Jalloy S-110
JLX-60W
Man-ten
Maraging
MLX-60
NAX steels
Pitt-ten X-60W
Republic X-60W
T-1
Tri-ten
YSW-60
Nine (9%) nickel
   (ASTM-A-353)
Yoloy

Copper, Brass and Alloys

Copper and Alloys

Oxygen free electrolytic
   copper
Oxygen free copper
Electrolytic tough pitch
   copper
Phosphorus deoxidized
   copper
Commercial bronze (90%)

Red brass (85%)
Low brass (80%)
Cartridge brass (70%)
High brass (68.5%)
Yellow brass (66%)
Yellow brass (65%)
Yellow brass (63%)
Muntz metal (60%)
Light leaded brass
   (62.2%)
Medium leaded brass
   (62%)
High leaded brass (62%)
Extra high leaded brass
Naval brass (63.5%)
Phosphor bronze A
Phosphor bronze C
Low silicon bronze B
High silicon bronze A
   (Ev-1010)
Leaded silicon bronze
   (Ev-1012)
Manganese brass
Cupro nickel
18% nickel silver
Nickel silver — 65-18
Nickel silver — 65-15
Nickel silver — 65-12

Stainless, Heat and Corrosion Resisting Alloys

Unstabilized Types

201	304
202	304-L
211	305
300	310
301	311
302	

Stabilized Types

316
316-L
321
347

Straight Chromium Types

405
409
410
430
430 Modified
442
446 Modified
Nickel Base
Monel 400
Monel 403

Superstrength Steels

SAE-4130/4340
Modified H-11
Vascojet-1000
Unimach II
D6AC

High Temperature — High Strength Alloys Iron Base

Unimach II
H-11, H-13
Vascojet 1000
Potomac M
D6AC
Carpenter 20Cb-3

Cobalt Base

Haynes Alloy 25

Nickel Base

Hastelloy-B
Hastelloy-C
Hastelloy-X
Inconel-600
Inconel-702
Inconel-718
Inconel-722
Inconel-X-750
Illium-G
Illium-R
Nichrome-V
Nickel-200
Nickel-201
Nimonic-75
Rene-41

Stainless Alloys High Strength — Iron Base

AM-350
N-155
PH-15-7 MO
17-4 PH
17-7 PH
Incoloy-800
Incoloy-801
A-286
18-8LN
19-9 DL
19-9 DX

Refractory Metals and Other Basic Metals

Gold	Silver
Lead	Invar
Magnesium	Kovar
Molybdenum	MU-Metal
Nickel	Pewter
Platinum	Zinc
Tantalum (STA-1000)	
222 Metal (Tantalum-Tungsten)	

Titanium — commercially
   pure
Titanium-6AL-4V
Titanium-5AL-2.5Sn
Titanium-8AL-1Mo-1v
Tungsten-2% thoria
Zirconium (GR-12)

Common Steels (SAE-1010-1020), Special and Coated Steels

Cold rolled deep drawing
   quality
Cold rolled — soft temper
   — low carbon
Vitreous enameling deep
   drawing & spinning
   quality
Hot rolled pickled &
   oiled — low carbon
Hot rolled — low carbon
Hot rolled — copper
   bearing
Lead coated (long terne)
Galvannealed (zinc
   coated)
Galvanized
Aluminized steel
Alphatized steel

FIGURE 9.6   Materials that have been spun. *(American Metal Stamping Assoc., Richmond Heights, OH)*

A comparison between two identical parts, one made by spinning and the other by press forming, illustrates the steps saved by spinning (Figure 9.7). To make the cone illustrated, spinning requires three steps, but press forming requires eight steps.

Spinning is ideal when production quantities do not justify draw-press tooling. This is especially true when dealing with large-diameter parts.

Consider the three basic shapes: cone, hemisphere, and straight-sided cylinder. The cone is a simple shape for spinning, but it is difficult for press forming. The hemisphere is more difficult to spin, but it is still easily done. Spinning a sharp angle exposes the metal to a great strain and requires more skill. Consequently, the straight-sided cylinder is the most difficult of the three shapes to spin and requires a more-skilled operator. Figure 9.8 shows the basic forms that can be easily spun.

Sometimes parts are spun in preparation for a machine finish and there are times when parts are made by other processes, then finished by spinning. For instance, it is possible to

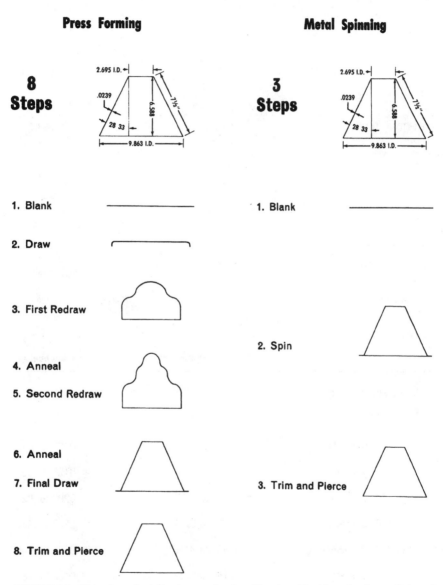

FIGURE 9.7 Comparison of spinning to press forming. *(American Metal Stamping Assoc., Richmond Heights, OH)*

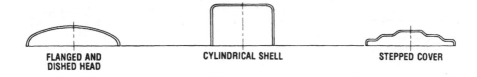

FLANGED AND
DISHED HEAD   CYLINDRICAL SHELL   STEPPED COVER

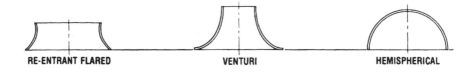

RE-ENTRANT FLARED   VENTURI   HEMISPHERICAL

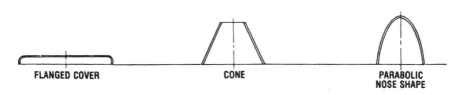

FLANGED COVER   CONE   PARABOLIC
NOSE SHAPE

FIGURE 9.8   Basic metal-spun shapes. *(American Metal Stamping Assoc., Richmond Heights, OH)*

	Commercial Applications	**Special Applications**
Up to 24″ Diameter	± 1/64″ to 1/32″	± .001″ to .005″
25″ to 36″ Diameter	± 1/32″ to 3/64″	± .005″ to .010″
37″ to 48″ Diameter	± 3/64″ to 1/16″	± .010″ to .015″
49″ to 72″ Diameter	± 1/16″ to 3/32″	± .015″ to .020″
73″ to 96″ Diameter	± 3/32″ to 1/8″	± .020″ to .025″
97″ to 120″ Diameter	± 1/8″ to 5/32″	± .025″ to .030″
121″ to 210″ Diameter	± 5/32″ to 3/16″	± .030″ to .040″
211″ to 260″ Diameter	± 3/16″ to 5/16″	± .040″ to .050″
261″ to 312″ Diameter	± 5/16″ to 1/2″	± .050″ to .060″

FIGURE 9.9   Spinning process tolerances. *(American Metal Stamping Assoc., Richmond Heights, OH)*

achieve lower cost by rolling sheet-metal shapes and seam welding them into cylinders. Then, the parts could be completed with a spinning operation. A multitude of parts could be made by spinning. You will have no trouble holding the tolerances listed in Figure 9.9.

The message from spinning vendors is short and to the point. If you have a small quantity of symmetrical sheet metal parts to make, consider spinning them. In fact, it might pay to consult with a spinning vendor for a moderate quantity—especially if you have a tight schedule.

CHAPTER 10
SHEARFORMING

BACKGROUND

History records the fact that the Egyptians began the art of spinning and practiced a very simple form of hollow metal fabrication. As a more serious production process, the art of metal spinning originated in China around 900 A.D. and was brought to America about 900 years later. In the beginning, only soft, pliable metals were worked. At first, the equipment was rudimentary and the operators had to be craftsmen to produce even the simplest parts. Because the preceding chapter in this book covers that subject of metal spinning, that subject is not included here.

However, as time passed, operators wanted to work thicker and stronger materials. This created a demand for mechanical advantage to replace the brute strength of the operator and relieve some of his fatigue. This led to power spinning and the introduction of air and hydraulic power to provide greater force. This modern variation of the spinning process is known as *shearforming* (also known as *flow turning* and *floturning*)—a sophisticated version of metal spinning. It has the ability to produce appreciable differences in wall thickness, and close tolerances are easily obtainable by the operator using a machine controller. See Figure 10.1 for a model of shearforming machine that was marketed around 1978.

PROCESS OVERVIEW

A typical shearforming machine is lathe-like in appearance with a low profile and a very heavy, rigid frame. Its motor-driven, heavy-duty, precision spindle holds a mandrel contoured to fay with the internal configuration of the workpiece. The workpiece blank is held forcefully against the front face of the mandrel by a small-diameter, rotating push rod, supported by a hydraulically actuated tailstock.

A massive, hydraulically actuated cross-slide supports a hardened-steel roller, which could be any diameter from 4 to 12". The roller is tipped toward the mandrel at 5 to 15 degrees to avoid hitting the tailstock push rod. The edge of the roller is radiused ¼" or more

FIGURE 10.1 Early autospin model #AS12.30-540CNC from around 1978. *(Autospin, Inc., Carson, CA)*

so that it can gently create and move a wave of metal on the workpiece. The roller can be angled to permit burnishing of the workpiece (Figure 10.2).

When shearforming, the roller is brought up to the rotating blank and moved out radially from the front face of the mandrel, forcing the workpiece against the mandrel in a stiff, servo-controlled path. The roller is microprocessor guided to form the shape desired. This path can be in quadratic equation form, for instance, like a copper EFP (explosive-forged penetrator), which is shearformed in 60 seconds (compared to an hour using other processes).

Metal flow is achieved by the shear forming roller, which applies a pressure on the blank against the supporting mandrel. Thus, conical or contoured components can be produced in a one-step operation. The rollers move the free material parallel to the axis of the mandrel—either in the same direction that the roller is traversing or opposite to it. The second instance is called *backflowing*. The remaining portion of the blank, which does not take part in the actual deformation, remains always at right angles, with respect to the axis of rotation, and does not change its external dimensions. Thus, circular, square, or other blanks can be shearformed.

Originally, large circular components for the aircraft gas-turbine industry had been weldments of several smaller parts, which were then machined at great cost. Shearforming can produce these components in one piece (thus eliminating welding), providing more consistent metallurgical characteristics at lower cost.

During the reduction in thickness, very great stresses are created in the unsupported area of metal, away from the confines of the mandrel. These stresses act in the same way as those encountered during deep drawing operations on a press: they tend to cause wrinkling around the rim. It takes skill and experience on the part of the operator to avoid this problem. Her judgment too, is required to know when an anneal is necessary. Her knowledge of the "feel" of the metal and how it is moving, tells her when that intermediate anneal is required. It is also her experience that dictates the number of steps to the spinning operation, the type of roller to use, and even the material for the chuck.

The material flow occurs in the axial direction, while the wall of the workpiece is produced from the reduction in blank thickness. The thickness requirement of the blank is determined by the wall thickness desired and the angle formed by the wall and the component's axis. It would require too much time to unravel some of the plasticity phenomena

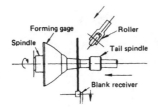

Place the material on the blank receiver and then switch on. The tail spindle moves forward automatically to chuck the material. The receiver then moves downward while the spinning roller goes ahead.

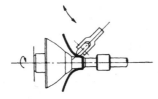

The spinning roller reciprocates along the rotary forming gage to from the material into the rough shape.

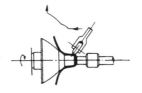

After rough forming, the spinning roller starts the copying job along the finishing gage.

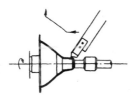

When a high surface finish precision is required as in aluminum goods, etc., change tools (e.g. roller to forming tool) by using the tool change unit to carry out surface finishing.

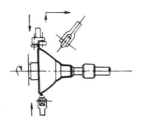

At the same time as the spinning roller moves backward, the trimming unit starts to cut the product to the required size. The curling unit just starts bending the edges.

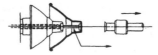

The tail spindle goes backward and then the knockout unit works to take out the finished product.

FIGURE 10.2 A complete shearforming cycle. *(Nihon Spindle Co., Japan)*

and the intricacies involved in the roller to blank relationship. This section does, however, include the sine law, which is followed, for instance, to determine the blank thickness for the specific angle of a cone of desired wall thickness. The shear spinning of any curvilinear shape will always follow the sine law.

Machine speeds and feeds are dependent to a great extent on the material being processed. Often, they must be determined by actual tests, but the machine manufacturers distribute information cards that recommend feeds and speeds for various materials and for different diameters. The surface speed increases dramatically as the workpiece diameters get larger. Much depends on operator experience. For instance, it is an established fact that stainless steels will provide finishes at high speeds superior to mild steel. As a general rule, speeds below 200 RPM are rarely used and speeds up to 1000 RPM are frequently used for medium-sized products. A good starting point when unsure of a new job would be a speed in this area coupled with a feed of about ⅟₁₆" per revolution..

The roller feed will likewise vary for different materials and hardnesses. The workpiece diameter and the percent reduction from blank to finished product also must be considered. The most practical method of specifying feed rates is linear travel per revolution of the spindle. If the feed rate is too small, it might develop "flattening" of the roller at the point of contact. This, in turn, causes "double working" of the metal at the instantaneous point of contact, and that local material might have a tendency to flow faster than the metal outside the influence of the roller. A slightly wavy condition could result, which is particularly noticeable on cylindrical work. The remedy is to increase the rate of feed slightly to reach the correct balance of speed, feed, and elongation of the material being worked.

In shearforming, the forming rolls are controlled by strong, hydraulic servo systems. The machine's structure is extremely rigid and massive. In fact, if the proper size of machine is used, the major limitation in shearforming is the physical properties of the blank, not the machine.

Whereas the earlier shearforming machines required the rollers to travel in a straight path, the newer models incorporate hydraulic tracer units, which allow shapes other than cylinders or cones to be formed. The newest models create parabolic or curvilinear shapes, provided that the initial blank is selected on the basis of the sine law, which is described shortly.

The word *blank* is applied to any part or form of metal secured to the chuck, preparatory to shearforming. The following are some of the starting blanks that have been used in flow forming: flat, round, square, hexagonal, and octagonal blanks, or almost any other shape, depending on the article required. It could be cylindrical blanks machined from forgings or castings. Normally, we think of a blank as being a piece of flat sheetmetal—either round, square, or some other shape, depending on the finished contour required—But it could be a cupped blank or cylinder produced as described by spinning, pressing, forging, or casting. It could also be a part that started as plate; it was rolled, welded, and rough machined before being shearformed.

A very popular product for shearforming is a conical shape. There are two limiting factors when considering conical work. First is the amount of reduction that the material selected will take without requiring annealing. Plain carbon steel, some steel alloys, annealed aluminum, and some stainless steels (such as the 300 and 400 series) will endure a 75% reduction. As is covered further along in this chapter, the amount of reduction is strictly governed by the sine function of the side angle of the cone. The thickness of the wall is equal to the starting thickness of the blank times the sine of that side angle. The sine of 30 degrees is 0.5000 and the sine of 15 degrees is 0.2588. So, when starting from a flat blank, if the cone's side angle is 30 degrees (60 degrees included), the reduction would be 50%. If the angle were 15 degrees, a 75% reduction could occur. Some of the high-strength alloys cannot stand such a large reduction in one stroke. In fact, some materials require heat when being floturned.

The second limiting factor is the capability (the size) of the machine to form the material selected. This limitation dictates the maximum blank thickness that a specific machine can form, as well as a maximum reduction. For example, even if a material can withstand a 75% reduction, if a blank, too thick for a machine, is positioned on it, the machine is not powerful enough to do a 75% reduction in one pass. The machine could probably perform the task in two passes.

Results obtained from flow forming will vary according to the work-hardening characteristics of the material and consequently, the magnitude of the residual internal stresses in the formed part. Correctly controlled stress relief annealing restores the desired properties of the cold worked components. It doesn't have to affect the dimensional accuracy of the completed part. That means parallelism, roundness, and size should be within a 0.005" tolerance on components up to about an 18" diameter, and if necessary, some dimensions can be held to half that tolerance.

Sometimes it is not possible to produce a completely shearformed part to the required finished size on all faces and surfaces because of some critical feature. It would still be practical to shearform the part, then machine the critical surface.

COMPARISON

In metal spinning, a flat blank of metal is forced by an operator, leaning on a long-handled forming tool, to conform to a convex mandrel. In the process, the wall thickness of the spun part is held reasonably constant, except for some stretching that occurs where the blank is bent. This change in thickness is not sought. In fact, it is normally undesired. It is accepted as an inconvenience of the process.

Shearforming, on the other hand, permits the creation of thicknesses that vary from point to point by as much as 100%. Another constraint of spinning involves blank thickness and radii. In spinning, blank thickness is generally ⅛" (or less) and most workpiece radii are at least 30 times the blank thickness. Shearforming can handle much greater thicknesses and bend any radius that the material itself can endure. Of course, this statement depends somewhat on machine size. Both processes can, under the proper circumstances, form steel plate up to 1" thick.

Although the process of spinning has been accelerated by several innovations, it has a few disadvantages that have been overcome by shearforming, and the technical advances of that process.

FORMULAS

Certain basic formulas are used in the calculations that are done preparatory to performing the process. These formulas must be used to achieve efficiency and quality in production. However, many occasions will arise for the use of initiative and ingenuity when new or unusual manufacturing opportunities present themselves.

The cone is the basic shape on which the shearforming process is founded. A formula based on the sine law is derived from the cone. The sine law presents a relationship between the thickness of the blank from which the part is to be formed, the included angle of the cone, and the wall thickness of the finished part. To satisfy a mathematical law, the cone is the basic shape adopted for the calculation of either the thickness of the starting blank or the wall thickness of the desired part.

The formula can be expressed two ways. It can be the sine function of the angle subtended by the cone centerline and its wall, or it can be the cosine function of the complainant of this angle. With reference to Figure 10.3, the formulas are written

$$t = T \text{ sine } a \text{ or } t = T \text{ cosine } B$$

where

t = Wall thickness of cone
T = Thickness of required blank
a = Angle subtended by centerline and cone wall
B = Angle subtended by cone wall and base line extended.

These formulas state that the depth of any conical part made by shearforming is not dependent on the diameter of the blank alone, as would be the case in spinning or deep drawing, but on the blank thickness also. The diameter of the blank is, in turn, dictated by the major diameter of the finished cone.

As the sine angle of the cone decreases, the thickness of the required blank increases. When the sine angle is 15 degrees, making the total cone angle 30 degrees, a second shearforming pass should be planned. A different formula is used to calculate the wall thickness when the included angle of a cone is less than 30 degrees. This formula is also based on the sine law. The cone angle and the thickness of the blank for any wall thickness are interdependent and altering one will vary the other. Whenever it becomes necessary to make two or more passes to form a cone, it is necessary to use a different mandrel for each pass (Figures 10.4 and 10.5).

Figure 10.5 indicates how a shallow-angled, straight-sided cone could be produced from a drawn blank. Two floturn steps were used so that the depth of the blank could be held to a minimum. The component shown was made from 6061 aluminum and the blank itself was made from annealed stock. After the first floturn operation, the material was solution treated (T4), then the second pass was made. This method takes advantage of the

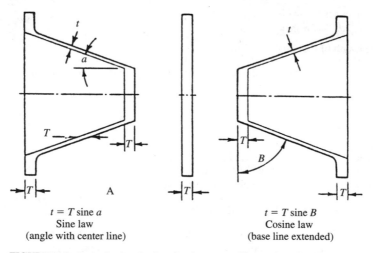

$t = T$ sine a
Sine law
(angle with center line)

$t = T$ sine B
Cosine law
(base line extended)

FIGURE 10.3 Basic sine law for shearforming cones. *(Floturn, Inc., Cincinnati, OH)*

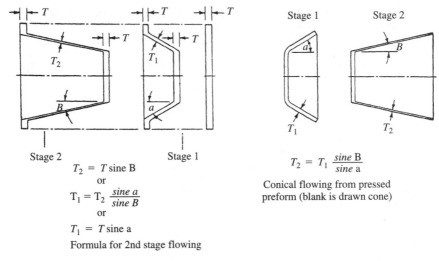

$$T_2 = T \, sine \, B$$

or

$$T_1 = T_2 \, \frac{sine \, a}{sine \, B}$$

or

$$T_1 = T \, sine \, a$$

Formula for 2nd stage flowing

$$T_2 = T_1 \, \frac{sine \, B}{sine \, a}$$

Conical flowing from pressed
preform (blank is drawn cone)

FIGURE 10.4 Additional sine laws for shearforming cones *(Floturn, Inc., Cincinnati, OH)*

delayed hardening characteristic of this particular material (age hardening) so that the second forming step was performed after the solution treating, thereby avoiding any problems associated with heat distortion. The distortion encountered during the solution treating was removed in the second floturn operation. Final operations involved trimming both ends of the part.

The sine law cannot be used to form cylindrical parts. Those parts could be formed accurately and satisfactorily by first forming a shallow cup by some other process (for instance, spinning or by using a press). Then, it could be shearformed using the formula for cylindrical components. At this time, the thickness of the blank depends on three things:

1. The thickness of the final wall required.

2. The thickness of the base required.

3. The length of the finished cylinder.

The formula generally used to calculate the thickness of the starting blank cup is

$$L = \frac{l \times t}{T}$$

where
    L = Required length of starting blank cup.
    l = Required length of finished cylinder.
    d = Inside diameter of both blank and finished cylinder.
    T = Thickness of blank cylinder wall.
    t = Thickness of finished cylinder wall.

When considering large-diameter cylinders with thin walls, which constitute the majority of cases, no factor of correction needs to be applied. In rare cases, the value of L should be multiplied by a correction factor (dt over dT).

The following three sketches (Figures 10.6 to 10.8) are used by the Floturn manufacturer, Lodge and Shipley, to illustrate the function of their machine. These sketches also facilitate understanding of the sine law. To comprehend conical part machining, you must understand the sine law. Figure 10.6 illustrates the basic elements of machining a cone.

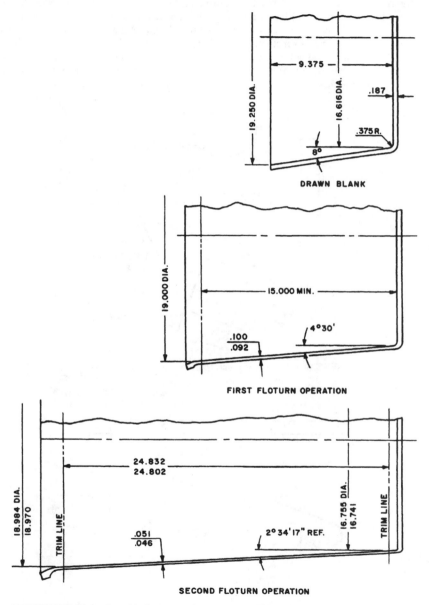

FIGURE 10.5 Floturning a blank to a cone in two passes. *(Floturn, Inc., Cincinnati, OH)*

Figure 10.7 explains the axial displacement of material and shows that the original blank thickness is transposed parallel to the centerline, and any increment of material that is at some distance x from the centerline, remains at the same distance (x) in the completed cone. The normal thickness of the cone is reduced by a percentage, as determined by the sine law.

Figure 10.8 illustrates how a blank is floturned into a cylinder. It illustrates a common type of cylindrical part made from a drawn cup and also provides the generally used method of calculating blank depth by comparing cross-sectional areas. A detailed comparison of the parts would indicate that this method of calculating blank depth is not quite accurate, and that equivalent volumes should be compared. But for most practical applications, this simplified method is satisfactory. The slight error introduced is on the safe side because more material is provided in the blank than volumetric calculations would demand.

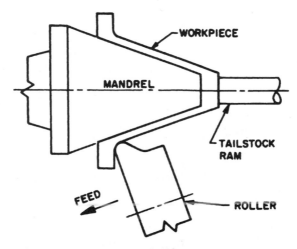

FIGURE 10.6 Machining a cone. *(Floturn, Inc., Cincinnati, OH)*

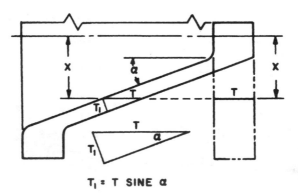

$$T_I = T \text{ SINE } \alpha$$

FIGURE 10.7 Example of sine law in floturning. *(Floturn, Inc., Cincinnati, OH)*

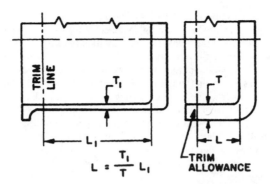

FIGURE 10.8 Floturning a blank into a cylinder. *(Floturn, Inc., Cincinnati, OH)*

$$L = \frac{T_1}{T} L_1$$

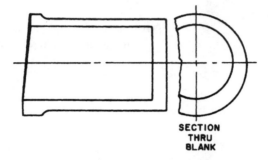

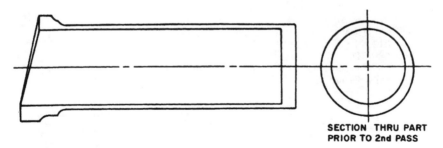

FIGURE 10.9 Avoid eccentricity of cylindrical blanks. *(Floturn, Inc., Cincinnati, OH)*

MISCELLANEOUS

When shearforming cylindrical work from a pressed blank, it is sometimes necessary or advisable to lightly machine both the ID and OD of the blank prior to shearforming. Otherwise, the result might be a varying wall thickness (Figure 10.9).

The concentricity of cylindrical blanks is of prime importance. Any time that a tubular blank is used as a first step to shearforming, it is necessary to consider lightly machining both the inside and outside of the blank. Figure 10.9 illustrates the condition that could exist after a first pass, if this step is not taken.

The production of cylindrical parts is generally a multi-pass process in which the roller makes multiple passes down the length of the component, making it thinner and longer with each pass. As a general rule, a single-roller machine can take a 20% reduction per pass, and machines with two or three rollers are capable of a 40% reduction per pass.

Tube spinning is a rotary point extrusion process performed on a shearforming machine, but not controlled by the sine law. Because the half angle of a tube is zero, the tube-spinning process follows a purely volumetric relationship with limitations related to a practical limit of deformation that materials will withstand without intermediate anneals (Figure 10.10).

The tube-spinning process has two categories, forward and backward. In forward spinning, the material flows in the same direction as the roller (toward the machine headstock). In backward tube spinning, the material flows in a direction opposite to the roller travel (toward the tailstock). Backward spinning has a few advantages. No clamping or holding

| | Shear spinning | | |
Material	Cone	Hemisphere	Tube spinning
4130	75	50	75
6434	70	50	75
4340	65	50	75
D6AC	70	50	75
Rene 41	40	35	60
A286	70	55	70
Waspaloy	40	35	60
18% Ni Steel	65	50	75
321 S.S.	75	50	75
17–7 PH S.S.	65	45	65
347 S.S.	75	50	75
410 S.S.	60	50	65
H 11 Tool Steel	50	35	60
6–4 Titanium*	55	—	75
B120VCA TI.*	30	—	30
6–6–4 TI.*	50	—	70
Commercially Pure TI.*	45	—	65
Molybdenum*	60	45	60
Pure Beryllium*	35	—	—
Tungsten*	45	—	—
2014 AL.	50	40	70
2024 AL.	50	—	70
5256 AL.	50	35	75
5086 AL.	65	50	60
6061 AL.	75	50	75
7075 AL.	65	50	75

*Spun hot

These percentages of reduction, exceeded under specific conditions, are intended as a guide which, if followed, will reduce development time.

FIGURE 10.10 Percentage of reduction for shear or tube spinning without intermediate anneals. *(Floturn, Inc., Cincinnati, OH)*

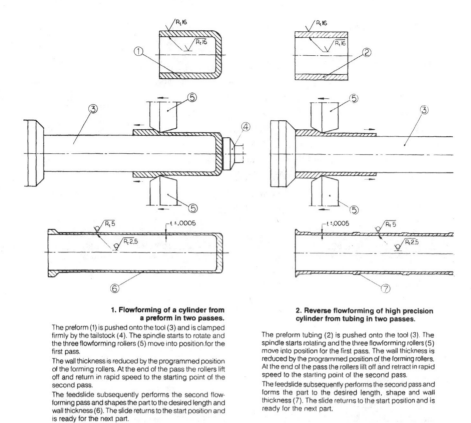

1. Flowforming of a cylinder from a preform in two passes.

The preform (1) is pushed onto the tool (3) and is clamped firmly by the tailstock (4). The spindle starts to rotate and the three flowforming rollers (5) move into position for the first pass.

The wall thickness is reduced by the programmed position of the forming rollers. At the end of the pass the rollers lift off and return in rapid speed to the starting point of the second pass.

The feedslide subsequently performs the second flow-forming pass and shapes the part to the desired length and wall thickness (6). The slide returns to the start position and is ready for the next part.

2. Reverse flowforming of high precision cylinder from tubing in two passes.

The preform tubing (2) is pushed onto the tool (3). The spindle starts rotating and the three flowforming rollers (5) move into position for the first pass. The wall thickness is reduced by the programmed position of the forming rollers. At the end of the pass the rollers lift off and retract in rapid speed to the starting point of the second pass.

The feedslide subsequently performs the second pass and forms the part to the desired length, shape and wall thickness (7). The slide returns to the start position and is ready for the next part.

FIGURE 10.11 Forward and reverse flow forming. *(Autospin, Inc., Carson, CA)*

by the tailstock is required. This method can also produce tubular sections longer than the normal length capacity of the machine. The major disadvantage occurs when the first section of material spun has to travel the greatest distance and is therefore subject to being out of plane. This, in itself, is not too significant in a straight tube (Figure 10.11). But some parts, such as a ballistic missile fuel case (which has substantial weld sculptures), cause a problem.

Once the procedure for shearforming a part is established, the process is quite simple. For any subsequent run, the setup is:

1. Mount the mandrel on the spindle faceplate or chuck and true it up.

2. Mount the selected (specific for that job) roller on its spindle.

3. Adjust the slide to the correct angle, relative to the mandrel, and position the roller at the correct distance from the mandrel.

4. Set the microprocessor for the calculated rate of speed and feed and all the travel limits.

5. Finally, turn the selector switch for the desired cycle: manual, semi-automatic, or fully automatic.

When those steps have been taken, a typical working cycle would be:

1. Load the blank into the machine. Sometimes a centering device is used to assist this step.
2. Use rapid advance to bring the tailstock to the clamp position.
3. Press the operating button, which turns the mandrel, supplies coolant, and begins the cross-slide movement to the mandrel.
4. The cross-slide's saddle feeds automatically for a predetermined distance to form the part.
5. The cross-slide retracts rapidly to the start position.
6. The spindle stops turning and coolant/lubricant stops flowing.
7. Tailstock goes into rapid retraction and the finished part is pushed off the mandrel by an ejector rod, which protrudes through the mandrel.

The results of the shearforming process vary, according to the work-hardening characteristics of the material and the magnitude of the residual stresses in the finished part. The quality of the finish will vary also, depending on the combination of blank size and machine feed and speed. If any significant features of operation are not properly adjusted, the materials will react in a manner similar to the results of deep drawing.

MACHINES

In Sweden, the Hallarydsvaken company built the first spinning-type machine that we call *shearforming*. Then, in Germany, the Leifeld, Bohner, and Kohle, and the Kellinghaus companies produced machines for sale. The Swiss Jenney Pressen Company and the American firms of Lodge and Shipley, Cincinnati Milling, and Autospin produced machines. Machines are also built in England and Japan.

All major parts of the machines are generally made of a good grade of cast iron. An important characteristic of cast iron is that it absorbs vibrations, thus machine builders use it generously. All the machines have several design features in common. The headstocks always have large roller bearings (some tapered) to withstand the high axial and radial loads imposed. Hydraulic power is generally used to operate the tailstocks and cross-slides. The main parts of the machines are extremely robust to withstand severe vibrations (Figures 10.12 and 10.13).

Wherever cast iron is not used, the machine manufacturers use heavy steel plates welded in an egg box structure—especially for the bed. A series of these weldments 1 or 2 feet square provide the necessary rigidity. The top plate of these squares is probably another steel plate, about 3" thick.

The first shearforming machines produced had only one roller, yet the work output was tremendous. Because the machine was computer-controlled, the tolerances held by this machine were heretofore unseen in the metal spinning arena. Although many products can be made in a single-slide machine, most machines now use multiple slides. Twin slides are helpful in avoiding axial and radial deflections. Three- and four-slide machines are also made and they improve production of specific products. Vertical machines are also manufactured to add to the versatility of the process.

The transmissions might have low- and high-speed ranges, from perhaps 10 to 3000 RPM, depending on the size of the machines and their duty rolls. Because there sometimes

FIGURE 10.12 Multi-spindle shearforming machine. *(Autospin, Inc., Carson, CA)*

is difficulty removing parts from the mandrel, an ejector rod is actuated hydraulically in the hollow spindle to push parts off.

The stopping distance of the roller from the chuck had been maintained by a micrometer dead stop operated by a handwheel. Although this has performed consistently successfully, this assignment is now generally handled by a microprocessor.

The process generates considerable heat, so cooling is of major significance. An important responsibility is the selection and maintenance of the fluid, which not only cools, but lubricates. One highly successful shearformer insists that his treatment of the fluid used for this purpose is instrumental in his success. Naturally, this treatment is proprietary and certainly is much more than filtration.

The earlier shearforming machines derive much of their functional advantage from the use of templates. Once the optimum operating adjustments were made, the movements were consistent and production was good. Now, of course, the microprocessor is set and reset until the ideal conditions are present, precluding the necessity of using templates.

Those early machines depended on microswitches, trip dogs, and solenoid valves to control the movements and operations of the roller slides, carriage, tailstock, and templates. These devices regulated the flow of hydraulic fluid to the rams and cylinders that caused the movements. The newer machines use microprocessors to control these movements.

When highly accurate parts are needed, care must be taken to ensure that the mandrel is turning absolutely concentric with the chuck. Sometimes this is ensured by finish grinding the mandrel on the machine with a portable grinding unit.

The machine manufacturers realized that too much floor space was required for the machines that were contemplated for some very large parts being designed. To work these large components, it was decided to change the largest machines from horizontal to vertical, just as the machine tool industry had done with lathes many years before. See

Application:
–Production of thick-walled parts which are difficult to
 form, such as wheel rims, wheel hubs, parts for jet engines,
 space applications, and components for chemical and
 agricultural equipment, etc.
–Spinning of aluminium, steel, special corrosion and acid
 resistant grades of steel, alloyed steels and titanium.

FIGURE 10.13   A two-roller shearforming machine. *(Autospin, Inc., Carson, CA)*

Figure 10.14 for a view of a vertical machine. This one is fairly small, but they are sometimes built to accommodate blanks as large as 10 feet in diameter. The builders retained
their insistence on extreme rigidity and strength. This is absolutely necessary when you
consider that as much as 250,000 pounds of force can be exerted by a machine powered by
as much as a 350-horsepower motor.

These machines have adapters that permit additional mechanical devices to be attached
to perform specific tasks, which leads to higher production. One device centers the blank
or workpiece. One loads and unloads workpieces. You can buy a trimming and also a beading attachment. There are heaters, both gas and induction, for necking high-pressure bottles and forming hard materials that require heating before and during processing. See
Figure 10.15 for such a machine and Figure 10.16 for such a process.

FIGURE 10.14 Vertical twin spindle shearforming machine. *(Autospin, Inc., Carson, CA)*

Application:
–Production of gas bottles, high-pressure bottles, accumulators etc., made of pipes, respectively pressed semi-finished shapes.
–Sealing of bottoms and neck shapes.
–Hot and cold forming of steel and aluminium

Design features:
–Welded construction, made of heavy steel plates.
–Guides are hardened, provided with automatic central lubrication and completely covered.
–Stop and ejector are located inside the hollow spindle and may be remotely controlled.

–DC drive for main spindle.
–Traversing of longitudinal and swivelling supports via electrohydraulic linear amplifier. Precise positioning and control of feed rate.
–Simultaneous traversing of longitudinal and swivelling supports permit the programming of most contours.
–Forming roller with hydraulic axial adjustment.
–Equipment for automatic feed and removal to facilitate simultaneous charging of a three-coil induction heating device.
–Burner unit for post-heating device.
–Short set-up times because of programs stored from previously processed workpieces.
–Heavy duty machine designed for continuous operation.

FIGURE 10.15   A neckforming operation. *(Autospin, Inc., Carson, CA)*

MATERIALS AND PRODUCTS

Many different materials have been processed by the shearforming technique. Some of these materials, which have been completed in one pass of the roller, would have required fairly expensive tooling and a number of cycles in a power press.

Several aluminum alloys and types of copper perform quite well. Normally, these are processed in the annealed condition. Another metal commonly used is Fortiweld, a boronized steel with a tensile strength of 40 tons PSI (hence its name). Because this strength is obtainable in the as-welded condition, it is a particularly useful material when weldments are to be shearformed. Several grades of stainless steel are easily worked. Many high-strength alloys have been shearformed, although they are difficult to machine by conventional means. Figure 10.10, once again, is a table displaying a compilation of materials and the amount of reduction possible in one pass. This is only a partial list of metals capable of being shearformed.

All kinds of military components, such as armor-piercing projectiles, components for guided missiles, rocket-booster tubes, and nose cones have been successfully shearformed. Many types of cooking utensils (especially large ones for hotels, restaurants, and

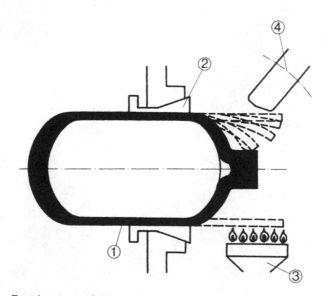

Forming example:

Neckforming on compressed gas cylinder.

A heated cylinder ① (tube, forged or drawn preform) is loaded into the machine and clamped in a collet chuck ②. The spindle starts rotating and the heating torch ③ is ignited by the pre-heated part. The roller ④ shapes the neck as per the computer program. The machine stops after completion of forming cycle and the finished part is ejected.

Important additional equipment:

- Induction and gas heaters
- Loading and unloading equipment
- Optical pyrometer to control even heating of part

FIGURE 10.16 Neckforming with multi-nozzle torch. *(Autospin, Inc., Carson, CA)*

hospitals) are easily made. The internal finish can be superfine so that no further polishing is required for utensils that the food-producing industries, such as canning and dairy, must use. Smaller sizes of these cooking utensils for domestic use are also produced (Figures 10.17 and 10.18).

FIGURE 10.17 Some parts made on a twin slide shearforming machine. *(Autospin, Inc., Carson, CA)*

Stainless steel "shims" are part of a family of parts, formed with a spherical radius and up to 42" in diameter. Some parts may start with a ⅜" thick stainless alloy blank which is reduced to 3/16" thick in forming.

"Packaged Power" is this shaped charge explosive canister assembled from three Floturn parts and filled with explosive. Used for underwater digging, demolishing buildings, etc., it provides (in various sizes) a standardized, directable explosive force.

Aircraft nose cone is produced from various aluminum alloys; process assures uniform wall thickness, seamless construction and precision dimensional tolerances.

Seamless hoppers, produced from stainless steel or other alloys, are made in a variety of sizes and thicknesses; feature an ultra-smooth, hard surface that minimizes chances for contamination, makes clean-up easy.

Printing drums, widely used in photocopiers, are produced from aluminum to close tolerances, then diamond turned to a 6 micro-inch finish.

The Floturn Process is ideal for production of conical, cylindrical and contoured parts. It can be used to work most metals and alloys, from .010" to 1½" thick, from 1" to 60" in diameter.

The Floturn Process (shear forming) is but one of a number of unusual metalforming processes available from Floturn, Inc. Others include: power spinning, deep drawing and ironing, tube necking and closing, warm orbital forging, etc.

Floturn offers a unique consulting service in metalforming . . . unusual equipment, methods, and production . . . plus unbiased recommendation of the right way to go!

Check your needs with Floturn . . . phone for an exploratory consultation . . . 513/671-0210.

FIGURE 10.18 Shearformed parts. *(Autospin, Inc., Carson, CA)*

ROLLERS

A massive hydraulically actuated cross-slide supports a roller that could be any suitable diameter from 4 to 12". The roller, which is vertical, turning about its center, is often tipped back up to 15 degrees from the spindle axis to clear the tailstock push rod. The leading edge of the roller is radiused in a very specific manner to gently create and move a wave of metal on the workpiece. In addition, this gentle metal movement is an act of burnishing. It improves the finish on the piece. For satisfactory metal flowing, there must be no tendency for the fibers to be subjected to either a shearing force or a stress that would lift the workpiece from intimate contact with the mandrel. The importance of the roller edge radius is shown in Figure 10.19.

If the roller edge radius coincides with the thickness of the metal blank or is smaller (as in Figure 10.19B), the radial force tends to hold the workpiece in contact with the mandrel. But, as the contact point is followed around the radius in an axial direction, there is no such

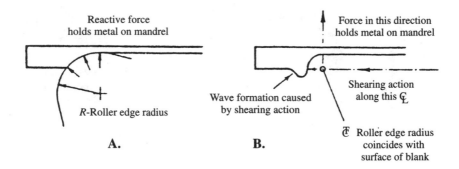

Roller edge radius relative to the wave formation

FIGURE 10.19   The importance of roller-edge radius. *(From a report by John M. Grainger, England)*

holding force. In fact, at this spot, there is the beginning of a shearing force. As the shearforming proceeds, the magnitude of the shear force increases and can spoil the finish, and leave the part in a stressed condition. Figure 10.19A shows a properly radiused roller edge.

Experience indicates that the minimum roller edge radius should not be less than 1.5 times the reduction in metal thickness. And, for a more satisfactory result, the radius should be two to three times the reduction. It is better to err on the side of a larger radius than on one too small.

Builders of shearforming machines generally supply a standard set of rollers with each machine. The light-duty rollers might have reflector rings on their front face, which bears a large portion of the forming load. Normally, rollers are made from high-alloy steel that contains molybdenum, vanadium, tungsten, and chromium. Metallurgists have discovered that mixing a small percentage of several metals produces a synergistic effect. This type of alloy can be hardened to a higher Rockwell (65 to 70) than the mandrel and is excellent when encountering heat during the manufacturing process.

In any metal-spinning operation, the form tool is brought to the center of the rotating blank, close to the tailstock, and it works outward toward the operator, forcing the blank against the mandrel. In shearforming, the process is really the same. The roller is brought up to the rotating blank, close to the tailstock push rod. It is then moved out from the center of the blank in a stiff, servo-controlled path.

The roller is guided by a microprocessor to form the part whose shape can even be in the form of a mathematical equation. One such equation, a quadratic, is used to roll the outer contour of a copper, tank-piercing projectile. The inner contour is determined by the mandrel. This ogive has a wall thickness that varies according to the formula, which is critical.

COMPUTER CONTROL SYSTEM

Computer numerical control (CNC) is used in shearforming for the same reasons that it is used on numerous metal-removing machines. It improves accuracy and repeatability, and reduces human errors. It also increases productivity by allowing those quick setups that permit small-lot production economies. All of the technical knowledge assembled by the engineering department can be inserted into a data base.

The process can be simulated and studied in the form of charts and graphs—even before actual execution. Complex geometries can be formed under CNC control, which could not be processed in any other manner. Can you imagine any other way to control two or four rollers working simultaneously on a part?

In addition to controlling a gamut of machine functions, CNC monitors several online devices, such as force- or electrical-measuring instruments, optical pyrometers, heat sources and pressure transducers. Papers, which usually accompany parts traveling through a shop, are no longer required because traceability of parts can be an automatic product of a CNC system.

The following are some additional advantages of a CNC system:

1. Each shearforming system sold is provided complete interactive software.

2. The computer executes the software controlling all attached devices and indicates prompts and messages for the operator.

3. The operator enters numeric data (such as machine functions, time delays, feeds, and spindle speeds) via a keyboard or numeric keypad.

4. Contours can be optionally entered from a scale drawing via the electronic digitizing tablet.

5. Program data can be printed out on one of several available printers.

6. All program changes and editing can be made at the machine.

7. An unlimited number of spinning passes can be programmed.

8. The system can control any roller or combination of rollers over all necessary spinning contours required to reach the final part configuration.

9. The computer monitors critical machine functions and conditions to maintain optimum safety and maintenance situations. Machine malfunctions will appear immediately on the screen.

10. The computer-interfaced servo-actuators control position, speed, acceleration, and thrust of the spinning axes to complete the program. Yet, the simple design permits years of trouble-free operation.

11. User (customer) program data is stored on the data disk for recall at any time for execution on the spinning machine.

Apparently, many more products could use the shearforming process to economize.

CHAPTER 11
ELECTRICAL-DISCHARGE MACHINING

HISTORY

The electrical discharge machining (EDM) phenomenon was first noticed about the year 1700. Soon after this, Benjamin Franklin wrote of witnessing "the actual removal of metal by electrical sparks." In 1881, Meritens first used arcs for welding. However, it was not until around 1948, that the Lazarenkos, a Russian husband and wife team, first applied the principle to a machine for metal stock removal.

The popularity of this machining method has grown by leaps and bounds since 1970. Lately, its growth rate has been about 30% annually. Machine power, speed of stock removal, and types of jobs EDM can do better than any other machining method have increased to the point that many jobs must now be done by EDM (conventional or wire) in order to be competitive (Figure 11.1).

PROCESS

The hardness of the workpiece has no effect on the process. The material could be hardened tool steel or even carbide. Rather than machine a part before heat treating it, EDM permits the machining to be done after hardening. This eliminates risk of distortion or any other damage. Graphite, copper, or tungsten are generally used to make electrodes. The electrode is always made slightly smaller than the cavity desired because the erosive action progressing outward from the electrode always produces a cavity slightly larger than the electrode. This size difference is called *overcut*. Once established, overcut is predictable.

EDM is a precision metal-removal process using an accurately controlled electrical discharge (spark) to erode metal. This process will machine any electrically conductive metal, regardless of its hardness. To visualize the process, picture one electric spark passing from

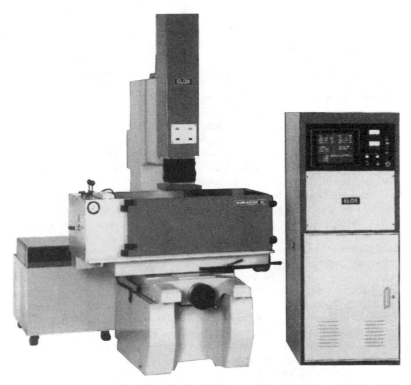

FIGURE 11.1   A typical die-sinking electrical discharge machine. *(Elox Corp., Davidson, NC)*

a negatively charged (–) electrode to a positively charged (+) workpiece, both of them immersed in the same bath of dielectric oil. The energy of the spark brings particles of the workpiece to a vaporized state. These particles immediately resolidify into small spheres and are flushed away by the dielectric oil, leaving a small pocket eroded in the workpiece. This cycle, repeated thousands of times each second, erodes material from the workpiece until a reverse image of the electrode is formed in the workpiece.

The current EDM process is used in a range from drilling small holes to machining huge 50-ton dies. The advent of CNC wire-cut systems has dramatically expanded both the quantity and sophistication of EDM applications. These new applications have improved the versatility and profitability of tool-making operations. In fact, they have also made EDM the logical choice for many production activities as well.

A good analogy is the comparison of EDM to a thunderstorm. The thunderstorm has a negatively charged cloud, a positively charged cloud, and wind movement. The EDM circuit has an anode, a cathode, and a switch with a servo system. The air gap between the two clouds acts as resistance, similar to the dielectric oil of EDM.

As the wind blows the clouds toward each other, the potential energy overcomes the air-gap resistance and electrons will jump the space. The electrons that ionize the gap, seeking unbalanced atoms, generate the tremendous energy (lightning) that causes havoc. EDM uses the same principle on a controlled basis to vaporize conductive metal.

There has been a simple evolution to the EDM process. It began with an RC circuit (resistor-capacitor), which provided a random discharge; some small, some large. If debris found its way into the gap and ionized, a chain reaction would result, causing random or uncontrolled timing of the discharge firing. This resulted in irregular surface finishes, slow cutting, different overcuts, and excessive electrode erosion.

Figure 11.2 illustrates an elementary EDM machine. It shows a workpiece sitting in a tank of dielectric oil, which covers both the workpiece and the electrode, and a source of electric power is connected through a switch. The switch can be turned on and off to provide an interrupted flow of electrical power to the electrode. Not shown is a servo mechanism that is needed to advance the electrode into the workpiece.

If you substitute a transistor for the switch, the on/off function can be achieved thousands of times per second. Another advantage of the transistor is that there is no build-up prior to turning on. The new type of power supplies have further improved control by assigning independent values to "on time" and "off time." The "on time" creates the spark crater and the amount of electrical energy determines the crater size.

The EDM is composed of four components: a machine tool, a power supply, a servo-computer, and a quill servo mechanism. The machine tool positions the electrode in relation to a workpiece to erode a cavity of some type. The power supply produces a high-frequency series of electrical arc discharges between the electrode and the workpiece, which erodes metal from the workpiece. This cutting power supply provides the required strength (amperes) and presets the gap voltage. It turns the power on and off.

"On time" is the actual cutting time when workpiece disintegration is occurring. "Off time" is the time provided to clear the disintegrated particles from the gap between the electrode and the workpiece. Both "on time" and "off time" are important. The "on time" is set together with the amperes to establish metal-removal rate, overburn, and surface finish. The "off time" (and the flush) keep the cut clean, which is necessary for efficient stock removal.

The servo computer constantly monitors and analyzes the gap conditions and, accordingly, modifies the on/off time settings and the up-down electrode movements. This computer recognizes four conditions: open gap, shorted gap, arc build-up, and proper EDM cut. The quill servo mechanism carries the electrode in and out of the cut position. The sensitivity of this unit establishes the cutting efficiency, the electrode wear, and the surface finish.

It is easy to check the servo system's response. Give the "full-down" manual command to the servo and note what happens when the electrode runs down to make "contact" with the workpiece. The electrode must dither on top of the workpiece with a total up-down movement of not more than 0.003" and there must not be any damage to either the workpiece or the electrode. A dither of less than 0.003" movement is an indication of a good-quality machine.

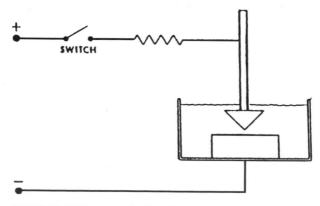

FIGURE 11.2 Elementary electrical discharge machine.

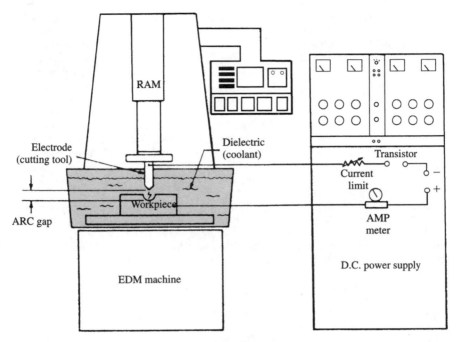

FIGURE 11.3 The electrical discharge machine. The electrode and workpiece are held by the machine tool (left), which also contains the dielectric system. The power supply (right) controls the electrical discharges and the movement of the electrode in relation to the workpiece.

Figure 11.3 illustrates the components of an EDM system. The electrode is attached to the ram of the machine tool. An hydraulic cylinder or dc servo unit moves the ram (and electrode) in a vertical plane to position the electrode in relation to the workpiece. This positioning is done automatically by the servo driven by the power supply. During normal machining, the electrode does not touch the workpiece. They are separated by a small gap.

Both the electrode and the workpiece are immersed in dielectric oil. The same oil acts as a coolant and is pumped through the gap to flush away eroded particles (swarf). In operation, the ram moves the electrode toward the workpiece until the space between them is such that the voltage in the gap can ionize the dielectric and allow a discharge to occur.

During the "off-time," the oil regains its insulating properties. It remains in this state until it is reionized by the next pulse. This process repeats thousands of cycles per second. Each discharge melts a small area of the workpiece. The molten metal cools, solidifies, and is washed away by the flushing action of the dielectric oil.

APPLICATION RECOMMENDATIONS

High machining rates are proportional to high current. However, high amperage generally requires large machining areas, and provides rough surface finishes. Fifty amps per square inch can normally be applied to graphite electrodes when roughing areas larger than 0.5 square inches. Copper electrodes will take as much as sixty amps per square inch. Electrode wear is at a minimum under heavy stock-removal conditions.

When considering finish-removal rates, the amperage, and, consequently, the stock-removal rate, is limited by the required finish. Fine finishes are possible only at low amps, which creates faster electrode wear. So, the finer the required finish, the slower the machining rate and the greater the electrode wear. That is why it is often more desirable and economical to finish and polish the workpiece off the EDM machine, rather than try to achieve a high finish through the processes. It is advised that you avoid extra-fine finishes in the process unless specifications and shape demand it.

When it is convenient to premachine a workpiece by sawing, drilling, or milling, it should be done (unless, of course, that isn't economical). Cost, as well as convenience, should be your guide. A significant consideration is how long you can run the job without stopping. For instance, an EDM job can be run continuously for up to 100 hours. One man could start and keep an eye on 3 or 4 EDM jobs simultaneously. It is common to load several EDM machines and run them in three shifts around the clock. Machine availability and the bottom line should be the deciding factors.

In general, difficult shapes are as readily machined as simple ones. The electrode of the complex shape is more difficult to prepare, but the actual EDM machining is no more difficult. It is no problem to make either sharp or round corners; deep, narrow pockets; thin ribs; twisted forms; holes; and multiple indentations. Blind cavities and through holes demand different techniques, but both can be handled by EDM. See Figure 11.4 for samples of work done by EDM.

Most dies and molds for any use can be made inexpensively by EDM. That includes (but is not restricted to): molds for the plastic industry, powder metallurgy, coining, stamping, forging, cold heading, extrusions, or die casting. Generally, EDMing these dies and molds will save time and money. In fact, in some situations, the use of wire EDM can produce male and female die components with one cut. (More about this later.)

Heat-treatment distortions are precluded by EDMing after heat treatment, rather than before. Because the electrode doesn't even touch the workpiece, there is little requirement for heavy clamping. In fact, the clamping can be feather-light for delicate workpieces. EDMed surface characteristics are very different from standard machine finishes. The surface is somewhat like one that has been shot-peened. The texture is nondirectional, covered with small pockets that retain lubricant. This feature makes the finish ideal for forging, stamping, and drawing tools.

Today, you do not avoid the EDM process. Instead the question is, "How should we EDM this job? Should we use a CNC machine? What kind of power supply? Will this job take pulse, vacuum, or pressure flushing? What is the proper technology necessary to minimize electrode wear, maximize metal-removal rates, and obtain a good finish?" Work that was once called "unconventional" is now a matter of course. In this process, the operator is the key to obtaining peak EDM performance. All machine manufacturers hold classes for operators. Material suppliers (like Poco) also have 3- to 5-day sessions that are very worthwhile.

It is best to get a user-friendly machine and software package. A package that a knowledgeable programmer can use to input a program that determines which roughing and finishing electrodes to use and the appropriate power-supply settings for the application. Today, a computer-controlled machine allows the worktable to be moved, as well as the electrode. This makes orbiting possible in all three planes. There are several distinct advantages to orbiting:

1. By distributing wear evenly over the entire cutting surface of the electrode, orbiting keeps corners and edges from being prematurely worn down.
2. If unexpected wear to the electrode occurs, with a CNC EDM, the operator would simply increase the size of the orbit and continue using the same electrode, and still obtain the proper dimensions in the cavity.

FIGURE 11.4 Products of electrical discharge machining. *(The Netw Corp., New Britain, CT)*

3. Moving both the workpiece and electrode results in superior flushing, which, in turn, means faster metal removal and finer finishes.

4. The money saved in electrodes justifies the use of CNC.

EDM can machine to a tolerance of 0.001" consistently. That includes the repeatability required, going from one electrode to another. The major roadblock to accuracy is the electrode. If the electrode is not accurate, the workpiece will reflect the inaccuracy.

POWER SUPPLY

The power supply controls the amount of energy consumed. First, it has a time-control function that controls the length of time that current flows each pulse. This is called *on time*. Then, it controls the amount of current allowed to flow during each pulse. These pulses are of very short duration and are measured in microseconds. There is a handy rule to determine the amount of current a particular size of electrode should use. It says that for an efficient removal rate, each square inch of electrode calls for 50 amps. Low current levels for large electrodes will extend overall machine time unnecessarily. Conversely, too heavy of a current load can damage workpiece or electrode.

The impact of each pulse is confined to a small area, the location of which is determined by the shape and position of the electrode. The arc always travels the shortest distance, across the narrowest gap to the closest "high spot" on the workpiece. After each "high spot" is removed, the impact goes to the next highest, and so on, until the machining is complete.

Figure 11.5 (principle I) shows how metal removal increases as energy input increases. The total energy depends on the number of sparks each second and the amount of energy in each spark. The energy is measured in amperes. Figure 11.5 (principle II) shows that the surface finish improves as the spark energy decreases. The amount of metal removed is normally proportional to the energy used.

Surface finish is important on many jobs. It is a function of two things: "on time" and peak current, both of which are settings of the power supply. Another rule is that long "on time" and/or peak current produces a rough finish; the reverse, short "on time" and/or low peak current produces a fine finish. When a finish better than 100 RMS is required, it should be polished manually—off the EDM. Fine-machine finishing can be achieved by using a tungsten electrode, but that would slow down the process.

ELECTRODES

One of the first decisions the EDM user must make concerns the choice of electrode material. With the exception of tungsten and graphite, most electrode materials melt near 1000 degrees C. That means they will lose material during the machining process. Graphite sublimates (changes directly from a solid to a vapor) at 3350 to 5000 degrees C, depending on the type of graphite.

Graphite is composed of particles and pores. The smaller particle size has better detail when machined, shows less wear when used in die sinking, and generally cuts faster if you wire EDM it into electrode shape. However, this type of graphite costs more and the resulting electrode removes material at a slower rate. The larger particle materials are less expensive, usually wear faster, and are slower to machine with wire EDM. Consequently, the selection of graphite material poses problems that experience alone can solve.

EDM electrodes are made from a variety of materials, depending on the workpiece material and the type of cut to be made. Some electrode materials are: graphite, copper, copper graphite, copper tungsten, tungsten, brass, and steel. Graphite electrode material is offered in many grades, which vary widely in price and quality. Some are useful for roughing only.

When roughing steel under average conditions, a stock-removal rate of 2 to 3 cubic inches per hour can be expected. This would be on large jobs that are running nonstop, perhaps overnight, using 100 amps when machining areas in excess of 6 square inches.

The EDM process will not erode graphite as fast as metal. An electrode's ability to resist wear has a significant effect on the surface finish of a machined part. An important feature of electrode material selection is its machinability. Except for tungsten, all common electrode materials machine fairly easily, but graphite combines good machinability with good EDM performance.

The cost of making electrodes can sometimes be discouraging. At least two machines on the market eliminate this problem. At one company, Easco-Sparcatron of Whitmore

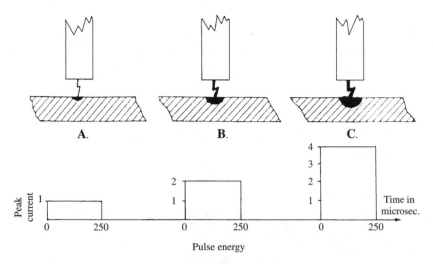

Principle I: Metal removal rates increase with the amount of energy per spark.

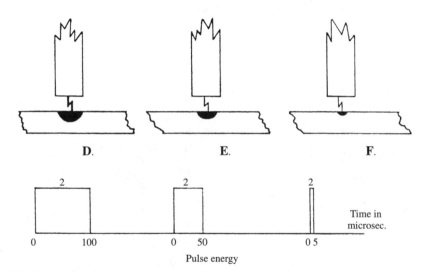

Principle II: Surface finish improves as the spark energy decreases. Energy may be increased or decreased by changing the peak current and/or on-time.

FIGURE 11.5 The principles of metal removal.

Lake, MI, has developed the "Total Form Machine." This is, in effect, a carbon copier, which turns out copies from a master. This master produces copies in minutes that would take hours if performed by conventional machining.

A second machine has a similar purpose. The Extrude-Hone Corp. of Irwin, PA, markets the OrbiTEX, a machine for the manufacture of 3-D graphite electrodes to be used on EDM machines. With this equipment, the uncut graphite block is mounted to the oscillating lower platen of the machine and an abrasive cutting master is fixed to the upper platen. Under the control of a microprocessor, the cutting master is lowered onto the oscillating graphite block and it imposes its shape into the graphite, totally forming the electrode to final dimensions. This eliminates the costly, time-consuming labor of single-point machining the electrode. Moreover, dozens of electrodes can be orbitally machined from one cutting master. After an electrode has been used for roughing, it can be redressed on the OrbiTEX, returned to the EDM machine, and used as a finishing electrode.

Resistance to wear is the electrode material characteristic that has the greatest effect on the metal-removal rate. If an electrode is wearing, electrical energy is being wasted and eroded electrode material will contribute to instability in the work gap. Both of these conditions are undesirable and will eventually inhibit the metal-removal rate.

Electrode material is usually selected because of:

- Metal-removal rate
- Resistance to wear
- Surface finish obtainable
- Expense
- Machinability

Electrode wear is exhibited in four ways:

- Corner wear
- End wear
- Side wear
- Volumetric wear

The most significant of these is corner wear because it determines the accuracy of the final cut contour. Any electrode material must be able to withstand the rigors of the process while maintaining its intended shape and fine detail. The wear is greatly affected by the material's melting point. The temperature at the point of spark discharge has been estimated to be in the range of 5000 degrees C.

FLUSHING

Three functions are required of the dielectric oil in electrical-discharge machining:

1. It must insulate the electrode/workpiece gap until the required voltage conditions are achieved.
2. It must cool the work, the electrode, and the molten particles.
3. It must flush the metal particles out of the gap.

As an insulator, the dielectric prevents a spark from occurring until both the gap and voltage are correct. When they are correct, the oil ionizes and permits the discharge to occur. Very high temperatures are present at the arc gap. The spark has enough energy to dissolve steel and carbide. Without a coolant, the electrode and workpiece could become dangerously hot. Sustained high-amperage machining will raise the temperature of the oil and make it difficult to handle the work. At 100 degrees F, it begins to lose efficiency; at 165 degrees F or hotter, it is unsafe to continue the operation until the oil cools.

What has been called the single most important factor in successful EDM work is flushing—the removal of metal particles from the gap. Good flushing provides good machining conditions. If the flushing is poor, metal will be removed erratically, time to machine will increase, and costs will increase. Figure 11.6 shows four common methods of flushing—two are pressure flushes and two are suction flushes. Sometimes a combination pressure/vacuum flush will be used. The diagrams show the most efficient methods to flush the swarf using internal pressure and vacuum (suction) patterns and external dielectric flow.

There is no substitute for coolant under pressure at the gap, but notice that coolant volume, not excessive pressure is the key to efficient flushing. Excessive pressure can accelerate electrode wear, create turbulence in the cavity, and possibly coagulate areas of conflicting flow patterns.

Pressure flushing through the workpiece is used in most through holes, such as those in stamping dies, cold headers, or any application where it is easier to provide flushing holes in the work, rather than in the electrode. Vibration can be used when the flushing is not as

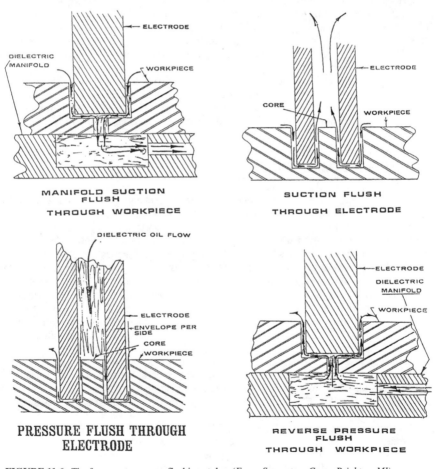

FIGURE 11.6 The four most common flushing styles. *(Easco Sparcatron Corp., Brighton, MI)*

good as you would like. The agitation created by this action tends to keep particles in suspension, where they can be flushed out by a jet of oil. It might be necessary when such problems occur to withdraw the electrode at intervals and allow the jet to purge the work area.

Vacuum flushing is always an alternative to pressure flushing. This method is useful when overcuts are 0.0015" or smaller. In this method, particles are pulled away from the cutting arc and are not permitted to remain between the electrode and the workpiece. Best results are obtained with relatively clean oil because the gap is flushed with oil already in the work pan, rather than filtered oil from the reservoir, as in pressure flushing.

Flushing is rarely a problem during roughing cuts because the overcut is large enough to provide good coolant flow around the electrode. As overcuts decrease, the problem of removing metal particles increases. Large electrodes allow, and sometimes require, multiple flush holes to make certain that flushing is effective across the entire surface of the electrode.

Flushing swarf (eroded particles) from the cavity is one of the most important features of efficient EDM machining. The electric spark does not differentiate between eroded particles floating in the dielectric oil and the workpiece material. Thus, it will react on the closest oppositely charged particle. If the spark gap between the electrode and particle is less than the predetermined optimum gap, which has been preset, it will cause the servo to respond, thus removing the electrode from the cavity. This situation is also conducive to the creation of a hotspot.

Oil that appears black in the workpan is not necessarily dirty in the gap. During pressure flushing, the oil in the gap comes directly from the filter, and the oil in the pan has been used. Most machines have, at the very least, an adequate oil-circulation system. Several companies market a good grade of dielectric oil, made especially for EDM work.

SMALL-HOLE DRILLING

Small-hole drilling deserves special notice. Small holes, 0.005 to 0.020" in diameter, can be drilled in hardened steel by EDM. Traditionally, this work has been done by a skilled machinist using a sensitive drill. Broken drills and scrapped parts have made this an expensive operation. Many parts in instrumentation, fuel injectors, and turbine blades demand very small holes. In some cases, small holes with other-than-round shapes are required.

EDM equipment in general use is not able to perform such work without serious problems, such as recast layers (surface EDMed, melted, and reformed) and spatter. That is because most EDM work is large (such as die jobs) and small-hole drilling requires a different technique. Amperages, spark frequencies, and overcuts that are proper for machining dies and molds, are not suitable for drilling fine holes with accuracy and a good finish. Holes under 0.015" diameter use tungsten wire for electrodes, and straight rods are used for larger holes.

However, it is possible to make and use special low-power equipment as the Raycon Company of Ann Arbor, MI, has done. With proper controls, the accuracy and repeatability of small-hole EDMing can be accomplished successfully and inexpensively. An operator could do small-hole EDMing instead of using a skilled machinist and a conventional drill.

Raycon has designed a small, table-top EDM machine with specialized power supply, electrode material, servo system, and dielectric. It requires a high degree of precise control and consumes very little power. This equipment can maintain a tolerance of 0.0001" in holes up to 0.010" in diameter. Once the job is set up, it can be continued in an automatic mode by an operator instead of a skilled machinist. In the absence of heat or mechanical force, there will be no distortion and delicate components can be EDMed. Multi-electrode machining has led to greater economies.

Whenever a need is created in manufacturing, some company will find a way to answer it. Currently, Charmilles Technology Corporation is marketing what they call a micro EDM unit, which operates on a desk top and machines small holes. In fact, by this time, other EDM manufacturers might be in this segment of the market also.

WIRE (EDM) ELECTRIC DISCHARGE MACHINING

Die-sinking EDM equipment achieved its first popularity around 1950, and wire EDM about 20 years later. The first wires used were copper and were capable of movement only in the XY plane at a feed rate of 1½ square inches per hour. The technology of wire EDM has advanced considerably since then.

Wire EDM is identical to die-sinking EDM, which we've been calling *conventional*. Die sinking uses a plunging electrode, and wire EDM uses its traveling wire. But each can be used in place of the other. The principles of wire EDM are essentially the same as vertical, die-sinking EDM. Metal is eroded from the workpiece by electrical sparks and the activity is protected from the environment by a dielectric. The dielectric used by the Elox Corp. of Davidson, NC, is deionized water. This is produced by circulating ordinary tap water through a deionizing system in the cooling module. Deionized water is an efficient insulator, but untreated water is a conductor and cannot be used.

The electrode, in this situation, is a vertically traveling wire, 0.002 to 0.012" in diameter, which travels through the workpiece like a bandsaw. The wire, like the solid graphite electrode of the conventional die-sinking EDM, wears as it cuts, and must be continuously replaced as it moves horizontally into the cut. It does not touch the workpiece. Its path is controlled by a computer-generated program. Any contour within the maximum worktable movements, including tapers up to 20 degrees, can be cut with great accuracy.

When considering the acquisition of a wire EDM machine, it is a good idea to get as many features as possible that would improve efficiency. One, for instance, is automatic wire threading. One operator can handle several machines, both die sinking and wire, so it is important to have convenience features. The proper grade of wire is also significant. High conductivity in the wire is desirable. A wire's thermal properties are determined by either the mix of alloying elements or by the base core material. A desirable melting point allows the wire to wear, promoting an open gap, which improves flushing.

Today, many molds and dies are produced by wire EDM because this technique saves time and money. Wire EDM is simply a traveling, thin wire, electrode, which cuts through a workpiece as it discharges. It is started through a hole (which has been drilled or EDMed), then it is fed by a microprocessor to cut out any shape required. The wire, generally brass, is held in top and bottom guides as it moves through the workpiece (Figure 11.7).

Since it was introduced around 1970, wire EDM has increased in popularity much faster than any other machine tool. Its feed rates have increased from 1.5 cubic inches per hour to about 23. As power-supply technology improved, surface finishes also improved.

The wire eliminates the need for elaborate and precision electrodes, which are necessary for conventional EDM work. The precision is in the memory of the CPU and you can make a new die or tool a month or a year later because the die's shape is always available in memory. A computer numerically controls the wire (electrode) movement so that the workpiece is made with exceptional accuracy.

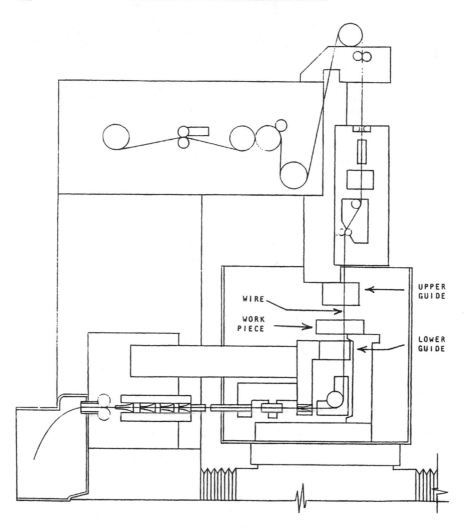

FIGURE 11.7 The schematic of typical wire EDM machine.

Punch and dies can be machined by wire, holding tolerances well within "001." One cut with wire can produce both punch and die from one block of tool steel (Figure 11.8). Generally, roughing cuts are within 0.004" of the finish size. When the slug is removed, and power levels are reduced, it normally takes two or three finishing cuts to get the desired dimensions and the proper finish. The total time for finishing cuts is about twice the time for roughing. Turn-around times are fast and finishes of 10 RMS can be commonplace.

Conductive ceramic can be cut by both die-sinking EDM and wire EDM, but you must be certain that the ceramic is truly conductive. If it were not conductive at some spot, the servo might interpret the lack of conductivity as an open-circuit condition and move the electrode into the workpiece causing a crash. This kind of situation is less significant for wire EDM because in that case, the wire would simply break.

CASES

Many companies offer contract machining to fill idle time on expensive equipment. Some become so successful at it that the "filler work" creates growth opportunities in new directions. This situation of idle machines is considered normal in many shops, but an idle machine is unprofitable. Companies that purchased EDMs to manufacture their dies competitively, soon learned that they could be run 24 hours a day and picked up the challenge to fill the empty working hours.

One company reported that they had a part that required 3½ hours to mill after heat-treating. In addition, the finish was never as good as they desired. But they owned a wire EDM, which was idle between die jobs. At first, the shop manager met stiff resistance to his suggestion that he try to machine the part on the wire EDM. In desperation, the company gave him permission and he wire EDMed the part in one hour with an excellent finish, exactly to size. The shop manager spoke of the difficulty getting engineers to think beyond the familiar, traditional machining methods. He said, "There's a learning curve to climb before you start thinking about EDM as a production process."

The Forma Tool and Mold Company produces molds for the plastic industry. Vice president Jack Browder says that the winning combination of automatic wire feeder and the capability of providing finishes better than 12 microinches and tolerances within 0.0001" has won considerable work from the competition. One application concerned a 3.000" × 7.000" steel plate 0.300" thick, which required 1522 holes, each 0.0550" in diameter, with a 1-degree taper per side on 0.0501" centers. By means of CNC programming, the machine

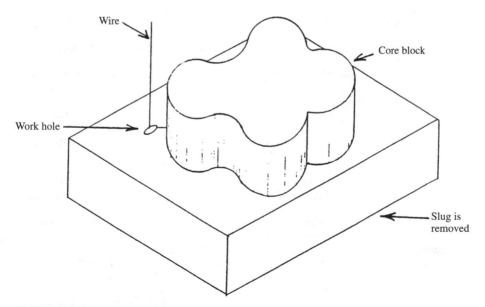

FIGURE 11.8 Making punch and die with one cut.

automatically positioned itself for each diameter to be wire cut using pickup points on the part centerline. Browder said that the setup and programming took about 8 hours and the machining, which included two roughing and two finishing passes, took about 106 hours. Tolerances on this part were held to ±0.0001". Browder then listed several jobs, requiring a total tolerance of only 0.0001", which could not have been done in any other method.

Wire EDM is a versatile machine that has a broad, growing range of applications in the area of production. It is no longer a machine solely for the purpose of making tools, molds, and dies.

GLOSSARY OF EDM TERMINOLOGY

amperes The measurement of electrical current.

arc gap The distance between the workpiece and the electrode when EDMing.

capacitor A component that stores a charge.

coolant Same as dielectric.

cycle Made up of the on-time and off-time and expressed in microseconds. Time associated with the discharge of one EDM spark.

dielectric Oil used in EDM system as a coolant, to flush away eroded metal particles and as a barrier between the electrode and the workpiece.

electrode A formed part made of electrically conductive material, used to machine the workpiece. It can be male or female in shape.

flushing Forcing dielectric through the arc gap to remove eroded particles.

frequency The number of sparks per second.

gap current The number of amperes flowing between the workpiece and the electrode. It can be read on the ammeter.

gap voltage The changing voltage, monitored on the voltmeter, which appears across the electrode-workpiece gap.

ionization Phenomenon by which dielectric oil breaks down and becomes electrically conductive.

machine tool The mechanical system that holds and positions the workpiece and guides the electrode.

microsecond One millionth of a second. The time unit used to measure on and off times in the EDM cycle.

on time The time during which current flows. Expressed in microseconds.

off time The time of cycle during which no current flows. This time allows the molten metal to re-solidify and get flushed from the gap.

overcut The clearance per side between the electrode and the workpiece after the machining operation.

peak current The amount of current that flows during on time.

power supply The electronic system that generates and controls the electrical discharges.

servo The system that converts electrical signals into mechanical motion to move the electrode.

short circuit This occurs when the electrode and the work piece are in direct contact. This causes the machine to stop and the ram to retract.

stepped electrode An electrode constructed in such a manner as to permit the roughing and finishing of a through hole or cavity in a single setup. The smaller front section is used to rough out the cavity and the larger rear portion is used for finishing.

striking voltage The voltage at which the dielectric coolant ionizes and allows the electrical discharge to occur between the electrode and workpiece. This is determined by the spark gap distance, in the gap, and the type of dielectric fluid.

CHAPTER 12

ABRASIVE-WATERJET CUTTING (AWC)

HISTORY

Abrasive waterjets were originally used to clean metal surfaces prior to surface treatments of various types. Although abrasive waterjet machining is relatively new, compared to most metalworking processes, it was first developed in 1974. Currently, the process is well established worldwide. In fact, a local distributor of abrasive waterjet cutting equipment was recently questioned by a company in India. The management there wanted to purchase one system that would be able to economically process many different materials.

Dr. Gerin Sylvia, professor of industrial engineering at the University of Rhode Island, began testing an automated abrasive waterjet system in 1985. Around 1990, the Southwest Research Institute of San Antonio Texas, opened a laboratory for studying precision cutting with abrasive waterjets. U.S. government facilities have been experimenting with this new cutting method as a means of expediting the manufacture of parts made from advanced composites (Chapter 46). Abrasive waterjet cutting is another manufacturing process whose use and popularity has grown dramatically as engineers find AWC solutions to problems that have been plaguing them.

One of the attractions of this process is the easy maneuverability of workpieces. Although saws cut mainly in straight lines, waterjets can be guided to cut curves, tapered or beveled holes, and complex shapes. Another application for this process is bridge inspection and the inspection of any concrete structure. A portable AWC machine can slice through thick slabs of reinforced concrete to expose the bridge's underlying steel framework. This permits easy evaluation of the framework's condition. The former method of using jackhammers and saws required more extensive cleanup and was much more expensive.

PROCESS DESCRIPTION

Waterjet cutting is a very effective cutting solution to a variety of industrial needs. In installations throughout the world, this process solves manufacturing challenges with waterjet technology. The method uses filtered water compressed to as high as 60,000 psi and directed through a sapphire (or some other hard material) nozzle. The result is a coherent jet stream that cuts cleanly through virtually any soft material without shredding or crushing. Materials that can be cut with water alone are plastics, foods, rubber, insulation, cardboard, automotive carpeting and headliners, and most woven and nonwoven textiles (Figure 12.1).

This method, a compressive, shearing action, produces a jet velocity of close to 3000 feet per second; a velocity about three times that of a pistol bullet. Any soft material in its path will simply be removed. Harder materials (such as glass, metals, superalloys, ceramics, concrete, and tough composites) are cut by adding an abrasive to the jet. See the abrasive waterjet cutting in Figures 12.2A and 12.2B. The cutting speeds for these hard materials range from 2–18 inches per minute. Cutting these materials, the waterjet drives the abrasive and is not the primary force any longer. Garnet is the most commonly used abrasive in this operation, but many different types are available.

In fact, foundry sand is most often used in foundry applications, such as cutting off gates and risers, even though garnet is 30% more effective than sand. That is simply because there are tons of sand available in all foundries, making sand more convenient. When castings are stripped this way, they should be dried to avoid corrosion and operators should wear hearing protection when using this method.

Three major components make up a waterjet cutting system: intensifier, nozzle, and receiver. The intensifier houses a pump that generates water pressure of 30,000 to 60,000 psi. The receiver catches the blast of abrasive-laden water and directs it to a sorting tank. The nozzles have been a problem since the start of the process (Figure 12.3). The nozzle is in the cutting head.

At first, the nozzles achieved only a 20-minute life. Then, they could take up to two hours of use. The lifetime was gradually increased to 5, then 10 hours. Now some manufacturers are claiming a nozzle life for as much as 100 hours of constant use. Nozzles with

FIGURE 12.1 These parts of rubberized cork, felt, urethane, and silicone were cut with a standard waterjet system in minutes. *(Jet Edge, Inc., Minneapolis, MN)*

B.

A.

FIGURE 12.2 Materials being cut: A. ½" aluminum, ¾" Kevlar, ½" glass, and ½" phenolic. B. 1" glass. *(A. Ingersoll-Rand, Baxter Springs, KS. B. Jet Edge, Inc., Minneapolis, MN)*

a small internal diameter make a deeper and higher-quality cut than those with a larger internal diameter. At present, a good nozzle can cut 0.066 thick gray cast iron at 36" per minute and a 0.88" thickness at 4" per minute. Nozzle materials have run the gamut with all kinds of hard, abrasive-resistant materials being tried. Carbides, ceramics, rubies, and sapphires have been used.

Ingersoll-Rand (I-R) invented the waterjet intensifier and sold their first commercial waterjet system in 1971. Their current models come equipped with calibrated rupture discs for safe venting of overpressure. They also have emergency stop, high-pressure bleed-down valves as standard equipment. Like all AWC producers, their intensifiers come in several horsepower ranges. The catcher tank comes equipped with sacrificial steel plates for effective dissipation of spent energy from the streams of water and the abrasives. An important consideration is that a single intensifier can supply high-pressure water for multiple applications in separate locations.

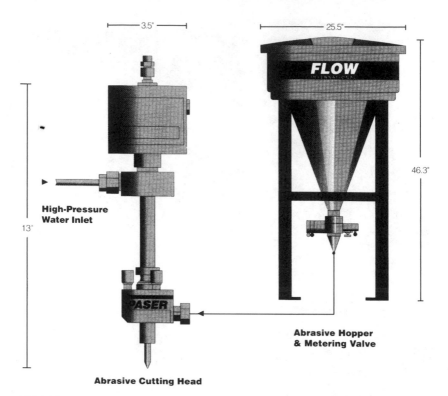

High-Pressure
Water Inlet

Abrasive Hopper
& Metering Valve

Abrasive Cutting Head

FIGURE 12.3   The route of abrasive in the AWJ cutting system, from the holding tank to the nozzle. *(Flow, Inc., Kent, WA)*

There's a lot of optional equipment offered by each AWC manufacturer. For instance, Ingersoll-Rand has a special waterjet slitting system for continuous slitting of web materials. There is a manual cutting station for freehand or template following. I-R uses a polymer mixer that meters and mixes a small percent (0.25%) of a nontoxic, long-chain polymer into the water. This improves the coherence of the high-pressure cutting stream and is useful in maintaining a narrow kerf when cutting thick materials.

All the companies mentioned in this chapter as manufacturers of AWC equipment are reliable, world-class, and absolutely trustworthy. Mention is made in this chapter of various pieces of equipment, both standard and optional, which are available to assist the purchaser and help determine questions to ask each individual AWC producer.

The waterjet cutting action occurs as a result of the high-pressure water being forced through an orifice as small as 0.003" diameter. The pressurized water exits the orifice as a coherent waterjet stream, which produces a clean cut with minimal kerf. The waterjet cuts cleanly taking kerf material with it. When coupled with a suitable motion-control system, the process provides accurate cuts with a high degree of repeatability. Incidentally, the process water is filtered to any desirable level of purity before being discharged as waste.

Abrasive waterjets can cut through 14" slabs of concrete or a 3" thick tool steel plate at a rate of 1.5" per minute and accomplish this in a single pass. The freshly cut edge will have a surface roughness of 150 to 250 microinches. In some situations, the plasma arcs can cut faster than AWC, but the heat-affected zone might make its use prohibitive.

Although the high-pressure intensifier and the receiver are heavy and large, only the cutting head needs to be manipulated for cutting parts into the desired shape (Figure 12.4). The receiver might require a little design work to integrate it into either a manual or automated system. In the least-sophisticated manual system, a hand-held wand is moved around the workpiece. But this requires much skill because of the high water pressure exiting the wand, which makes it difficult to control.

Various companies have integrated special holding fixtures or robots to guide the wand. These additions maintain a constant nozzle-to-workpiece distance and a constant feedrate of abrasive. A number of companies offer a turnkey system, including an articulating arm, gantry, robotics, and CNC tooling for accurate positioning. These parts, the articulating mechanism, the robot, and the tooling must be protected from the abrasive-laden spray.

FIGURE 12.4 Mass producing parts. *(Flow, Inc., Kent, WA)*

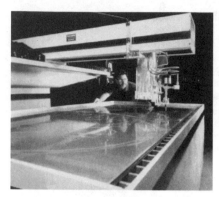

- *PASER 3* abrasive waterjet system
- 60,000 psi pump
- Machine accuracy of +/- 0.015 inch
- Choice of two large work envelopes to accommodate standard stock size material
- CNC controller
- Rugged cantilever-style X-Y configuration allows accessibility from three sides of work envelope
- Automated material handling equipment and dual cutting heads available to increase production capabilities and minimize set-up time

FIGURE 12.5   The AD series Flying Bridge. *(Flow, Inc., Kent, WA)*

- Machine accuracy of +/- 0.005 inch
- 5 foot by 8 foot work envelope with self-cleaning catcher tank (larger, field proven systems available)
- *PASER 3* abrasive waterjet system
- 60,000 psi pump
- FLOWPro software
- User friendly Allen Bradley Series 9 CNC controller
- Three-dimensional, offline programming packages available

FIGURE 12.6   The AF-5800. *(Flow, Inc., Kent, WA)*

A key to the successful use of AWC is the development of rotary joints and high-pressure water lines. Swivels were required that would stand up to the severe environment of AWC. A manufacturer of both laser and AWC systems estimated that the laser was more cost-effective at cutting metal thicknesses less than 0.375", but the AWC was better for thicker metals. Of course, this is a ball-park approach, but it is a starting point for process consideration. This might be the main difference between lasers and AWC, as far as that manufacturer is concerned, but for the aircraft industry, the difference is the ability of AWC to cut advanced composites. Military aircraft (especially) are using more and more advanced composites because that's the material for the next generation of airplanes.

The accuracy of this process depends on the type of machine used. The larger size, generally, a gantry type would have an accuracy of ±0.015" (Figure 12.5). The intermediate-sized machine, with an 8- × 5-foot envelope, could maintain ±0.005" (Figure 12.6). The very high-tech machines have an advertised accuracy of ±0.0025".

MAINTENANCE

The Omax Corporation provides the following precautions. Most nozzle difficulties from excessive wear to defective cutting are caused by problems with the jewel (nozzle). A flawed or poorly seated jewel can arise from a chip on the edge of the jewel or from dirt— either on the jewel seat or around the edge of the hole in the jewel. The result of this imperfection is that water issuing from the jewel might not pass directly down the center of the mixing tube. And a jet that grazes the mixing tube wall will cause rapid mixing tube wear and defective cutting.

This alignment can be easily checked. Remove the abrasive feed tube and stop air flow into the nozzle (a finger can sometimes do this). Now, examine the waterjet stream from the mixing tube. If the jet comes out in a solid stream for ½" or more before it becomes fuzzy, the alignment is alright. Otherwise, the cause of this misalignment must be investigated. An inferior jet will cost money, both in nozzle wear and poor parts.

Misalignments are easy to fix. Disassemble the nozzle and clean it and the seat while examining the jewel. Examine the collet for enlargement of the hole directly beneath the jewel. If recirculation of abrasives has opened the hole (listed at 0.030" diameter) and you find that it is 0.040" or larger, the jewel might lose its footing and tilt out of alignment. In this case, the collet must be replaced.

If a clogged nozzle doesn't open after conventional methods (freeing the clog with a small wire), try installing the mixing tube upside down. An inferior type of abrasive can create clogs because some suppliers do not guarantee that there are no larger particles than called for in their product. Omax provides an option to permit users to buy less-expensive garnet. It is a vibrating sleeve attachment that fits on top of the abrasive dispenser.

Each of these well-known AWC manufacturers should supply the purchaser with a maintenance manual that will allow the user to avoid most system problems. For instance, the operator must know exactly what turning on the machine does and what the computer does for him.

CASES

A manufacturer of spin windows was making his 9" diameter × ⅛" thick windows out of tempered glass because higher-quality laminated glass could not be cut using conventional methods. The tempered glass warped out of tolerance and created a high scrap rate. Changing the cutting process to AWC permitted the use of laminated glass without the scrap problem of the tempered glass.

Today's advanced composites can be as hard and rigid as steel. The same properties that make these materials so tough, also make them difficult to cut without deteriorating their quality. Composition technologists continue to introduce new material combinations that defy the capabilities of traditional machining methods. The nature of metal, ceramics, and carbon matrix composites slows down cutting speeds and rapidly dulls conventional cutting tools.

Until recently, conventional cutting methods were used to cut these new materials. Because of the composition and the fiber orientation of these composites, they were often damaged by the heat of the operation or by having their edges frayed or delaminated. Much of this damage could not be determined before assembly and use. AWC cuts 10 times faster and generally without damage.

An Ingersoll-Rand hydroabrasive nozzle was used by North American Aircraft to cut components for the B-1B bomber. The parts were cut from titanium sheet 0.125" thick. The operating pressure was 50,000 psi using 2 lbs. per minute of #60 grit red garnet abrasive and one gallon per minute of water. The garnet was introduced into the mixing chamber by vacuum created by the moving jet. This is the same technique used for mixing weed killer or fertilizer into a garden hose waterjet. One of the main reasons for using AWC on these titanium sheets is that the cuts were almost burr-free. This helped raise production by a factor of 10 to 1 in favor of AWC over the previous conventional method.

In 1989, Lockheed Aeronautical Systems of Georgia began using an AWC system that combined abrasive waterjet technology and computer-aided design. The system was used to cut production parts for the C-17 and the C1-130 Hercules. The waterjet workstation used an Ingersoll-Rand cutting nozzle and a Cimcorp controller.

The operator secured the material to be cut on a work table, then the nozzle was automatically guided. The guidance was achieved by the operator selecting instructions from the computer menu and a 6-axis robot-controlled waterjet delivered a mixture of abrasives and water through the nozzle.

The Huffman Corporation of South Carolina has a working relationship with Ingersoll-Rand that is unique. This company modifies the basic AWC machine so that it can cut under water. This greatly reduces the noise created by the waterjet in operation. It also preserves the accuracy of the cutting bed. After much use, the cutting bed shows the wear of bombardment by abrasives. If the bed is lowered and positioned under water, much of the erosion will be prevented.

Pratt and Whitney created a versatile waterjet cutting center to cut complex shapes from space-age materials, as well as to remove coatings from parts. One example of their work is a titanium flange that is 40" × ½" thick. Conventional milling of this part would take 12 times as long. Plasma-sprayed coatings used to be removed from engine components by lathe turning or by a chemical process. AWC reduced the time for that work by 90%.

Flow Corp. has joined with ASI, a developer of advanced gantry robots, to install waterjet systems for Boeing, McDonnell Douglas, Lockheed, and others. These projects were to provide the capability of cutting very large workpieces. As a contrast, disposable diapers comprise Flow's largest market segment. An unusual service provided by Flow is its patented HydroMilling process to remove concrete from bridge decks, parking garages, airport runways, and other concrete structures

The NLB Corp. of Wixam, MI, manufactures waterjet cleaning systems. Their machinery harnesses the power of water for cleaning industrial plants. Whatever the buildup, rust, scale, resins, chemical residues, paint, or epoxies, an NLB waterblast system is much faster than manual cleaning.

LAI is a job shop specializing in waterjet and laser processing. A job they undertook recently was to drill 194 holes, each with a diameter of 6.000" ± .002", in a ⅝" thick × 101" diameter plate of Inconel metal. It was done successfully by waterjet technology, whereas the previous vendor had wasted $17,000 of stock trying to do the job on a large boring mill (Figure 12.7).

Jet Edge, Inc., developed a waterjet system for machining forged titanium aircraft turbine blades. The previously used machining method had created a lot of warpage, and required grinding and deburring before the parts were done. Switching to AWC improved critical tolerances and increased production six-fold. The heart of the system is a six-station indexing table. Waterjet cutting heads are mounted on a three-axis servo-driven table with encoder feedback. This system uses measurement feedback from the cutting of the previ-

FIGURE 12.7 The stock is inconel ⅜" thick by 101" diameter. The holes are 6.000" (±0.002") diameter. *(LAI Corp., Westminster, MD)*

ous blade to offset the cutting head for the current blade. This system meets the engine manufacturer's specification of ±0.001".

Rockwell reports cutting 0.063" thick titanium using 1.5 lb/min. of garnet with a speed of 12" per min. Avco Aerostructures-Textron is using AWC to modify wing structures of the C-5 transport planes. Using garnet as abrasive, Avco cuts through aluminum and titanium up to 2.5" thick and through tough graphite epoxy up to 3" thick.

EQUIPMENT

AWC is a reliable and accurate tool for specialty users, such as abrasive jet job shops, laboratories, development centers, and a multitude of various manufacturers in a variety of industries. A new development in the technology has enabled AWC manufacturers to advertise that tolerances of ±0.005" can now be held. Repeatability of ±0.0015", squareness of 0.0015" per foot, and straightness of 0.002" per axis can be expected. These advances should preclude the necessity of many secondary operations.

The Omax Corporation's AWC is a complete system. Its AWC computer loads a CAD drawing from another system and changes it to one for the Omax system. The computer then determines the starting and stopping points and the sequence of the cut. Then the operator enters the type of material, its thickness, and tool offset data. The computer then calculates the feed rate and makes the part. The machine itself is hard anodized aluminum castings and frame. The ball screw, bearings, and motors are enclosed in bellows for complete protection. Each manufacturer has to protect his machinery somehow from the abrasive environment of the process.

Some AWC manufacturers provide "turnkey" systems that operate with a modem and CAD/CAM capabilities that permit transfer from CATIA, AUTOCLAD, IGES, and DXF formats. Their computer runs a program that determines in seconds how to minimize waste when cutting from blocks or plates.

Ingersoll-Rand has a new concept for its intensifiers. Instead of using an accumulator to flatten surges in pressure from the intensifier, I-R uses a programmable controller to operate twin cylinders, one discharges high-pressure water while the other recharges. In this way, without using an accumulator, deadband is eliminated and pressure fluctuations are reduced significantly.

All manufacturers have something that is proprietary and different in their product. The Flow Corporation's machine includes an abrasive cutting head and an abrasive metering system. Figure 12.8 shows twin "Pasers" at work on nested parts and Figure 12.9 shows another "Paser" in a fairly intricate cut. The ability to nest parts saves material, as does the ability to cut kerfs under 0.06".

FIGURE 12.8 Twin nozzles and stacking multiple sheets of material increases production. *(Flow, Inc., Kent, WA)*

FIGURE 12.9　Flow Corporation's patented "PASER."
(Flow, Inc., Kent, WA)

The biggest improvement in AWC is tool-life (nozzles). One manufacturer has claimed that theirs now is good for 100 hours of operation. Ingersoll-Rand has a diamond nozzle that they claim is good for 500 hours. Always consider the parameters under which these claims are made. For instance, the rate of abrasive use and the water pressure.

All machines have some type of computer-controlled motion system with an X-Y table. They are capable of machining a large variety of materials and thicknesses. Cut edges are usually smooth and satiny in appearance.

Punching is still the fastest way to produce holes or cutouts in any part that is punchable. After all, a high-speed press operates at hundreds of strokes per minute. Yet press manufacturers are now producing combination presses. That is, if the work can be punched, punch it. But if requirements preclude that operation, the work can be performed in an auxiliary mode. In some cases, that would mean by laser, in other cases by plasma arc. If the preponderance of work is thin sheetmetal, 0.060 to 0.135" thick, laser would be the choice. If the work is thicker, up to 1", plasma arc would be the choice. Yet many shops are buying AWC systems because there are so many situations where AWC would be preferred. See Figure 12.10 for a variety of work that can be done by AWC.

Both lasers and plasma arc equipment produce heat that could spoil the workpiece. This same heat could generate toxic fumes. These are not problems with AWC. In addition, AWC carries the kerf material into a catcher tank, significantly reducing airborne dust. Many processes have been tried to cut advanced composites, but nothing does the job as well as abrasive waterjet cutting. The point is that there is a need for and a place for each process. Management must contemplate the big picture. Which process or combination of processes is the most economical for the projected workload?

Abrasive waterjet cutting offers an alternative to the traditional machining methods. This type of cutting has no heat-affected zones, no recast layers, no toxic fumes, no micro-fractures, and (generally) no secondary operations. The elimination of friction,

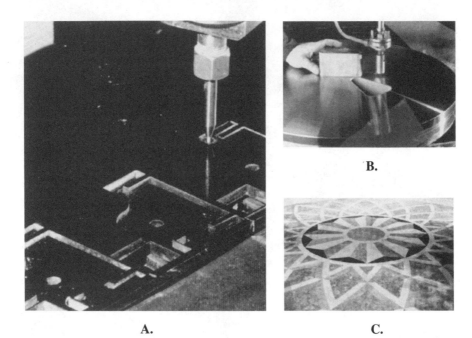

FIGURE 12.10 A variety of work can be performed by AWJ cutting: A. Minimal kerf width allows tight nesting and common line cutting of parts. B. Cutting thick metals. C. Cut stone and marble. *(Flow, Inc., Kent, WA)*

which normally occurs in tool-to-part contact, avoids the thermal damage that can adversely affect metallurgical properties in materials being cut. The USDA has approved the process for the sanitary cutting of food products. Job shops have found that they can increase cutting speeds through thick materials. Sheetmetal fabricators find AWC useful for a wide range of materials and thicknesses from prototypes to volume production. Management, which never before considered this process, should reappraise it to remain competitive.

CHAPTER 13

LAPPING, POLISHING, AND HONING

Lapping, polishing, and honing are tried and true forms of surface finishing. They were formerly very expensive and designers avoided them like the plague. But I recently witnessed the solution of a very serious production problem by substituting lapping for the current process, double disc grinding, and I was determined to include the process of lapping in this book.

An aluminum computer part, about 3" × 1" × 0.125" thick, full of holes, was being made by fine blanking, then was finished by double-disc grinding. Ordinarily, double-disc grinding is a good process for making a workpiece flat. But these parts were consistently just enough out of flat to require a straightening operation. That part is now being lapped in a twin-wheel lapping machine. The bottom line is that the new process is producing flat pieces and saving money. Lapping, polishing, and honing in modern machines is so fast that these processes are being reintroduced to American industry.

The demand for improved quality of manufactured products has challenged traditional machining processes. This chapter is about abrasive machining processes, which are used to produce extremely close tolerances for flatness, parallelism, and thickness with minimum operator judgment. These processes are improvements over grinding and are presently used by manufacturers of all sizes of bearings, hydraulic pumps, seals, cutting tools, computer components, various glass products, ferrites, carbides, ceramics, measuring tools, semiconductor materials, and automotive parts, to name a few.

LAPPING

The Process

Lapping is a low-velocity abrasive machining process that uses loose-rolling abrasives in a liquid or paste vehicle. A slight pressure and a relatively slow motion are applied by the lap to the surface to be machined. The workpieces are free to align themselves against the lap. The process permits an accurate, though nonreflective, surface finish (Figures 13.1 and 13.2).

Lapping occurs when abrasive grains in a liquid vehicle, often known as a *slurry*, are guided across the surface to be lapped and backed by a lapping plate. Abrasive grains used for lapping have sharp, irregular shapes. When pressure and motion are applied, the grains are forced into the workpiece and cause the material to chip away in microscopic particles.

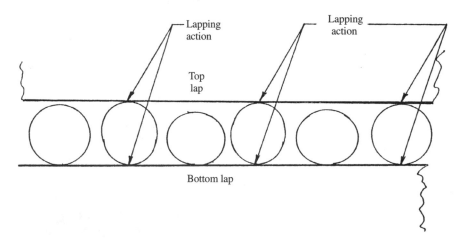

FIGURE 13.1 Cylindrical lapping between flat laps. Workpieces roll between flat laps. The top lap is free floating and self aligning. The largest workpieces and the largest part of each workpiece receives the lapping action of the top lap while smaller or oval workpieces receive no lapping action on the smaller diameters. *(P. Wolters, Inc. Plainville, MA)*

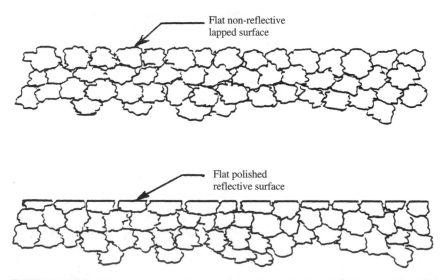

FIGURE 13.2 The difference between reflective and nonreflective finishes. *(P. Wolters, Inc. Plainville, MA)*

The Procedure

Typical operational procedures have been established. Parts are loaded into the slotted plate type workholder and a small amount of lapping compound is applied to the top of the parts or to the bottom of the upper lap. The upper lap is positioned above the lower lap on the same centerline and is locked. The upper lap is then lowered down on top of the parts until the programmed starting pressure is supported by the parts.

The machine is started, rotating the lower lap in a counter-clockwise direction, causing the parts to roll and push the workholder around its centerline. After a predetermined time (approximately 30 seconds), the automatic cycle timer stops the machine. The operator raises the upper lap and swings it out of the way. She checks the size of a few parts and rearranges the workload by transposing every other part with one directly opposite in the load. She might also turn the parts end for end as they are transposed. A small amount of lapping oil is added, the upper lap is lowered again, and another 30-second cycle begins (Figure 13.3).

At the end of the second cycle, the upper lap is raised and a few more parts checked for size to confirm a time/stock-removal ratio. Each part is then turned end for end. A small amount of lapping oil is added and the upper lap is lowered again. This short-cycle procedure is repeated until the desired part size is reached. The lap pressure is adjusted to provide the best lapping action.

Liquid Vehicle

With regard to the movement of the abrasive grains, special importance must be assigned to the liquid vehicle. The grains are rarely shaped evenly round and are much more likely to be odd shaped, perhaps as slivers. Activity occurs within the slurry. The workpieces move, the adherent liquid moves with it, carrying the abrasives. Vortices develop in the liquid, which force the abrasives to work.

Water-based vehicles are commonly used for double-side lapping although oil vehicles can and are being used. A few objections to oil are: its tendency to penetrate the pores of some materials (ceramics), its expense, and the greater difficulty of disposing of it. The purpose of the vehicle is to carry the abrasives and position them to work most efficiently. It also lubricates the surfaces, carries away the abraded material removed from the workpieces, and also removes some of the heat of the operation. Inhibitors should be added to water vehicles to prevent rust. In some cases, suspension agents are added to prevent settling of the abrasives.

Surface Finish

The surface finish produced by lapping is almost always nonreflective and has a gray matte or frosted glass appearance. The finish texture depends on the composition of the abrasive. For example, a steel part lapped with a 12-micron silicon carbide abrasive would probably produce a surface finish of 5 to 6 microinches, whereas the same steel part lapped with a 30-micron silicon carbide abrasive would have a 12- to 13-microinch finish.

The finish texture is unique in that each microscopic indentation is separate from its neighbors. Thus, it provides an ideal sealing surface. A ground or honed surface that consists of machine marks in linear form can be a source of leaks. Although some parts that have been honed and others that have been lapped could have the same microfinish, the lapped surfaces would be less likely to leak.

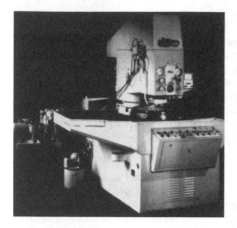

A.

B.

C.

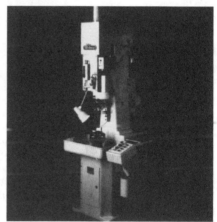

D.

FIGURE 13.3 A few models of lapping and honing machines: A. Twin-wheel lapping machine, Microlap AL 2 L. B. Twin-wheel edge-honing machine, Microlap AL 00 K. C. One-wheel flat lapping machine, Microlap FL 8. D. Cylindrical internal lapping machine, Microlap IL 12. *(P. Wolters, Inc. Plainville, MA)*

When the dimensional tolerance is less than 0.000050", time plays a very significant role. The operator handling the machine soon develops an understanding of the time/stock-removal ratio. The cycle is stopped frequently to turn the parts end for end. At this time, the operator measures a part and records the size. The first cycle always removes the most stock because the previously machined finish has many peaks and valleys and the peaks lap off easily. Also, the lapping compound is sharpest when the process first starts.

It is possible to attain accuracy and straightness to 5 millionths of 1" by lapping. The lap flatness and the operational procedure are the two factors that determine the machine capability. With the proper abrasives and lubrication selection, finishes of 0.5 microinch are obtainable on hard, dense metal. Finishes of 1 microinch and less are being routinely achieved on plug gauges and valve parts. Although these super finishes have been achieved by expert operators, finishes under ½ a light wave (5.8 millionths") are difficult and require more care and time.

A lap finish is exceptional and cannot be duplicated in any other way. The finish pattern is completely multidirectional on parts finished by flat lapping. Any other finish provides directional patterns. Experience indicates that a multidirectional finish has a longer life. Parts that have been ground between dead-centers prior to lapping can be flat lapped round to 0.000005". That same part, centerless ground, could not achieve that accuracy.

Abrasives

Abrasives are available in a variety of forms: soft to hard, strong to brittle, coarse to fine, uniform to irregular. The most common abrasives for machining softer materials are aluminum oxide, garnet, and cerium oxide. Harder materials are machined with silicon carbide, boron carbide, and diamond (Table 13.1).

TABLE 13.1 Hardness Chart.

Abrasive	Mohs	Knoop
Zirconia	8	1160
Garnet	8	1360
Aluminum Oxide (Alumina)	9	2100
Silicon Carbide	9 to 10	2480
Boron Carbide	9 to 10	2750
Cubic Boron Nitride	10	4500
Diamond	10	7000

Stock Removal

The material-removal rate is proportional to the following influences: lapping speed, wheel pressure, grit type and size, type of lapping liquid, and proportion of grit per unit of liquid. The volume of stock removed naturally increases as the speed of the wheel increases. Because the depth of penetration of the grit does not change with a change in velocity, the surface finish is normally unaffected by the speed of lapping. However, it can be easily seen that the depth of penetration does change with increased pressure.

Normally, the amount of stock removed in lapping is between 0.00015" and 0.0005". If a decent grinding job has been done prior to the lapping, the lapping should remove imperfections in the part. If a finish of less than 1 microinch is desired, fine abrasive should be used and the stock removed would be about 0.00015". If the required finish is 2 microinches, then the stock removed would be about 0.0003".

Workpieces roll between the flat laps. The top lap is free floating and self aligning. The largest workpieces receive the lapping action, and smaller or oval workpieces receive little or no lapping action.

Advantages of Double-sided Lapping

- Machine two sides of workpieces in the same time as one.
- No problem holding nonmagnetic materials.
- Machine any kind of stable material from plastic to diamond.
- Best method to produce close accuracies of flatness, parallelism, and size.
- Removing stock from both sides simultaneously helps relieve internal stresses of workpieces, thus making it easier to maintain flatness.
- Permits simple workholder design without clamps or anything to hold workpieces rigidly, thus improving accuracies.
- Free-floating top wheel achieves accuracies. This precludes critical machine alignment, precision high-speed bearings, or accurate sliding ways.

Production Rates

These machines can handle about 100 parts per hour, varying with the tolerance required, part size, and machine size. The smaller the part, the more parts can be worked per hour. Two or three loads per hour can be lapped with an average of 40 parts per load. Of course, this includes handling, checking, and lap-conditioning time.

Materials That Can be Lapped

Theoretically, any material that can withstand the lapping pressure can be lapped. Hard, stable materials (such as glass, hard steels, and certain ceramics) are usually lapped with conventional abrasives and cast iron laps. Carbides are lapped with very hard abrasives and a hard, sintered, metal-bonded, abrasive lap. Otherwise, there would be excessive lap wear.

Lapping Abrasives

A variety of abrasives is utilized for cylindrical lapping hard alumina, 2- to 3-micron size, suspended in an oily paste, is generally selected for a finish of 1 microinch on hard steel. A corundum of 800 grit is often used if much stock is to be removed and finish requirements are broader. Harder materials, such as carbides and certain alumina ceramics, can be lapped with boron carbides or diamond abrasives.

The correct choice of abrasive for lapping or polishing depends on various factors (e.g., material, method of premachining, amount of stock removal, required surface quality, and the necessary accuracy of the geometry). Another significant factor in achieving consistent results and to avoid scratches is the consistent quality of the abrasive and the smallest variation in bandwidth of the grain size, as well as an equal distribution in the carrier liquid.

In any comprehensive program for lapping and polishing, experience is very important. For that reason alone, it is important to consult with an expert: the machine manufacturer or a reliable vendor. The Peter Wolters Company (among others) markets a wide variety of reliable abrasives. They include silicon carbide, alumina, and boron lapping compounds, and several for special jobs, such as bore lapping. They have polishing agents, diamond paste, and a number of additives for the lapping and polishing slurries. These include antifoaming agents, which are sometimes required, depending on water hardness, and rinsing and cleaning agents.

Type of Parts

The most common cylindrical parts lapped between flat laps are hydraulic valve spools, plug gauges, piston pins, diesel injector parts, and special armatures. More and more ceramic parts are currently being used. When flatness or thickness is critical, there is no better way to achieve those goals than by lapping. Water-based vehicles are generally used with ceramics because of the tendency of oil to penetrate the pores. See the display of lapped parts in Figure 13.4.

FIGURE 13.4 Products that were lapped. *(P. Wolters, Inc. Plainville, MA)*

FIGURE 13.5   A sprocket-type workholder for double-sided lapping. *(P. Wolters, Inc. Plainville, MA)*

Machine Design

Two annular laps, each mounted on a vertical spindle are used on cylindrical lapping machines. Depending on the machine model, one or both of the laps rotate. Workpieces are positioned in a workholder, which guides them between the lap faces. The workholder is disc-shaped and is thinner than the workpieces. It is guided in the center by a pin that can move the workpiece eccentrically to the middle of the lower lap.

The workpieces are placed in slots that are tangent (not radial) to the center of the workholder. The rolling action of the parts drives the workholder. In some machines, the lower lap rotates while the upper lap is stationary, but self-aligning. See the double-side lapping machine with the workholder sprockets in Figure 13.5. The sprockets in this picture are for flat, round components .

FLAT HONING

Unlike grinding, flat honing is a cool, gentle process. There are no sparks and only enough pressure is applied to cause the abrasive grains to penetrate and cut. Slow wheel speeds are used and a flood of filtered honing fluid is circulated around the parts to carry away material honed off and to stabilize the temperature. Heat-sensitive components are often better honed than ground.

This is a high-precision, low-velocity abrasive machining process using bonded abrasives, called *flat honing plates* (wheels). These honing wheels are, in reality, fine-grit abrasive grinding wheels. Pressure and motion are applied between the honing plates and the workpieces, which are free to align themselves. In other words, the workpieces are guided between two wheels. Honing oil is used to lubricate, clean, and cool the parts and equipment. Modern machines work so fast and to such close tolerances, that they come equipped with heat exchangers to help maintain a uniform temperature. No loose abrasive is used in honing: it is fixed abrasive in the form of a wheel.

This type of honing, around since about 1940, has been used to obtain flat, parallel surfaces, which could not be achieved by grinding alone. Flat surfaces can easily be obtained by lapping or honing in any manner. However, parallelism can be achieved only by using twin lapping or honing machines. When equal amounts of stock are removed from both sides simultaneously, stresses are relieved at the same time that mirror finishes are left on the surfaces. The honing wheel is really a fine-grit grinding wheel run horizontally at low speeds (Figure 13.6).

The type of parts that gravitate to this process are those requiring precision surfaces. Examples are valve plates, ball bearing races, seals, pump components, hardened wear surfaces, and hydraulic valve parts.

Advantages of Flat Honing

1. This process produces flat, clean, parallel, reflective surfaces.
2. Most materials no harder than Rc 62 can be flat honed. When there is a bore that must be perpendicular to the honed surface, be sure that it is perpendicular in the preceding machine operation. If it square before honing, it will be square afterwards. However, honing will not improve an out-of-square condition.

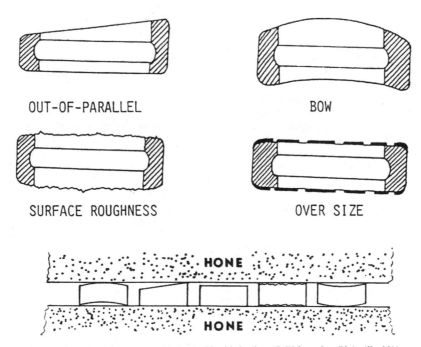

FIGURE 13.6 Conditions correctable by double-side honing. *(P. Wolters, Inc. Plainville, MA)*

3. Magnetic or any other type of fixturing is not necessary. Only the simple holding disc is used. This allows work on nonferrous parts.

4. The gentle operation results in little surface damage. Stresses are released equally from both sides. This is less likely to cause out-of-flat conditions. In fact, flat honing should correct them.

5. Finished parts come clean from this operation with only a thin film of mineral oil coating. Elaborate cleaning or rustproofing is not normally required.

Material hardness must be considered. Very soft material, such as carbon seals, cut too easily, resulting in loss of control of size and parallelism. Such parts are better processed by lapping when close tolerances are required. When materials are very hard, like alumina ceramics, tungsten carbide, or hardened tool steels, they do not hone well, either. The traditional abrasives used in flat honing just do not cut these materials.

Wheels

In the flat-honing process, silicon carbide or aluminum oxide wheels are used. Although the silicon carbide is more versatile, aluminum oxide is used wherever possible because of its lower cost. The wheels are often a vitrified bond with hardness rated between I and K and a structure of 8 or 9. The grits are generally 150 to 400. A vitrified wheel is the most economical.

The more-expensive resin-bonded wheels cut faster because they break down under load faster than vitrified bonded wheels. Because of their polishing effect, they produce a more-reflective surface. Resin-bonded wheels are more suited to honing harder steels than nonferrous materials.

Ideally, the honing wheel should have a cutting surface that will stay sharp and not load up and lose its flatness. A small amount of wheel breakdown is desirable to expose fresh, sharp abrasive grains for continuous and uniform honing.

Normally, honing wheels are solid, full-faced. When honing harder, solid-surface workpieces, perforated wheel faces are used. The holes in the face of the wheel sometimes permit better abrasive penetration and coolant circulation.

The modern two-wheel honing machine provides fast, consistent production. The honing action doesn't change all day because of temperature control. Water-cooled wheels, temperature-controlled hydraulics and cutting fluid aid in producing precision parts. An automatic electronic-measuring device controls workpiece thickness within 0.0002".

A self-aligning, free-floating top wheel provides the ultimate in accuracy without costly, periodic alignment of the spindle. The machine is heavily constructed to minimize deflection. The workholder guides the workpieces in an ever-changing pattern between the revolving wheels to provide even honing action and uniform wheel wear. There is a steady flow of filtered honing oil that cleans, lubricates, and cools the process. A rigid, diamond trueing device reconditions and trues the wheels. The top wheel can be raised and lowered under power and it can be moved aside for access in loading and unloading at the end of cycles. The wheels and workholders have independent drives that permit reversing direction of rotation.

The Honing Process

During the honing process, a coolant must be applied copiously:

1. It prevents thermal distortion by cooling the workpieces during honing, and cools the trueing diamond when the honing wheel is trued up.

2. It prevents wheel face loading by flushing the wheels.

3. It flushes away honed workpiece material and abrasive grains broken off of the hone wheel.

4. It lubricates both the workpieces and the wheels, reducing friction.

The lubricant should be introduced at the wheel center and allowed to flood over the wheel faces and workpieces by centrifugal action. A thin mineral seal oil seems to work best when flat honing with silicon-carbide wheels. Synthetic coolants or soluble oil/water mixtures are less desirable. Insufficient lubrication leads to wheel loading and poor finish.

Filtration down to a maximum of 5 microns in particle size should be continuously conducted. Honing softer materials (such as iron, aluminum, and soft steel) is not generally a problem. When honing harder materials, however, heat develops because of higher pressures and increased friction. Sometimes it is necessary to add a heat exchanger to the system. If heat is allowed to build up, the cutting action slows down.

Hardened workpieces with broad surfaces are more difficult to hone because of the load they present. Hone pressure varies, depending on the material and shape. Heavy, solid parts can withstand higher pressures than thin parts. Pressures of 5 to 10 psi are common. Although too little pressure is usually harmless, except for lengthening the job, too much pressure will cause excessive heat, dulling of the grains, and glazing of the surface being honed.

General Principles of Honing

Parts are presented to the flat honing machine in such a condition as to do as little honing as possible. The principles of flat honing are as follows:

1. Usually, two flat honing wheels are used.

2. Hone wheel faces are maintained in precision flatness.

3. Workpieces are guided in workholders between flat hone surfaces.

4. Workpieces are always free to align themselves during honing. They are never clamped, held rigid, or tightly fixtured.

5. Usually, a light oil (coolant) is used to lubricate and cool, as well as keep the hone cutting surface clean.

6. The hones apply pressure to the workpieces to provide the proper cutting action.

Flat honing is designed to correct out-of-flatness and out-of-parallelism, as well as to improve the surface finish and produce a desired thickness. Typical surface conditions prior to flat honing are: oversized, out-of-parallel, bowed, and rough. Usually, a lapped or honed surface is a base line from which important dimensions are taken. There are places where 0.0001" out-of-flatness or parallelism could ruin an assembly. A hydraulic seal or ball-bearing race are examples of this.

Batch-mode Processing

The machine is loaded with as many components as can be fit in the workholders—the more the better. Batch-mode processing allows a number of workpieces to be honed at the same time, and each piece will be honed to the same thickness. Basically, doing many pieces at one time actually improves uniformity of production.

In this mode, two wheel machines are used. The best-possible flatness and parallelism can be obtained by this method for several reasons. Because the flatness of the wheel faces are maintained precisely, the surfaces being generated actually mirror this flatness. If any internal stresses are present in a workpiece, by removing stock from both sides simultaneously, it helps to normalize the part and maintain close tolerances. If a part has internal stresses, machining one side at a time will create a bow. When working to tolerances of less than one micron (0.000040"), such stresses are a factor that must be considered.

Most honing is batch mode, but there is through-feed mode as well. This is good for such parts as small ball-bearing races under ½" diameter. Parts are generally fed through the revolving hones once, but if more stock needs to be removed, multiple passes are necessary.

Cylindrical Honing

This process applies a superfinish to the outside diameter of components, such as valve spools. The polishing action occurs between two flat revolving wheels of the same configuration as for flat honing. A quantity of workpieces are nested in each slotted workholder, where each individual part is free to revolve in its slot. The workpieces roll between the rotating wheels, which are turning in opposite directions. The slots in the workholders are designed with the centerline tangent to a circle in the center. In this way, the workpieces slightly scuff as they roll, producing the desired superfinish. The coolant used is normally water, the wheel bond is resin, and the abrasive is silicon carbide fine grit (400). The high points left from grinding require about a 5-minute cycle to improve the surface finish to acceptable.

Edge Honing

This process has a resilient pad bonded to the lower wheel. The pad is saturated with abrasives. Pressure is applied by the upper wheel. The workpieces are partially submerged in the (hard) resilient pad. This causes the part edges to be honed or deburred in a clean and dry process.

Internal Honing

This low-velocity abrasive machining process utilizes bonded abrasive hones to do the cutting. Either the workpiece or the hone is free to align itself. Pressure and motion is applied between the workpiece and the hone. Presently, the Sunnen Corp. of St. Louis, MO sells a fixture that allows outside diameters also to be finished on the standard Sunnen (internal) honing machine. Their standard internal honing machine, the MBB-1660, can hone any hole diameter between 0.060 and 6.500" in regular or odd-shaped parts. See Figure 13.7 for a photograph of the popular Sunnen internal honing machine that can be found in so many machine shops.

This machine will hone bores with keyways or splines. Specially designed stones and mandrels make this easy (Figure 13.8). Adapters are available that permit honing to the bottom of blind holes. Because no clamps are used to hold the workpiece, thin wall, and delicate parts can be internally honed (Figure 13.9). Ten common hole inaccuracies can be corrected by internal honing (Figure 13.10).

Bores in most materials can be honed: Alnico, aluminum, carbide, ceramic, ferrite, fiberglass, gold, nylatron, plexiglass, porcelain, quartz, silicon, stainless steel, tantalum, zircaloy, and just about any other material.

SUNNEN

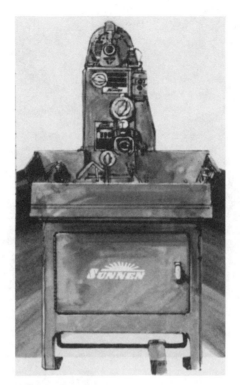

FIGURE 13.7 The standard Sunnen internal honing machine. *(Sunnen Products, Co., St. Louis, MO)*

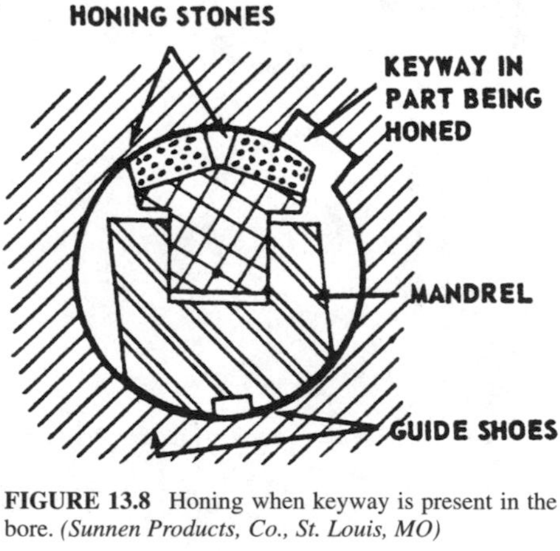
HONING STONES
KEYWAY IN PART BEING HONED
MANDREL
GUIDE SHOES

FIGURE 13.8 Honing when keyway is present in the bore. *(Sunnen Products, Co., St. Louis, MO)*

FIGURE 13.9   Bores in a variety of products are honed. *(Sunnen Products, Co., St. Louis, MO)*

FIGURE 13.10   Ten common bore inaccuracies. *(Sunnen Products, Co., St. Louis, MO)*

Procedure of Internal Honing

1. Measure the bore.
2. Set honing dial for amount of stock to be removed.
3. Hone until the dial reads zero.
4. Be sure that a satisfactory supply of honing oil reaches the stones and mandrel—especially when working blind holes.

Setups don't generally take much longer than 10 minutes because all fixtures and special tools are with the machine and indexing, and critical tool adjustments require only moments. Easy-to-read charts and tables show clearly which type of tooling, cutting stones, speeds, and cutting pressures to use.

The Sunnen mandrel design with one stone and two unevenly spaced guide shoes permits the machining of round bores with the least removal of stock (Figure 13.11). The company claims that their machine can consistently hold a 0.0001" tolerance and deliver a 2-microinch finish.

Generally, a bore is considered short when its bore length is less than the bore diameter. This makes it difficult to maintain the bore centerline and it requires the use of short bore techniques. If the outside diameter of a part is much greater than the bore to be honed, a short-bore technique should also be used. Again, this is true if the bore to be honed is not centered, with respect to the weight of the part.

Three basic techniques are used for honing short bores:

1. *Stacking—clamped or loose* In stacking, short parts are lined up by their bores, end clamped, and honed as one long bore. This is the preferred method of honing short bores because it provides high production rates. There are two preconditions for stacking. The pieces must have flat, parallel end surfaces and the bores must be close to the same diameter.
2. *Honing free-hand—one or two at a time* When honing free-hand, the operator should avoid side thrusts by honing with both hands, keeping fingers as close to the bore as possible.
3. *Honing against a faceplate* If the short-bore part has one face that is flat and perpendicular to the bore, install a face plate on the machine and hone while pressing the flat surface against the faceplate.

Select a mandrel with the shortest stone and guide shoe length when honing short parts. When honing stacked parts that are clamped, the stone length should be between ⅔ and 1½ times the stack length. Actually, the longer length is preferred. It pays to become familiar

| Integral steel shoe honing tool | Double length stone honing tool | Honing tool with adjustable guide shoes |

FIGURE 13.11 Mandrels with their stones. *(Sunnen Products, Co., St. Louis, MO)*

with the printed advice of the machine manufacturer as to how to handle various materials and jobs. There are many types of tools, mandrels, and stones that have a place in honing internal bores satisfactorily.

Many designs call for blind or shouldered holes. A hole that has a bottom, a shoulder, or any kind of obstruction that would prevent a mandrel from going all the way through the hole is termed a *blind hole*. Designers would be of great help to the manufacturing process if they would at least provide a relief in the bottom of the hole.

FLAT PRECISION POLISHING

Hard workpieces (such as ferrites, ceramics, and tungsten carbides) are polished in a manner different from soft workpieces (such as aluminum, silica, quartz, and silicon). A high-precision, low-velocity abrasive machining process that utilizes abrasives embedded in a hard polishing pad is used for the hard materials. Pressure and motion is applied between the polishing lap and the workpieces (Figure 13.12).

For the soft materials, a relatively soft pad adhered to the flat polishing wheel loads up with soft abrasive and improves the finish and flatness of the workpieces. Controlled lubrication is used in the polishing, which results in a reflective surface.

Silicon wafers are now being finished by lapping and polishing. The manner in which this is being done has been modified slightly and is also being used on germanium, sapphire, and a few other similar substances. In wafer production, an etching process preceded the polishing because it was believed that this would remove the subsurface damage, thus accelerating the polishing step. However, this seemed to cause a deterioration of geometry. Consequently, if any etching is done now, the step is short.

Wafer polishing results are influenced by the following factors:

- Type and quality of the polishing machine used and its control.
- Type and quality of the polishing cloths.
- Polishing compound and other chemicals used.
- Polishing temperature and pressure.

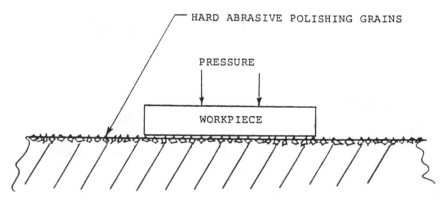

FIGURE 13.12   Hard abrasive polishing (pad shown on wheel). (*P. Wolters, Inc. Plainville, MA*)

Currently, a four-station, single-wheel polishing machine has been popularized for processing wafers.

Machine Design

Notice that in most new finishing machines, two wheels are used. They are required to achieve parallelism and they generally work turning in opposite directions, although each wheel's speed and direction is adjustable. The pressure used in most cases is about 2 to 4 psi. Most of these machines are computer controlled for automatic operation.

These machines have rigid support for the abrasive wheels and heavy design throughout to minimize deflection. They have adequate power to drive the wheels and workholders and supply lubricating/cooling oil. Provisions are made to raise and lower the top-honing wheel and to swing it out of the way to facilitate the loading and unloading of workpieces. There is a control of honing pressure whereby the cycle starts with light pressure on the top wheel to hone off high points and slight bows and to gradually increase as the workpieces become somewhat flat and uniform in thickness.

The abrasive grains begin to break down and cut more slowly so that each 30-second cycle removes less stock. The operator is able to machine to very close tolerances because of this reduction in stock-removal rate. For example, the first 30 seconds might remove 0.000050". The second cycle would remove 0.000030", the third 0.000025", etc. The final cycle could remove only 0.000003 to 0.000005" of stock.

Workholding

During the honing process, it is necessary to hold the workpieces. When honing both sides of the workpieces simultaneously, the workholder guides the parts between the wheels. The workholders must be thinner than the finished pieces. To ensure uniform wheel wear, it is good practice to design the workholders to traverse the parts a little beyond the OD and ID of the hone face. One or two pieces cannot be processed in this manner, unless, some of the vacant positions are filled with dummy parts.

A popular workholder design uses a set of four or more gear-shaped (sprockets) workholders driven by an inner gear. The workholders rotate inside a stationary ring of pins lining the outside edge of the machine table. This provides a positive drive and traverses the workpieces over the wheel face in a constantly changing pattern that is conducive to uniform wheel wear. This produces parts of excellent parallelism and uniform size. An independent motor drives the inner ring, which permits different speeds for the various job requirements. It also allows periodic reversals of rotation, which tends to maintain wheel flatness.

Numerous holes or cutouts are machined into the workholders to hold the workpieces. There should be sufficient clearance between the parts and their cutouts to permit the workpieces to rotate and align themselves easily during the honing process. The workpieces must not fit tightly in the cutouts. Workholder material is generally steel for thin carriers and plastic, or micarta for thicker carriers—weight being a factor.

Workholders are used to guide workpieces between the revolving hones, laps, or polishing wheels. They guide each part across the full face of the wheels to maintain uniform wheel wear. The parts should be able to turn freely in the workholder and align themselves between the wheels. Typically, workholders are sprocket carriers, driven in an epicyclic manner by a rotating inner ring of pins around a nonrotating outer pin ring. The speed and direction of the inner pin ring can be adjusted for optimum performance (Figure 13.5).

To manage the large variety of jobs, many workholders are available. When working parts to a low-microinch finish, the parts should not be allowed to rub against the workholder. Straight, solid pins are the easiest parts to lap. However, any cylindrical parts can be lapped between flat laps. When lapping minor diameters on shafts, the workholder is arranged with the larger diameter overhanging the OD and ID of the laps.

Wheel Trueing

The two faces of the annular shaped, bonded abrasive lapping and honing wheels must be absolutely flat, in order to machine flat to close tolerances. This is usually accomplished by using a heavy pivoting arm to traverse a diamond tool across the faces of the two wheels while they are rotating.

The hone wheel faces must be smooth to produce a low-microinch finish. This is accomplished by using a slow traverse speed when trueing with a single-face diamond. If a diamond cluster is used, the traverse speed can be increased considerably.

When honing harder materials with large surface areas, the honing action can be improved by trueing the wheels with a single-point diamond at a rate of 0.060" per revolution. There are methods to extend the time between wheel true-ups. For example, when the honing action starts to slow down, the operator can shut off the coolant while the honing continues. The wheel face will then break down and make the worn, dull grains fall out and expose new sharp grains. This should not be done more than once between conventional trueing operations.

Like many chapters in the *Advanced Machining Technology Handbook*, this subject also could completely fill a book by itself. There is so much more to it, and yet the purpose of the book should not be forgotten. It is to arouse the readers' curiosity and act as a catalyst to considering new manufacturing processes. I haven't mentioned the light-band measurement of flat surfaces or the method of restoring cast-iron lapping plate flatness. Vacuum chuck design or how the machines combine lapping and polishing hasn't been mentioned either. I haven't included the small lapping, honing, and polishing machines, which fill the need of smaller production quantities or just occasional use.

Many machine features are left to a live demonstration when you are investigating the subject and contact machine manufacturers. I have mentioned two machine manufacturers, The Peter Wolters Corp. and Sunnen. This is a good place to name a third manufacturer, Spifire, Inc. Chicago, IL. They make and sell products, competitive to the Wolters line. Some of these features are waxless wafer polishing, then the popular method using wax beds for wafer polishing. Many material-handling accessories are available. The sales representative will be happy to demonstrate anything that might be useful to your production.

CHAPTER 14
INDUSTRIAL SAWS

BACKGROUND

Every product or part of a product starts with a cutoff job. The raw material has to be cut. In machine shops all over the world, whether the job calls for one part or one thousand, the bar stock, tubing, sheetstock, pipe, or plate must be cut. It is the first step in production. There are four alternative methods for doing this: hacksawing, cold sawing, abrasive cutting, and band sawing.

The oldest form of power metal cutting is the power hacksaw. See Figure 14.1 for a view of two Keller gravity fed hack saws, a type that has been in service many years. Then there's Figure 14.2, a modernized Japanese Amada, doing the same type of job. It doesn't look much different, except for the electronics, which permits automation.

Around the middle of the 20th century, three other forms of power cutting became popular: power band saws, cold-cutting saws, and abrasive-cutoff saws. Today, there is considerable overlap because each type of saw can handle a wide range of cutoff applications. Each has advantages and disadvantages that have determined their niche in industry. All types have the basic ability of repetitively cutting stock into specific lengths. All types are available in two basic styles: manual and automatic. The four types have one other common feature: a replaceable cutting tool.

The major segment of power hacksaw cutting in the entire range of cutting capacities can be divided into two main categories; over-10" solids and under-10" solids. Formerly, there were critical mechanical and/or physical problems for each of the other types of saw in the under-10" capacity, However, advances in technology since 1975 have changed the technique of cutting.

A saw is one of the oldest types of machine tools to be found in any machine shop. In fact, it would be difficult to locate any shop that didn't have a saw of some type. Generally, the first machine a shop apprentice learns to use is a power saw. It is one of the oldest and the most basic and elementary tool in the shop; so why is it listed with modern manufacturing processes?

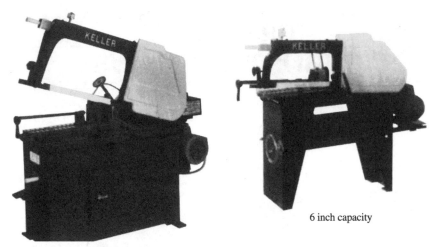

10 inch capacity

6 inch capacity

FIGURE 14.1 Keller gravity-fed hacksaws, a simple and reliable type: A. 10" capacity. B. 6" capacity. *Keller Industries, Hollandale, MN)*

FIGURE 14.2 High volume cut-off machine. *(Amada Power Hacksaws, Japan)*

Because, most manufactured products begin with cutting material, it is important to know what type of saw to use for each situation. The workhorse of the average shop for cutting bar stock used to be the power hacksaw. Sheetmetal was generally cut on a band saw. Plate stock involving a contour of any odd shape was also a candidate for band saw-cutting. Abrasive-cutoff saws are excellent for cutting angles and for cutting tubing or fairly thin, small pieces—especially of hard materials. But when it comes to cutting hard solids, the abrasive wheel wears down fast. Based on large volumes, the production of an abrasive wheel can cost twice as much as that of the hacksaw blade.

The question always arises as to what type of cutoff wheel to use; or what type of band-saw blade or what type of power hacksaw blade to use. Seek experience. Look for either someone with years of cutting behind him or a representative of the machine manufacturer. There are many types of abrasive wheels and many types of blades for hacksaws and band saws. They each have an area of effectiveness.

At one time, a shop was cutting sheets of copper in a bandsaw. Frustrated by the burr problem this method created, the operator experimented with other approaches. He was surprised to find one specific abrasive cutoff wheel that cut the copper sheet without any burrs. There is absolutely no better source of information than experience; yours or somebody else's.

It is possible to cut stock, such as steel, brass, aluminum, or copper, into precise lengths, consistently holding ±0.005". These days of severe competition have prodded machine tool manufacturers to do everything in their power to earn your loyalty. They will tell you what blade to use; its width, set, and thickness. They will tell you what speed and feed to use, then you will be able to hold tight tolerances in your cut pieces.

Parts that are going to be coined, orbital forged, or pressed in a closed die have to be carefully weighed, often in grams, because you want enough material to fill the die completely and yet, not squeeze out between die halves. It is simple to transform weight to length and thickness of bar stock and that is what's generally done. Accordingly, the slugs must be cut to a consistent size. Plenty of saws advertise the ability to hold close tolerances.

When operators can make accurate cuts, secondary operations (such as turning or grinding) to bring a cut part into tolerance, are greatly reduced. This saves valuable overhead and production costs. Because secondary operations are generally performed at much more costly stations, this difference in expense should not be overlooked in shop quotes.

HACKSAWS

Power hacksaws, on the other hand, have a fixed capacity rating, so whether the material being cut is large or small, the frame needs no adjustment. Another advantage is that the blade is held securely between two points, vertical to the workpiece and is not subject to the same stress that the band saw encounters as it travels around drive wheels and through alignment guides.

The "gravity" feed series of power hacksaws are simple, reliable, and durable. This is a straightforward, weight-of-the-saw-frame, feed system. These generally rugged, cast-iron machines are easy to maintain. Many frames come with adjustable gibes that are replaceable. They have horsepower-rated switches, automatic shutoffs, and quick-acting swivel vises.

Keller Industries, a long-time manufacturer of hacksaws, provides a chart with directions for blade selection, speed, and feeds. On the back of the chart, they have printed the advice contained in the next few paragraphs (Figure 14.3).

The proper tensioning of the power hacksaw blade is most important. Insufficiently tensioned blades wear rapidly, cut inaccurately, and deliver a blank with a poor finish. A blade tensioned too tightly breaks prematurely or pulls out at the pin hole. If chips are burned, the feed is too heavy. If chips are fine and powdery, the feed is too light. A free cut with nicely curling chips indicates ideal feeding pressure, fastest cutting time, and longest blade life.

TYPE OF MATERIALS		TEETH PER INCH		STROKES PER MIN	FEED		
		MATERIAL 2" and Under	MATERIAL Over 2"		RATE		POUNDS PRESSURE
ALUMINUM	Alloy	6	3	100-150	L	M	150
	Pure	6	3	100-135	L	M	150
BRASS	Free Machining	6-10	3-6	120-150	L	M	150
	Hard	6-10	6	100-135	L	M	150
	Tubing	10-14	10-14	120-150	L	M	150
BRONZE	Commercial	6-10	3-6	90-120	L	M	300
	Manganese	6-10	6	60-90	L	M	150
COPPER		6-10	3-6	90-120	M	H	300
HIGH DENSITY ALLOYS	A-286	4	3	50-75	M	H	400
	Discalloy	4	3	50-75	M	H	400
	Hasteloy	4	3	50-75	M	H	400
	Titanium	4	3	50-75	M	H	400
IRON	Cast	6-10	4-6	90-120	M	H	300
	Malleable	6-10	6	90	M	H	300
	Pipe	10-14	10-14	90-120	M	H	150-200
MAGNESIUM		6	3-4	120-150	M	H	150
NICKEL ALLOY	Inconel	6-10	3-6	50-80	L	M	150-300
	Monel	6-10	3-6	60-90	L	M	150-300
	Nickel	6-10	3-6	60-90	L	M	150-300
NICKEL-SILVER		6-10	6	60	L	M	300
STEEL	Alloy	6-10	3-6	60-120	L	H	150-300
	Carbon Tool	6-10	6	60-90	L	H	225
	Cold Rolled	6	3-6	100-135	L	H	275
	Hot Rolled	6	3-6	100-135	L	H	275
	High Speed	6-10	6	60-90	L	M	225
	Machinery	6-10	3-6	100-135	L	M	300
	Pipe and Tubing	10-14	6-14	90-135	L	M	100-150
	Stainless	6-10	3-6	60-90	L	H	150-300
	Structural	6-10	6-10	90-135	L	M	225
	Tool	6-10	6	60-90	L	H	225

Where more than one tooth specification is given, select the proper blade on the bases of the shape and size of the material. Use blade with coarser teeth for cutting large, solid stock; blades with finer teeth for smaller stock.

Blades should be wide enough to withstand required pressure...1-1/4" for light cutting (such as pipe or tubing) with short run blades...1-1/2" to 1-3/4" for heavy cutting with longer blades. For general purpose cutting a 1-1/2" x .075" x 6" tooth blade is recommended.

*When saw using medium to heavy feed rate, always use a coolant.

FIGURE 14.3 Blade selection, speed and feed chart. (*Keller Industries, Hollandale, MN*)

For most cutting jobs, the all-hard blade is first choice for straight, accurate cutting. The all-hard tungsten blade is unexcelled for retaining its sharp teeth. It handles work-hardening materials, abrasive materials, stainless steels, high-manganese steels, and the low-machinability bronzes. Molybdenum blades are good for fast, accurate cutting—especially on low- and medium-alloy steels, iron, and most nonferrous metals. See Figure 14.2 again; this Amada power hacksaw contains control electronics.

You gain more by selecting the coarsest tooth for the work. This is necessary for good chip appearance as more pressure can be applied for a better bite without clogging. Of course, the feed-pressure-per-tooth must be kept below the point of fracturing the teeth. Large sections and soft materials require coarse teeth. Thin sections and hard-to-machine materials require fine teeth.

Normally, you should set the feed pressure as heavy as possible without breaking the teeth or making the blade cut crooked. Excessive pressure and stroke speed increase the cutting rate at the expense of blade wear. When in doubt, keep pressure at maximum, but reduce the stroke speed. The heaviest practical pressure and the fastest reasonable stroke speed produce the most efficient cutting. A feed rate that is too light results in rubbing instead of cutting. Tooth points overheat, soften, and break down.

The optimum feed rate for hard, dense materials is heavy; for thin or soft material, it is light. For maximum production, you can increase feed by using coarse blades on soft materials. However, for straight, accurate work, moderate feed is required. Start the coolant flow before the cutting begins. Coolant is needed on all materials, except cast iron, copper, and some brasses, to reduce friction, blade wear, and chip clogging. Keep the coolant flowing until the job is completed and the blade has stopped.

The first practical cutoff machines were power hacksaws, and they are still well-established and in wide use today. The reciprocating power saws use a short, rigid blade, much like a hand hacksaw. The cutting occurs only on the forward stroke so that only one half the motion cuts. But when it comes to cutting solids, the abrasive wheel wears down fast. Based on large volumes, the production of an abrasive wheel can cost twice as much as that of the hacksaw cuts. However, the hacksaws now can come with a cam-operated back stroke so that the time lost from productive cutting is reduced. Nevertheless, it cannot match cutoff speeds of the other three models. Furthermore, the need to stop and reverse, then come up to speed, prohibits full efficiency. Also, the reciprocating action means the blade cannot be supported close to the cutting, as with a band saw. Therefore, it requires a thicker blade to provide comparable beam strength for cutting accuracy, which means more horsepower is required.

The popularity of the hacksaw results from its reliability, long service life, and the relatively low investment required. The simplicity of its design and small number of moving parts make it a long-distance performer. Plenty of these machines are still cutting reliably 30 to 40 years after purchase. The downside is simply that performance-wise, these old-timers can't keep pace with the "new boys on the block."

CIRCULAR OR COLD SAWS

Circular saws have been around for almost 200 years. They are large, massive machines, in relation to their work capacity, and their circular blades provide a continuous sawing action. The three basic configurations are horizontal, vertical, and tilt. Some models can use either circular saws or abrasive wheels. The blades are usually available from 12 to 55" in diameter, providing with allowances for the blade arbor, a capacity somewhat less than their radius. The kerf removed is quite large and the machine uses 2 to 10 times the horsepower required for band saws to cut the same capacities.

Like all blades, cold-saw blades are sensitive to differences in workpiece size. On all types of machines, for saws to be efficient, blades should be changed for different-sized workpieces and different materials. This blade-changing step is time-consuming—especially for the large-sized cold saws. Although they cut rapidly, blade cost, kerf removed, and initial machine investment can make them costly to operate. In general, cold saws are used where a high volume of nonferrous alloys or standard machinery steels must be cut. Considerable production work is required to justify the purchase of a large cold saw.

ABRASIVE SAWS

This type of saw operates much like the cold saw and ranges in capacities from the small "chop saws" used in many shops to the very large, sophisticated units that are similar in design and appearance to large cold-sawing machines. On the small capacity end, the chop saw has many effective applications; at the large capacity end, with big abrasive cutoff wheels, the competition is mainly from large cold saws for the high production of exotic metals.

Abrasive sawing is not to be confused with friction sawing. Friction sawing heats the metal by friction, then uses its teeth to scoop out metal softened by heat of friction. This cutting is effective in a comparatively small area. The abrasive wheel, on the other hand, removes metal the same way that a grinding wheel does. The abrasive wheel constantly breaks down to expose new, tiny, abrasive, cutting points as a grinding wheel does. So, wheel life typically is short. The kerf is large like a cold-cutting saw and, therefore, uses more horsepower. The most practical use for an abrasive wheel is for cutting hardened or nonmachinable metal.

BAND SAWS

Inherently, bandsaws twist the cutting blade 45 to 90 degrees around the guides of the band saw to permit the cutting process. This could lead to faster blade fatigue and premature failure. That's the reason it is so necessary to follow the manufacturer's recommendations and suggestions for speeds and feeds for the various blades doing sundry cutting jobs. The older band saws require constant adjustment of guides and blade tension throughout the life of the band-saw blade. The newer models need much less attention when cutting.

This type of cutting saw is probably the most commonly used at present. In modern metalworking, there are many situations where objectives and requirements often conflict. A shop might require a single blank in the morning and a large number of slugs cut off in the afternoon. There might be short, then medium runs all week, and then suddenly there is a need for extreme accuracy. The basic character of band saws accommodates all these needs. There have been continuous improvements in most areas of band-saw cutting, making it imperative to closely watch out for them. Never in the history of metalworking has it been so important to find ways to decrease costs while at the same time increasing productivity (Figure 14.4).

Recently, the trend has been to sawing systems built around more powerful and faster machines incorporating a whole new array of features that increase accuracy, reduce cycling time, improve material flow, and make the operator's task easier and safer. The most significant factor prompting those changes has been the development of exotic cutoff materials. The new materials were necessary to cut high-alloy, heat-resistant steels for space-age applications, which have "spun-off" for more commercial prod-

**2013-V &
3613-1**
General purpose
contour band
machines offer
maximum versatility for minimun cost.
Both machines have 13" workheight and
a tilting 26" × 26" table. Model 2013-V
has a 20" throat distance from band to
column; model 3613-1 has 35 ¹/₂"
clearance.

**2013-20 &
3613-20**
These versatile
contour band machines
feature a fixed table, DC
drive and electronic speed
control. Both feature 13" workheight
capacity, a built-in blade welder and a
chip blower. Model 2013-20 had a 20"
throat clearance and model 3613-20
features a 36" clearance.

**2012-2H &
3612-2H**
Contour-matic®
band machines
combine precision
DC drive with a hydraulically operated
worktable. A mist cooling system, blade welder,
and chip blower are standard. Both feature a 12"
workheight. Model 2012-2H has a 20" throat
and model 3612-2H offers a 36" throat capacity.

**2613-1H,
2612-1H3 & 2613-3**
These Contour-matic® band machines
feature a 26" throat with hydraulic
control of table feed. Model 2613-3
features a 10-Hp drive with 55 to 10,000
fpm band speed range and a 32" × 41"
table with 16" stroke, and powered chip
conveyor for high production sawing.

**2612-D15
Diamond
Band Machines**
This 26" throat capacity machine
features a 12" workheight and utilizes a
diamond edge band to cut optical glass,
unsintered carbide, ceramics and quartz.
It features infinitely variable band speeds
from 0 to 7,000 fpm and a DC drive table
system to provide the required feed pressure.

**Zephyr®
Friction Saws**
High velocity saw-
ing of 10,000
to 15,000 fpm, allows
Zephyr® saws to cut materials at rates
30 times faster than conventional machines.
Available in fixed, variable and wide-range
variable speed models, Zephyr® saws have
35-1/4" throat and 20" workheight capacity.

FIGURE 14.4 A variety of bandsaws. (*Doall Co., Des Plaines, IL*)

ucts. Some of these materials are so expensive that the size and amount of cutoff kerf
is an important consideration.

Efficient sawing systems are available for all types of high-production cutting. The
great majority of production cutoff jobs can be performed satisfactorily on standard, general-purpose, cutoff machines of the requisite power and capacity without any attachments
or modifications. There are, however, accessories that permit these standard machines to
handle volume production more efficiently (Figure 14.2 again).

In those special situations where you have high-volume jobs in a limited size and material range, or volume jobs that call for a routine function, such as making miter cuts in a
variety of angles, it might be worthwhile to consider purchasing a system designed specifically to handle them. These specialties could be where large quantities of extrusions or
forging blanks and slugs are required. There is a special machine for cutting large, curved
stock or plate stock of large thickness. Special machines exist for cutting turbine rotors,
connecting rods, heat exchangers, honeycomb, and for trimming. A maintenance schedule
exists for each machine and for each accessory. This information must be obtained and followed.

It is most important to select a blade for maximum life and resistance to corrosion. This
is when the manufacturer's advice should be sought. It doesn't take long to change blades

FIGURE 14.5 With the correct blade in the right bandsaw, you can cut almost anything. (*Doall Co., Des Plaines, IL*)

and it pays to use the correct one if any production at all is contemplated. In fact, there are times when a specific blade has to be installed because it would be folly to attempt the work with a blade of any other material.

Manufacturers not only give advice freely, but they provide operator training, which is invaluable. Teaching programs cover troubleshooting, blade and fluid selection, break-in procedures, speeds and cutting rates, machine adjustments, and more. They will supply handbooks, wall charts, application guides, and other technical educational material as needed. All this is generally free. There are many competitors in this field (Figure 14.6).

The DoAll power band saws use hydraulics to apply a tension of 30,000 lbs. per square inch to the band, imparting great stability to it. With the machine constructed very rigidly, cutoff tolerances of 0.002" per inch of work height can be maintained. If you are investigating a competitor's machine, look for these features.

It has been said that in 1981, band machining produced the lowest cost per cut for an estimated 70 percent of all cutoff applications in the United States. Band machining is not always the fastest cutting method, but its cost parameters and excellent performance characteristics most often best fulfill the varying volume and size cutoff requirements.

In high-production shops, one man can control five band saws. DoAll claims that with their latest equipment modifications, one man will be able to handle 10 machines. The control features on these latest models are so automatic that high production rates can be maintained with very little attention. There are all kinds of band machines to suit any requirement. Some jobs are better done on a horizontal machine, but others seem better for a vertical machine because this type, known as a *tilt frame*, can save secondary steps.

Selection of the most appropriate band tool for any given application is the key to cost-effective band machining and there are many choices. Begin with cutting-rate tables and job selectors to provide a starting point, but an understanding of blade characteristics is necessary for the best selection. The type of material and the cross section to be cut determine the blade characteristics required.

A properly adjusted machine, equipped with the correct blade and fluid, is the answer to the majority of band-sawing problems. The DoAll Company will send a specialist on a regularly scheduled preventive-maintenance basis if your company agrees to buy blades

FIGURE 14.6 Wilton, Inc. recently bought the rights to the Dake Shark 320SX. *(Dake, Inc., Grand Haven, MI)*

and fluid from them. He will also work with your operators to show them how to keep the equipment in top working condition. If DoAll does this, you should be sure that any other manufacturer whose product you are considering, offers the same opportunities.

Saw Blade Selection

The cutting tool consists of a welded loop of flexible steel saw band, carried in one direction on two wheels, one of which is the driving wheel. This permits full utilization of a cutting tool with hundreds of teeth. Because the entire band of teeth passes through the workpiece, tooth wear is distributed fairly equally over the entire length. The thinnest of all production cutoff tools for a band saw range from 0.058 to 0.080", with the most common being 0.062". This band cuts a narrow path, leaving the smallest kerf of any saws.

Three basic types of band-saw blades are used most: carbon alloy, solid or bimetal high-speed steel (h.s.s.), and tungsten carbide. Bimetal h.s.s. blades are appropriate for most applications involving virtually all ferrous cutoff jobs. Carbide-tipped blades might be required for harder or more abrasive materials, and carbon blades might be suitable for nonferrous materials. Diamond-coated blades are available for some unusual jobs, but normally these should be used only in special machines. The best blade for a given job is the one that will do it to the required specifications at the lowest cost. See Figure 14.7 for a special blade that can do what the "standards" cannot handle.

Tungsten carbide blades have a toothform all their own. The bimetal and carbon steel blades come in three toothforms: precision, buttress, and claw tooth.

- *Precision* This is the first blade developed for band saws. It has a deep gullet with a smooth radius at the bottom of the tooth. The rake angle was originally 0 degrees, but it can be obtained now with as much as a positive 7 degrees, and with a back clearance of about 30 degrees. The greatest selection of blade widths is available in this tooth form because it produces fine-finish cuts very accurately and has sufficient chip clearance for most sawing operations.

- *Buttress* This form is similar to a Precision, except that the teeth are more widely spaced to provide more chip clearance. The rake angle is 0 degrees and the back clear-

FIGURE 14.7 Bandsaws can cut soft, fibrous materials, but the correct blade must be selected. (*Doall Co., Des Plaines, IL*)

ance is the same as Precision. This has a coarse pitch on a narrower band width, which increases beam strength. The gullet is wider than a Precision. These bands are recommended for deep cuts on soft materials.

- *Claw tooth* This hook-type form has a 10-degree positive rake angle, which makes it possible to obtain faster cutting rates with reduced feeding pressures. The back angle is slightly less than Precision and Buttress. The gullet, which is a stress-proof design, is the widest of the three. This design should be used whenever possible because it keeps cutting costs at a minimum.

One standard chart shows the maximum blade width for any required radius. Blade width is measured from the back of the blade to the tips of the teeth. Always use the blade width specified for the desired radius. But for contour cuts, use the largest-width blade that will cut the smallest radius needed for the job.

The teeth on a band saw are offset to give clearance for the back of the band. Blade set is the overall amount that the teeth are offset to each side to provide this clearance for the body of the blade during sawing. This dimension usually determines the kerf (width of the cut) made in the workpiece. The kerf of a bandsaw blade is normally about ½ that of a power hacksaw blade and ¼ that of a circular-saw blade. The cost of materials saved by bandsawing can be significant when cutting high-priced alloys.

The three common blade sets are: raker set, wave set, and straight set.

- *Raker* The Raker set has one tooth offset to the right, the next offset to the left, and the third straight or unset. This is the pattern most often used probably because it takes a clean, even cut for a better finish.

- *Wave* Several teeth to the right, one unset, several to the left. This has a slightly wider set that helps prevent pinching of the band in work that tends to have internal stresses, such as pipe, structurals, and work with varying cross sections.

- *Straight* All teeth are set right or left with no unset teeth involved. This pattern has more teeth cutting and is used for nonferrous and nonmetallics, where the rudder effect is not needed and the extra cutting teeth survive abrasion for longer life.

The job-material specification, thickness, machinability, and type of cut desired dictate the choice of blade. Once the material is identified, locate it on "The Job Selector Dial" or in the "Saw Recommendation Tables." All modern DoAll band machines have a built-in Job Selector to guide the operator. Experienced band machinists modify these recommendations for specific jobs to get best results. The type of blade, pitch, and tooth form of the best blade for the job will be indicated on both Job Selector Dial and Sawing Recommendation Tables. An experienced machinist will be guided by this recommendation, but will vary it, based on personal experience.

Prior to the 1950s, the mainstay of sawing operations had been carbon steel blades. Now they have been bypassed technologically by newer types and are rarely used for production cutoff of ferrous metals. The single-metal h.s.s. blade's red heat hardness is more than twice that of plain carbon steel. On many cutting operations, a compromise must be struck between hardness at the cutting edge and spring tempering of the back. This results in less-than-optimal flex life and limited depth-of-tooth hardness. For this reason, it is not practical to resharpen single-metal h.s.s. blades.

Bimetal h.s.s. blades have overcome these problems and virtually made the single-metal h.s.s. blades obsolete. Machinists can now select tooth material for best cutting action and backing material for maximum flexibility. Bimetal blades can handle more tension than single-metal h.s.s. blades, cut 50% faster, and last two to four times longer. The heat-treating process allows the tooth material to achieve depth-hardness in the cutting-edge material that is four times that of single-metal h.s.s. blades, fostering longer life and permitting resharpening. If the blade is removed promptly for sharpening it probably will survive 2 resharpenings.

Cutting Fluids

The efficiency of almost all cutoff band-saw operations can be improved by using the proper cutting fluid. Correctly selected and properly used, cutting fluids permit faster cutting rates by providing a proper balance of lubrication and cooling that allows heavier feeds and/or faster speeds and promotes better finishes. To achieve maximum results, the cutting fluid must be selected for the machine tool, the material being cut, and the blade. There is a distinct interdependence between all members of the cutting team/operation.

Considering chemical composition, all cutting fluids can be classified in three broad categories:

1. Cutting oils with additives consist of mineral oil containing minor amounts of lubricity and extreme-pressure additives. The additives might be fatty oils, sulfurized materials, chlorinated materials, or all three. These additives are effective in preventing chip welding. (One precaution: never mix this oil with water.)

2. Soluble or emulsifiable oils consist of mineral oil to which emulsifiers have been added to permit mixing with water. The same additives that were listed in 1 (cutting oils with additives) can be added to the soluble oil. Technically, this is an emulsion, not a solution.

3. Synthetic or chemical cutting fluids are also mixed with water. They form true solutions; the concentrate is no longer recognizable in the mixture. For this reason, they do not readily separate, as will the soluble oils. They are also more tolerant of water with high mineral content than the soluble oils. The proper method of application and quantity of the oil is important. Naturally, experience is vital in this step, and if unavailable in your company, the machine manufacturer should be contacted.

ACCESSORIES

Power saw accessories and attachments are numerous and helpful for production. Some are needed to handle crooked stock or indexing to a length beyond the machine's standard stroke. Some determine the cutting rate and time automatically. There is an outboard vise for holding large cutoff pieces during cutting, and a stock-feed modification for continuous sawing. A hydraulically-powered roller stock table is helpful and so is the automatic chip conveyor. Even the polyurethane wear plates can be useful. Bandsaws can be used to expedite filing and polishing (Figure 14.8).

Heavy work clamps are used for stack sawing or contour cutting of heavy workpieces. Machine-speed clamps are available which clamp in T-slots on hydraulically powered tables. There are table extension bars, heavy work slides, chain-type workholding jaws, ratchet feeds, and hydraulic foot controls to move the table. There are blade shears and band welders. Most manufacturers will be glad to show you a complete line of their accessories, many of which are necessary for more efficient production.

Entire books have been written on the single subject of hacksaws, band saws, or blade selection. This chapter penetrates those subjects no further. There are so many pieces to the puzzle of how to cut stock that you will have to seek advice on specific jobs from either the machine manufacturer or an experienced machinist. How should you stack multiple bars of

Band filing that's seven to nine times faster than hand filing.

Band polishing for the quickest way to a smooth finish.

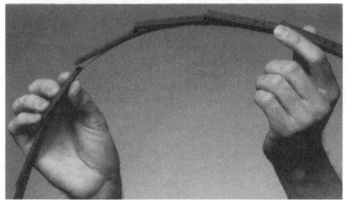

Precision file segments linked together produce a file band for continuous internal or external filing that is up to nine times faster than hand filing. File bands are mounted on the machine in the same manner as a saw blade and can be used on any low-velocity DoALL contour band machine. Special file guides are used so that band will track properly to give smooth, fast cutting action with precise accuracy.

Width inches (mm)	Shape and Cut	Teeth Per Inch (25.4 mm)	Length inches (mm)	Catalog Number*
1/4 (6)	Flat, Bastard, Medium	20	153 (3886)	311-266153
3/8 (10)	Flat, Bastard, Medium	20	153 (3886)	311-167153
3/8 (10)	Flat, Bastard, Medium Coarse	16	153 (3886)	311-142153
3/8 (10)	Flat, Bastard, Coarse	12	153 (3886)	311-126153
3/8 (10)	Half-Round, Bastard, Med-Coarse	16	153 (3886)	311-241153
3/8 (10)	Oval, Bastard, Medium-Coarse	14	153 (3886)	311-209153
1/2 (13)	Oval, Bastard, Medium-Coarse	14	153 (3886)	311-084153
1/2 (13)	Flat, Bastard, Medium-Coarse	14	153 (3886)	311-043153
1/2 (13)	Flat, Short Angle + +, Coarse	10	153 (3886)	311-027153

* Length shown for DoALL model 2013-V contour band machine. Other lengths available.   + Bastard cut for general use on all types of steel.   + + Short angle cut for use on aluminum, brass, cast iron, copper, zinc and other nonferrous metals.

FIGURE 14.8 A bandsaw can file and polish. (*Doall Co., Des Plaines, IL*)

stock for horizontal band sawing vs. vertical cutting? Is it cost effective to increase feeds, rather than speeds? What is the best combination of feed and speed for an accurate cut?

A large manufacturing supplier of military equipment assembled five experienced engineers as a group to review drawings, sit in at design reviews, and to investigate the best way to manufacture products. These men were all over 60 years old and each had an average of 40 years of experience. They saved the company much money and much time, which was often more significant than money. Many of their recommendations started with the cutting of raw stock, then went on to the best manufacturing process. Their value lay in the experience they brought to decision making.

CHAPTER 15

INDUSTRIAL PARTS CLEANING

Health, safety, and environmental regulations are redefining the workplace. Toxic chemicals and hazardous waste must be eliminated. Of major concern are traditional vapor degreasing and cleaning solvents. The potential effects to our health and the ecological consequences of these toxic solvents are well documented. The 1995 deadline for ending production of Class 1 ozone-depleting chemicals, such as 1,1,1-trichloroethane (TCA) and CFC-113, has already impacted the way that industry is handling the cleaning of parts and equipment. Safe alternatives have been produced and research is continuing to find other acceptable, safe products. Industries now notice that they are responsible for their hazardous wastes, including used solvents, forever.

Finding satisfactory cleaning processes has become a serious problem for manufacturers across the country, in many different businesses. Each company must evaluate the many alternatives available. This task, in itself, is enormous and expensive. There are options with trade-offs between cleaning effectiveness, equipment costs, operating and maintenance costs, and the costs of disposing effluents in a legal manner. Parts-cleaning issues are increasingly becoming complex. Rapidly expanding federal and state regulations governing discharge of effluent, water quality, waste disposal, and emissions to the atmosphere are causing consternation among business leaders.

At this time, the greatest need is for the advice of a cleaning expert. Understanding the particular requirements of your business, as it is affected by the current state and federal regulations, is the first step to resolving questions pertaining to solvent usage, handling, and disposal. Before buying any cleaning equipment that requires the use of solvents that are being phased out by the Resource Conservation and Recovery Act, consider the possible limitations and constraints. Determine if your spent solvents are recyclable and at what cost.

In selecting a component cleaning system, the size, shape, weight, and configuration of the parts should be considered. The types of contaminant to be removed must be considered. Blind holes, large surface areas, and the parts' materials must be taken into account. These physical variables should all be considered if proper cleaning is desired. Furthermore, there are several washing methods available today. These are: cavitation, mechanical agitation, solution turbulation, and high-pressure water. Each method has a special set of cleaning characteristics that can be used to solve specific problems.

VAPOR DEGREASING

Vapor degreasing has been the workhorse cleaning process for more than 50 years. On several occasions, the federal government has insisted that the solvent in use at that time be discontinued and replaced with less-harmful chemicals. Currently, the only cleaning chemicals available that pass OSHA regulations are very expensive or are not quite as efficient as the forbidden solvents. Several companies, including Exxon and DuPont, are busy researching other solvents and hopefully the situation will improve. Alkaline detergents with a pH of 10 to 14 are currently being used as a substitute cleaning medium.

For some time now, vapor degreasing and ultrasonic cleaning have been performed within one compact unit. The units have stainless-steel cleaning chambers with built-in refrigeration requiring only an AC connection. Cooling water at the top of the vapor zone condenses the solvent that has evaporated out of the boil chamber. See Figure 15.1 for a photograph of a unit produced by Sonicor Instrument Corp. of Copiague, NY.

Some units are equipped with casters for easy mobility. Each has two chambers; one for vapor generation and removal of gross contamination, and an ultrasonic chamber for precision cleaning. The full vapor zone allows vapor rinsing over both chambers. This equipment removes particles, wax, grease, oil, and other soluble contaminants from electronic components and both metal and plastic parts prior to assembly or further processing. Optional equipment includes a dessicant dryer, which makes possible the use of Freon TMS and TE (these are prohibited from most use). A further option provides for a manual spray of pure solvent.

ULTRASONIC CLEANING: CAVITATION

Ultrasonic cleaning is the rapid and complete removal of contaminants from objects by immersing them in a tank of liquid that is being flooded with ultra high-frequency sound waves. These nonaudible sound waves cause the liquid to act as its own scrub brush. This process is brought about by high-frequency electrical energy, produced by an ultrasonic generator. That energy is converted into high-frequency sound waves (inaudible ultrasonic energy) by means of transducers intimately bonded to the bottom of the tank. The ultrasonic energy enters the liquid in the tank and causes the rapid formation and collapse of minute bubbles—a phenomenon known as *cavitation*. The bubbles, traveling at high speed, implode on the parts immersed in the tank, thus cleaning both the surfaces and innermost recesses of intricately shaped parts (Figure 15.2).

The modern ultrasonic cleaners utilize solid-state circuitry in conjunction with lead-zirconate-titanate transducer elements. This cleaning technology is combined with aqueous and semi-aqueous chemistry to eliminate the undesirable solvents from the workplace. Certain models of the ultrasonic cleaners are computer programmable and feature robotic part-handling and monitoring of cleaning, rinsing, and drying.

The following is a list of applications where ultrasonic cleaners have done excellent work:

1. In the electronics industry.
2. In medical laboratories.
3. For jewelers.
4. Cleaning electron microscopes.
5. Cleaning glassware and laboratory equipment.

Thorough Vapor Degreasing And Ultrasonic Cleaning From One Compact Unit

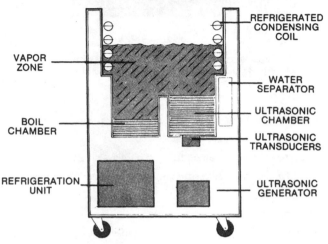

FEATURES:

- ☐ Minimum space-requirement
- ☐ Water separator, cover
- ☐ Drains each chamber
- ☐ Caster mounted for easy mobility

STAINLESS STEEL CONSTRUCTION:

- ☐ Chambers, condensing zone
- ☐ Condensing coil
- ☐ Water separator
- ☐ Cover & Cabinet

FRONT PANEL CONTROLS

- ☐ Main power switch & indicator
- ☐ Ultrasonic switch & indicator
- ☐ Start/reset button & indicator
- ☐ Run indicator

OPTIONS FOR MODEL UDR-30, UDR-50

- ☐ Spray lance/pump with reservoir.
  Add "SP" to model number.
- ☐ Desiccant dryer for solvents containing
  alcohols. Add "DD" to model number
- ☐ Filter System for ultrasonic chamber
  Add "FS" to model number

SAFETY CONTROLS
☐ High vapor level,
☐ High solvent boil temperature,
☐ Boil chamber low solvent,

FIGURE 15.1 Vapor degreasing and ultrasonic cleaning in one unit. *(Sonicor, Codiague, NY)*

6. Cleaning timing mechanisms.
7. Cleaning printed circuit boards.
8. Lens and substrate cleaning.
9. Radioactive decontamination.

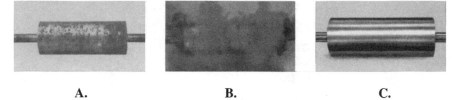

A.   B.   C.

FIGURE 15.2   An example of ultrasonic cleaning: A. Heavily soiled component prior to ultrasonic bath. B. High-speed photography captures the ultrasonic action. C. The cleaned part is ready for use. *(Sonicor, Codiague, NY)*

10. Cleaning business machines and typewriters.

11. Cleaning dental instruments and appliances.

12. Cleaning all types of nozzles: diesel fuel, oil burner, and all industrial types.

13. Cleaning metal and ceramic filters for aircraft and industrial equipment.

14. Cleaning surgical instruments.

15. Cleaning firearms and vending machine parts.

16. Removing dust and dirt, light oil and grease, buffing compound, tarnish, and corrosion dust from metal and plastic parts.

Ultrasonic companies offer a wide range of accessories that maximize the efficiency of these systems. The accessories include oil skimmers, digital timers, and hand-held spray wands, plus many types of solvents for removing such things as temporary cements, tartar and stain from dentures, and carbon deposits. It pays to discuss your contaminant problem with an ultrasonic expert—especially now that regulations prohibit or constrain the use of so many previously used solvents.

For a limited number of difficult-to-clean parts, the chlorinated solvent alternative most nearly like the forbidden 1, 1, 1-trichloroethane (TCA) involves either methylene chloride, perchloroethylene, and trichloroethylene. These three solvents share most of the advantages of TCA, but are not scheduled for phaseout under the Clean Air Act or the International Montreal Protocol. Nor are they subject to the excise taxes levied on ozone-depleting substances (ODS). Currently, they are listed by the U.S. Environmental Protection Agency (EPA) as acceptable alternatives to TCA.

Methylene chloride is a powerful and versatile chlorinated solvent. Its high solvency and low boiling point make it especially effective for cold cleaning and vapor degreasing of metal parts. Its ease of recoverability has led manufacturers to increase recycling efforts.

Perchloroethylene is a powerful and versatile solvent for waxes, oils, and greases, and it is compatible with most plastics, coatings, resins, elastomers, and rubbers. Because of its relatively high boiling point, it can be used in vapor degreasing of products with high melting points, wax or paraffin soils. In addition, the solvent's resistance to degradation by water makes it usable in wet applications.

Trichloroethylene (TCE) has been characterized as the ideal degreasing solvent because of its high solvency of oils, greases, waxes, tars, resins, lubricants, and coolants, which are generally found in metal-processing plants. In addition, TCE will not attack steel, copper, zinc, or other metals. It is highly stable in the presence of common chemical stabilizers, and its low boiling point permits vapor degreasing with low heat input and rapid handling following degreasing.

Generally, chlorinated solvents have been preferred by industry because they are practically nonflammable. They have no flash point, as determined by standard test methods. With the exception of perchloroethylene, however, they have flammable ranges when high concentrations are mixed with air and exposed to a high-energy ignition source. Consequently, all ignition sources must be eliminated in areas where high vapor concentrations could occur; only electrical equipment approved for use in explosive atmospheres should be used in these areas.

Each of these three alternative solvents brings with it significant health concerns. Methylene chloride and perchloroethylene are carcinogens. Trichloroethylene is suspected to be a carcinogen and is a threat to the liver. All three are poisons and suspected causes of cardiac dysfunction.

Hydrofluoroether (HFE) compounds was introduced by the 3M Company of St. Paul, MN as CFC/ODS replacements for parts cleaning before the end of 1996. HFEs are said to closely resemble, in performance, the ozone-depleting substances they will replace, but unlike CFCs, they have zero ozone-depleting potential, very short atmospheric lifetimes, and low global-warming potential.

Careful containment in vapor-degreasing machines is essential and mandated under standards for new and existing halogenated solvent-cleaning operations, which are governed by the National Emission Standards for Hazardous Air Pollutants. Procedures are available for both vapor degreasing and cold cleaning that greatly decrease vapor losses from existing equipment. Compared to open-top degreasers, totally enclosed degreaser designs can reduce emissions by more than 95%. Some enclosed degreasers are available to contain and reuse almost all the solvent vapors.

To get the greatest efficiency and longest life from an ultrasonic cleaner and understand what's going on, there are specific "dos and don'ts."

1. Most ultrasonic cleaners show high bursts of surface energy on initial activation—especially when room-temperature solvents are used.

2. Cold water will not normally cavitate well, although there might be an initial burst of surface energy.

3. Do not introduce excessively hot water into a cold ultrasonic tank because it will cause undesirable start-up conditions and thermal shock.

4. Do not use water solutions above 180 degrees F at any time.

5. Add chemicals to water prior to initial start-up and mix well—especially if the chemical is powdered material. It should be premixed before adding to the ultrasonic tank.

6. Allow a sufficient degassing period after the initial start-up and prior to introducing work into the cleaning tank. The degassing time will vary, depending on several factors, such as type of chemical liquid (i.e., solvent or aqueous), temperature, viscosity, etc.
   Note: Normally solvents degas much slower than aqueous solutions. Degassing can vary from a few minutes up to a few hours in large solvent cleaning tanks. For faster start-ups, solvents should be heated to accelerate degassing.

7. Be aware that surface activity might disappear completely after work is introduced because the ultrasonic energy is now being utilized by and on the work.

8. Be aware that certain frequencies provide different surface activity. Generally, higher frequencies will show more violent activity than lower frequencies. Surface activity is not necessarily related to cleaning power. For instance, a low-frequency system might show little surface activity and yet clean a part just as quickly and as thoroughly as a higher-frequency unit that produces violent surface activity, or vice versa.

9. Do not mechanically agitate parts or introduce air or fast-flowing liquid (i.e., recirculation filter system) into an ultrasonic cleaning tank.

Note: Use recirculation-filter systems preferably when ultrasonics are not being used, and be certain that the return line is beneath the liquid surface. Very slow mechanical agitation can be utilized.

10. Position parts so that air is not trapped in blind holes and cavities because no cleaning will occur in these areas. Liquid is required to carry the ultrasonic cavitation.

11. Do not use racks, trays, baskets, etc., fabricated from heavy and dense materials, because much of the ultrasonic energy will be absorbed by these devices and detract from the cleaning.

12. For the same reason as 11, do not use racks, trays, baskets, etc., fabricated from or coated with soft material (namely rubber, wood, fabric, or soft thermoplastics). In addition, soft materials could disintegrate.

13. Be aware that certain chemicals or excess amounts of these chemicals will not support ultrasonic cavitation. Check with the ultrasonic supplier if you are in doubt.

14. Most ultrasonic units are "liquid-level sensitive." That is, at various levels, they display more surface activity than at other levels. This is a function of wavelength and does not necessarily have any affect on the cleaning ability of the equipment.

15. Do not overload an ultrasonic cleaning tank with work. It is difficult to establish guidelines because of the wide variety of sizes and weights of parts encountered. It is wise to clean fewer parts quickly and thoroughly. Start with a few parts, add additional items, and check to ensure satisfactory cleaning results.

16. Do not place anything on or against the radiating surfaces of ultrasonic tanks or immersibles because this might dampen ultrasonic activity and cause equipment damage.

17. Do not permit cleaning solutions to become excessively contaminated or sludge to build up on radiating surfaces, causing a loss of cavitation power and possible damage to the equipment.

MECHANICAL AGITATION

To replace solvent degreasing, old habits and procedures must be changed. Traditional solvent cleaning is primarily a dipping process using aggressive chemicals to do most of the work. Modern detergent cleaning is a washing process wherein safe, mild chemicals rely on the performance and flexibility of the washing machine. The equipment must provide sufficient mechanical agitation, part orientation, and bath conditioners to facilitate the process. The Ramco Company of Hillside, NJ, makes the Migi-Kleen, an immersion parts washer.

This machine removes oil, chips, and dirt from a wide variety of machined components, castings, and housings. It accomplishes this by using a simple pneumatically powered reciprocating platform. This type of machine is also used for cleaning prior to welding, deburring, blasting, phosphating or assembly. It is effective for maintenance and rework, and it can remove rust, carbon, ink, and paint (Figure 15.3).

The operation is simple:

1. The operator loads contaminated parts into a basket, which is placed on an elevator.

2. The flip of a switch automatically lowers the elevator into the immersion chamber and it reciprocates, up and down, 240 strokes per minute.

3. Vigorous mechanical agitation swishes solution against all surfaces, in and around holes and recesses. Stubborn contaminants are washed away without hand labor.

MECHANICAL AGITATION
Cleaning Characteristics
1. complete immersion
2. recessed areas in contact with solution
3. parts moved up and down
4. vertical motion swirls solution against surfaces dispersing the
   contaminants.
This is the most common method of batch washing parts.

1. Load

2. Flip a switch

3. Unload

FIGURE 15.3 Two views of mechanical agitation. *(Ramco, Hillside, NJ)*

This basic machine is also manufactured so that the reciprocating platform is integrated with an ultrasonic station, making it a dual-mode machine. The scrubbing action of the ultrasonic addition provides a significant boost in chemical activity. Variable preset timers are used to control each cycle of the ultrasonic cleaning and agitation washing. The machine can be automated with a power-feed conveyor and a precision ball-screw linear-transfer mechanism.

TURBULATION SYSTEMS

Ramco, also makes a Vector washer, which provides an intense turbulence of massive volumes of solution to impact against parts and scrub away contamination. Heavy loads are no problem. The machine uses a Whirljet agitation system, which is both versatile and dependable (Figure 15.4).

A turbulation system can be added to the basic mechanical agitator machine. It uses high-volume pumps and closed-loop manifolds with multiple solution injectors directed at the work zone. The system provides an even flow of turbulence over, under, around, and through the contaminated components. This accelerates cleaning time and improves overall

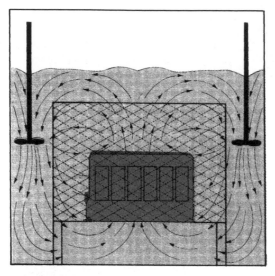

SOLUTION TURBULATION
Cleaning Characterisitcs

1. complete immersion
2. all recessed areas in contact with solution
3. high velocity agitation of solution around and through the parts.
4. shearing action strips away contaminants from the surfaces.

Ramco Vector Washers are designed for cleaning operations where violent solution movement is desired. This is the preferred method for immersion cleaning of massive components or where batch loads are routinely very heavy.

FIGURE 15.4   Solution turbulation. *(Ramco, Hillside, NJ)*

effectiveness in both cleaning and rinse cycles. See Figure 15.5 for a view of the turbulation system added to a standard agitator machine.

Check numbers on photograph with the following Migi-Kleen features:

1. Heavy-gauge steel construction with flange-up solution drip lip.
2. Sturdy transport elevator with no internal bearings.
3. Hinged splash guard cover with fusible safety link.
4. Safety guard to enclose moving parts.
5. Air line filter, regulator, and lubricator.
6. Single switch control.
7. Compact oil-removal unit skims off tramp oil that rises to the surface.
8. Electric heating system provides rapid heat-up and controlled solution temperatures up to 250 degrees F.
9. Preset timer precisely controls overall cleaning cycles.
10. Standard reciprocating platform provides vigorous mechanical agitation, up to 240 strokes per minute.

FIGURE 15.5 The "Migi-Kleen" machine combines agitation and turbulation. *(Ramco, Hillside, NJ)*

11. Dual stroke control; rapidly reciprocating short stroke (for general cleaning) or long stroke (in and out of solution, for flushing blind holes).

12. Turbulation manifolds with multiple injectors directed precisely at the work zone distribute a high volume of solution over, under, around, and through components.

A high-performance filtration unit in line with the turbulator removes suspended particulate. Mesh, cartridge, or bag filter elements are standard. Low-micron filtration is readily available. Oil-removal systems also are easily added to any machine.

ICE BLASTING

Today, two general ice blasting technologies are in use. The older one, dry ice blasting, uses solid pellets of carbon dioxide as a blasting medium for cleaning metal parts. This method reduces harmful air emissions and eliminates, or at least reduces, company liability associated with solvent chemical wash processes. The carbon dioxide pellets disintegrate upon impact and dissipate into the atmosphere. Consequently, there is no solvent-recycling problem.

The other technique uses water in the form of ice as the medium. The ice chips, driven by compressed air, removes surface grime by a scrubbing and flushing action. Both of these systems replace chemical methods by providing environmentally friendly degreasing and paint removal.

AQUEOUS CLEANING SYSTEMS

It's not hard to imagine that replacing chlorinated solvents for part cleaning would be very difficult. Unlike the open-top tanks or leaky degreasers used in the past, substitute solvents and government regulations require virtually emission-free cleaning devices. The switch from solvents to aqueous-based cleaning systems brings with it a whole new set of operating requirements. Many companies have been stockpiling the forbidden solvents even though they will have to pay an onerous excise tax to continue using them for metal-degreasing operations. In the meantime, totally enclosed emission-free vapor degreasers are essential, whether the machines are bought new or are retrofits of older systems.

Most observers believe that anywhere from 75% to 95% of metalworking cleaning can be handled by the right aqueous-based cleaning system. Aqueous cleaning, however, is not a simple "dunk and dry" process, like solvent cleaning. It is application-specific and requires commitment of capital and resources, as well as reconfiguration of manufacturing processes and time to analyze cleaning needs and waste-water management issues.

Several companies manufacture aqueous-based cleaning systems. They feel that there are three major reasons for the tremendous growth of their sales in the past five years.

1. *An increased attention to cleaning parts as part of a move toward better quality.* An example is the problem that an automobile manufacturer uncovered. His transmissions were giving trouble too consistently at about 60,000 miles. The competition's transmissions were surviving more than 100,000 miles. A thorough examination of his units found that over 80% of the particulates located in the fluid of the 60,000-mile transmissions were not metal shavings or chips caused by normal wear. They were dirt introduced in the manufacturing process. After the manufacturing cleaning process was improved, transmission life increased dramatically.

2. *Environmental concerns and government regulations of petroleum-based solvents and vapor degreasers.* There is sincere desire in industry to avoid anything that might harm personnel or the environment. This is because most industries really want to do the right thing and government regulations force the slackers among them to comply.

3. *The trend toward cellular manufacturing, which clusters production machines to build a particular part or product.* The last 10 years have witnessed a change in modern industries to cellular manufacturing. This new arrangement of machine clusters does not work well with centralized cleaning equipment, and many companies have purchased several portable cleaning machines to service the flexible manufacturing areas. The machines are rolled over to where they are needed and this saves much part-handling costs.

The Aqueous Cleaning Process

The simplest aqueous system consists of a wash cycle, a rinse cycle, and one or more drying cycles. A programmable controller moves parts (in baskets or racks) through the process stages. In the wash stage, precisely positioned nozzles thoroughly flush every surface. In the rinse stage, the nozzles completely rinse all part surfaces and hard-to-reach areas. In the dry stage, each blind hole and each surface receives a powerful blast of air that will blow off and dry any areas that collect water. Strategically placed baffles between wash, rinse, and dry stages help prevent solution "carryover."

The stainless steel wire baskets and racks expose hard-to-clean areas and facilitate complete drainage and drying. A common cleaning liquid would be a low-foaming alkaline cleaner formulated to remove oil and lubricants from aluminum and work as a general-purpose cleaner for ferrous materials in automatic washers and both spray and immersion applications.

Some stages of the aqueous cleaning process reach a temperature of 300 degrees F. Electric fans mounted above the unload areas permit almost immediate manual handling of the clean parts for packing in cartons. Particulate filtration removes fine particles to maximize the life of the cleaning solution, and differential pressure gauges indicate when filters require replacement or cleaning.

Wastewater treatment systems clean the process liquid so well that it might leave the plant through the public sewer system. A batch process might use chemical flocculation to remove suspended solids, heavy metals, and emulsified oils. The sludge from materials cleaned off the parts is a dry, white, nontoxic powder. Collected from an integral press, it goes directly to a local landfill without special handling.

The cleaning solutions are water-based and can contain a combination of water conditioners, detergents, and surfactants that promote better cleaning of metal parts. Special additives (such as pH buffers, inhibitors, saponifiers, emulsifiers, and deflocculants) are added to meet specific requirements. Tap water is usually sufficient for the early stages of aqueous cleaning, but should be followed with deionized water during rinse stages. Ions, such as calcium or magnesium in water can corrode or permit deposits to form on cleaned parts. Aqueous cleaning offers an effective alternative to solvent-based cleaning, but it does have some disadvantages.

It appears that water-based cleaning is the future for both U.S. and worldwide manufacturing. Before selecting a cleaner for your manufacturing operation, it is essential that a review of the Material Safety Data Sheet (MSDS) be conducted. First, be sure that the proper product name is listed in the MSDS in Section 1, Product Identification. The formula and the chemical family should be listed. The formula is a list of the ingredients in the product. The chemical family should state "Liquid Alkaline Cleaner" or other similar name. If solvent emulsion, solvent cleaner, or solvent appear under this section, it is probably not the right type of cleaner for your system.

Second, every MSDS sheet should list *all* hazardous ingredients contained in the product. The proper name and the Permissible Exposure Limit (PEL) and/or Threshold Limit Value (TLV) must be listed for each hazardous ingredient. Common ingredients in alkaline cleaners that might be found in MSDS lists are sodium hydroxide, potassium hydroxide, sodium metasillicate, and ethylene glycol monobutyl ether.

The exposure limit levels are the amount of a hazardous material that a worker can be exposed to without any expected health effects. Levels of potential exposure are measured in Parts Per Million (PPM) or in Milligrams of Particulate Per Cubic Meter of Air. Most industrial cleaning can be conducted without using cancer-causing agents and products that release hazardous fumes.

Finally, check the date of the MSDS sheet. Data on exposure limits are updated on an annual basis according to federal regulations. If the MSDS sheets' date is over 1 year old ask for a new sheet.

Aqueous Cleaning Machines

Everything from tiny needle bearings and stampings to large machine parts should have a cleaning machine selected especially for the particular work to be handled. Most cleaning-machine manufacturers make a number of different type and size units. Those described here are made by Hurricane Systems Division of Midbrook Products, Jackson, MI (Figure 15.6).

Rotary drum units are ideal for smaller parts that need to be immersed in the wash solution. The parts are gently moved through the spray and immersion steps cleaning all surfaces, including blind holes and threads. The drum is available with perforations down to $\frac{1}{16}$" for very small parts.

The belt system is more versatile than the rotary drum. It can handle a broader range of parts, from small to large and from delicate to iron castings. Variable speed control allows flexibility in cleaning cycles. Belts are available in stainless steel, plastic, and steel in a wide range of sizes.

The overhead conveyor system is designed for continuous, high-volume, heavy-duty operation. It is equally functionable as part of an in-line system or as a self-contained, continuous-loop system.

Multiple-stage custom systems can be easily modified to fit your exact production requirements. The drum, cabinet, fixtures, belt width, and speed, stage functions, quantity of spray nozzles and their location, and construction materials can all be selected for maximum effectiveness.

Dunnage cleaning systems clean the trays and baskets that hold your parts during cleaning. This is a useful machine when you have a number of different cleaning machines cleaning parts in continuous service.

The Cyclone tank washer works on the simple, time-proven method of agitation. The compact machine operates more efficiently than many larger units. It can be quickly transported from place to place on an optional caster cart. It has adjustable stroke lengths and speed and has self-contained heater controls with digital readout.

One additional piece of equipment that is often used in conjunction with cleaning machines is a tramp oil-removal system. It removes free-floating, dispersed, and tramp oils from metal-working fluids, thereby improving performance. Figure 15.7 shows Hurricane company's Extractor. An adjustable skimmer floats on the surface of the cleaning machine's reservoir and draws the top layer of fluid and/or tramp oil into the Extractor.

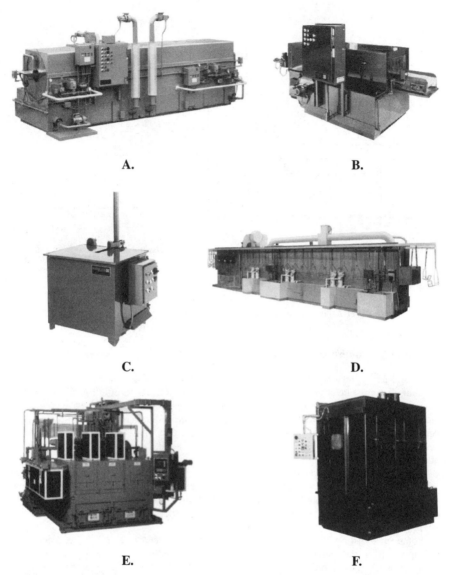

A.   **B.**

C.   **D.**

E.   **F.**

FIGURE 15.6 Hurricane cleaning machines: A. Multiple-stage rotary-drum system. B. Single-stage belt system. C. Agitation-tank washer system. D. Overhead conveyor system. E. Multiple-stage custom system. F. Dunnage/tote cleaning system. *(Hurricane Systems Div., Midbrook Products, Jackson, MI)*

The Extractor consists of a process tank with a series of chambers through which the fluid passes. The first chamber collects the contaminated fluid and houses a solids removal filter. Tramp oils begin to rise to the surface in this chamber. The second chamber houses a separation media pack, where dispersed and mechanically emulsified oils coalesce and

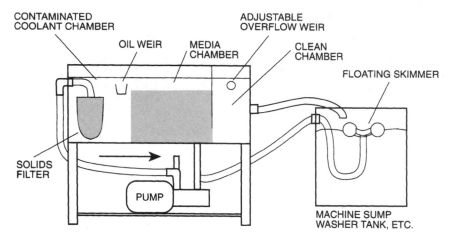

FIGURE 15.7   The Extractor oil remover. *(Hurricane Systems Div., Midbrook Products, Jackson, MI)*

float to the surface. The third chamber holds the clean fluid, which gravity feeds the oil-free fluid back into the original tank from which it came. A weir between the first and second chambers diverts the coalesced oil into a collection receptacle.

Rust and Corrosion Inhibitors

Water and air on ferrous metal oxidizes and creates rust. This is an element of aqueous cleaning that must be addressed and it usually means that a rust inhibitor is needed. If the next step in the manufacturing process is painting or high-precision welding, then a completely clean surface is desirable. If the parts go into storage between operations, a rust-preventing light oil might have to be applied. The reason is that water and alkaline cleaners are very effective. They remove everything down to bare metal, including any rust-prevention oils that were previously applied.

Zinc and aluminum parts, because of their sensitivity to corrosive alkaline cleaners, require different formulations than ferrous alloys. Aqueous cleaners for nonferrous alloys should be mixed with inhibitors that prevent the alkalies from attacking the sensitive metals. The inhibitors deposit a thin protective coat on the metals as soon as the contaminants are removed.

Sometimes rust inhibitors can safely be omitted from cleaning fluids. If the machining coolants used in the manufacturing process are water soluble and have rust inhibitors in their chemistry, those inhibitors can be omitted from the aqueous cleaning fluid. When rust inhibitors are used in the cleaning process, care must be taken to keep the concentration correct. If the concentration is too high, the parts will stick together. The chemistry of any cleaning solution must be controlled and audited regularly for best results.

Evaporation of Wastewater

An alternative for wastewater disposal is simply evaporation. *Evaporation* is the change of a liquid into a gas without boiling. On the other hand, forced evaporation is the change of a liquid into a vapor by using a heat source to the boil the liquid. If the liquid is forced to evaporate in a closed vessel, the space above the liquid becomes filled with vapor that begins to condense. A forced-evaporation unit has a vent stack to help the vapor escape with the help of a draft inducer and/or a vent fan. These fans move the gases across the liquid's surface to eliminate condensation.

Most evaporator manufacturers offer assistance in completing a Material Stream Analysis (MSA) of your company's wastewater. A certified laboratory will provide the data for air-shed authorities to determine if your evaporator poses any safety hazards. If your generator meets the requirements of federal, state, and local regulations relating to on-site waste treatment, you will be permitted to continue with the forced-evaporation method.

The reduction of weight and volume of a wastewater stream can reach 98%. The remaining 2% is a sludge of suspended solids, heavy metals, dirt, and other contaminants, and a small amount of water. Several types of wastewater evaporation systems are on the market today.

The five most common are:

1. Under-floor heating units that use convection to move heat from an insulated chamber through the wastewater.

2. Immersion-tube units act as a heat exchanger.

3. Submerged combustion units have small holes in the top of the heat tube, allowing bubbles of hot gases to directly contact the wastewater.

4. Direct-injection units spray wastewater into a natural gas or propane flame.

5. Thermal-oxidation units, also known as *incinerators*, burn, rather than evaporate, the wastewater. Operating temperatures can reach 1400 degrees F.

PLANNING AHEAD

Cleaning parts is a necessary extra expense in their manufacture, like deburring. It very often must be done to ensure quality and make sales. Consequently, plans should be made to understand the cleaning situation before decisions are made. As a start, answer the following questions:

1. How are you currently cleaning your parts and at what cost?

2. Describe the parts to be cleaned. What is the material? How complex and/or fragile are they?

3. What soils are put on your parts as they are processed?

4. What is the objective of the parts-cleaning project?

5. What are the biggest problems you are currently experiencing?

Now start discussing your cleaning project with a few manufacturers of cleaning equipment.

CHAPTER 16
HOT ISOSTATIC PRESSING (HIP)

BACKGROUND

Hot isostatic pressing (HIP) is a process of manufacturing that relatively recently acquired commercial production status. During the 1950s, it was only a laboratory procedure. At that time, the process was used to make synthesized industrial diamonds. Gradually, since about 1975, applications were found that required process improvements. Those improvements built one upon the other, until now, there are several applications where this process is outstanding.

It is curious that, like many technological advances, hot isostatic pressing of castings came about by accident, almost as an afterthought. One application of HIP that was being performed successfully was the compaction of metallic powders. It seemed that if it was good for powder metallurgy, it could also densify castings. So that was tried. Now the compaction of metal and ceramic powders and the elimination of porosity in castings are routinely performed (Figure 16.1).

A certain amount of porosity is present in any casting. Industry has lived with this condition for years. Hot Isostatic Pressing (HIP) has long been recognized as a process that can heal internal porosity in castings. HIPing is the simultaneous application of heat and inert gas (generally argon) pressure in thick-walled pressure vessels, such as that shown in Figures 16.2 and 16.3.

The gas pressure exerts the same force on all the castings within the pressure vessel, which causes internal defects (pores) to close, then seal by diffusion bonding at the process temperature. Conditions are selected for each alloy/material, which provide permanent healing of these internal pores without distortion of the major casting features. By eliminating porosity in castings, they are provided with properties that are normally associated with forgings.

The process has been used for the compaction of powders, both metal and ceramic, and, only more recently, the densification of castings and the elimination of micro-shrinkage

FIGURE 16.1 Powder billets formed by hot isostatic pressing. *(Industrial Materials Technology (IMT), Andover, MA)*

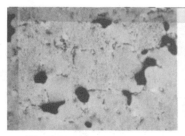

IN-100 Turbine Blade as
cast. Porosity evident
between grains (200x)

IN-100 Turbine Blade after
HIP at 2100°F/45 ksi/2 hrs.
Porosity eliminated (200x)

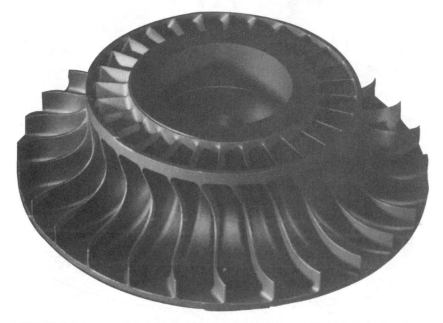

FIGURE 16.2 Scenes at Industrial Materials Technology. *(Industrial Materials Technology (IMT), Andover, MA)*

porosity. Howmet, Inc. of Whitehall, MI, is the first foundry in the country to establish their own production facility to HIP their castings. They have launched extensive research and development to establish themselves as leaders in casting and HIPing titanium and superalloy products—especially for the aerospace industry. Figure 16.3 shows one of their furnaces being lowered into a HIP pressure vessel.

The HIP process underwent much development at the Battelle Institute in 1955, and was first applied to casting densification in the early 1970s. At that time, it was very expensive because the HIP units were small and support equipment was coupled with each unit. In addition, controls were manual, the cycles long, tooling was cumbersome, and vessel loading was inefficient.

An IMT operator loads the isolation
chamber in a HiPIC unit.

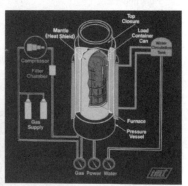

Schematic of a typical HIP unit.

An IMT operator loads the 60" HIP unit.

FIGURE 16.2 Scenes at Industrial Materials Technology. *(Industrial Materials Technology (IMT), Andover, MA) (Continued)*

Now, the HIP unit size is dramatically larger, the support equipment (such as vacuum pumps and gas storage) are shared by several HIP vessels, the former slow cooling is now fast because of the addition of mechanical cooling equipment, and the controls are computer operated. Tooling and handling equipment is flexible and easy to use.

Cut-away diagram of Howmet #3

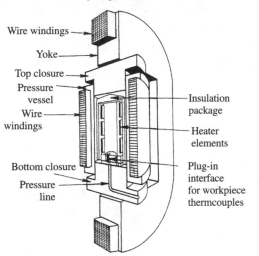

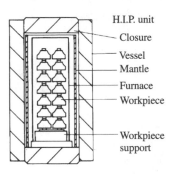

Howmet HIP capabilities

	#1	#2	#3
Work diameter	20"	10"	41"
Work height	69"	20"	96"
Temperature capability	2250F	2640F 3180F*	2300F
Pressure capability	30KSI	15KSI	15KSI
Heating element	Mo	Graphite	Mo
Workload T.C. capability	16	8	25

FIGURE 16.3 Howmet Foundry HIP equipment and diagrams. *(Howmet Turbine Components Corp., Whitehall, MI)*

Battelle Institute has been experimenting with HIP for several decades. A series of tests on as-cast vs. HIPed properties in copper alloy No. 903, and various aluminum alloys had dramatic results. The as-cast alloy showed widespread fine shrinkage defects. The porosity was highly interconnected and finish machining of this casting resulted in much leakage. The HIPed casting showed none of these defects, had no leakage, and showed a marked improvement in tensile properties (Figure 16.4).

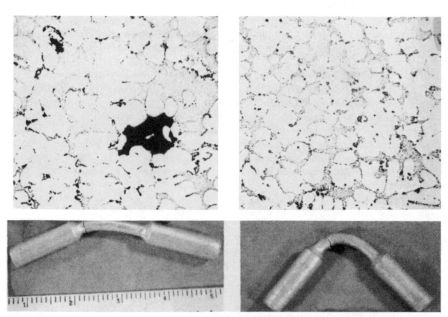

Aluminum Alloy Casting. Left: Before HIP. Right: After HIP.
Microstructures shown at 100X.

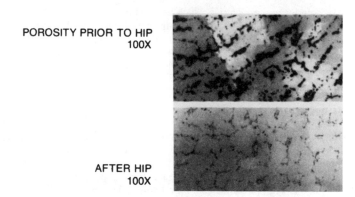

POROSITY PRIOR TO HIP
100X

AFTER HIP
100X

FIGURE 16.4 The beneficial effect of HIP on aluminum alloy casting: A. Before HIP. B. After HIP. Microstructures shown at 100X. C. Closing the internal porosity. *(Howmet Turbine Components Corp., Whitehall, MI)*

According to the Battelle Institute, the use of the term *healing* during the HIP process derives in part from some early work in which cemented carbide parts containing tear defects that had resulted from pressing and sintering problems, were upgraded to quality components by HIPing. The Battelle studies also indicated that both ferrous and nonferrous castings responded well to hot isostatic pressures.

You have to expect a certain amount of porosity in any casting. Industry has lived with it for many years. Since 1975, attention has been paid to HIP as a means of eliminating that porosity and providing the casting with properties that are normally associated with a forging. Despite the obstacles that HIP has caused to schedules and cost, industry must ask itself some key questions. "What is the reject rate? How much does the rework truly cost? How does this affect deliveries?" In some instances, HIP might solve these problems.

But it was not until Alcoa evaluated the effects of the process on cast aluminum alloys on a systematic, experimental basis, that the beneficial effects created by the elimination of porosity received proper consideration. Alcoa proved that HIPing improved the mechanical properties of commercial, aluminum alloys. In that work, high-strength aluminum alloys containing high concentrations of fine porosity were demonstrated to yield "water clear" radiographs following hot isostatic pressing. Figures 16.5 and 16.6 indicate success by Avco with aluminum, Howmet's success with titanium, and IMI's work with 17-4 PH stainless steel. Figure 16.7 shows large titanium investment castings being prepared for HIPing.

There are still technical drawbacks to the use of this process. Some castings have porosity that vents through the external skin. If the process gas were to enter the voids through a surface defect, there would be no closure or benefit to the casting. This type of castings have to be culled out and surface sealed in some way before HIPing. And some castings simply are too large for the current HIPing equipment. A third significant reason is that most castings are too inexpensive to engage in a HIP process until the process price per pound decreases somewhat.

TRADITIONAL HIP APPLICATIONS

High-performance materials (such as titanium alloys, nickel-base alloys, stainless steels, and cobalt-base alloys) have been the traditional work for HIP densification. Structural castings for jet engines (such as fan frames, compressor cases, and turbine frames) are major users of the titanium alloys and the iron base and nickel-base superalloys. Similarly, the fatigue-critical orthopedic implant devices made from cast titanium and cobalt alloys are routinely HIP densified.

Cast nickel-base superalloy turbine blades and vanes are frequency HIP densified to remove microporosity. The typical reason that HIP is specified for these critical, high-performance applications is the improvement in mechanical properties and fatigue life that results from healing internal porosity. In typical nickel-base alloys, ductility improves about 50% and fatigue life over 100%. The maximum-strength values are not improved, but scatter bands are narrowed.

Fine grain structure and superior mechanical properties result from HIPing, which requires very high pressures and temperatures. The range of pressures goes from 10,000 to 50,000 psi, depending on the circumstances. Usually, argon serves as the pressure medium, although sometimes nitrogen is used. The temperatures are always below the

One of the biggest users of HIP in the integral turbine rotor field has been the Avco Lycoming LTS101 engine where the axial compressor rotor (in Custom 450) and power turbine rotor (C101), shown together, and centrifugal impeller (also in Custom 450), shown below, are HIP processed. The advantages in strength and ductility achieved by the elimination of porosity through HIP'ing provide maximized ductility and low-cycle fatigue (LCF) life as shown in the diagram.

The ability of HIP to eliminate shrinkage porosity defects, even in the heavy cross section of integral rotor hubs, makes it particularly attractive in the small turbine engine industry. This class of casting is generally HIP processed with the same set of parameters as used for turbine blades, although lower temperatures have been found adequate for the stainless steels.

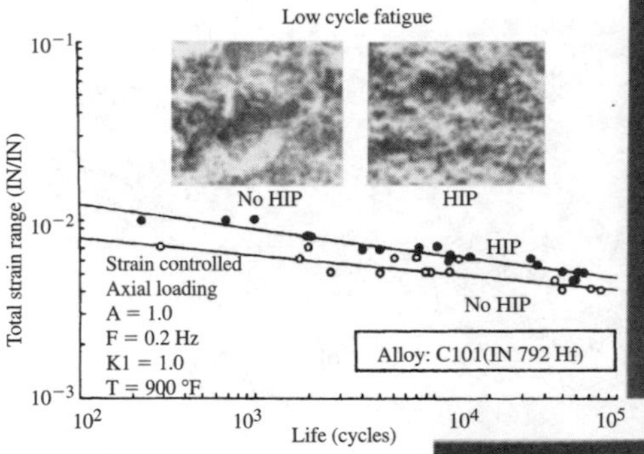

Low cycle fatigue

No HIP   HIP

Total strain range (IN/IN)

Strain controlled
Axial loading
A = 1.0
F = 0.2 Hz
K1 = 1.0
T = 900 °F.

HIP

No HIP

Alloy: C101(IN 792 Hf)

Life (cycles)

In-plant HIP processing has become bill-of-material for virtually all titanium castings produced by Howmet. The HIP processing reduces the frequency of weld repair, improves ductility and high cycle fatigue properties, enhances uniformity in tensile properties and, therefore, produces an improved reliability in the titanium cast parts. Titanium castings are HIP'ed in the 1550-1850F range at 15,000 PSI for 2-4 hours. The photo shows an array of investment cast, HIP'ed parts.

FIGURE 16.5   Traditional users of the HIP process. *(Howmet Turbine Components Corp., Whitehall, MI)*

melting point of the metals being processed. For aluminums, it would be around 1000°F, but the steels and their alloys could go as high as 2900°F.

When components must stand extremes of both heat and cold, the material generally considered is a carbon-carbon composite. Chapter 46 describes the carbon-carbon composite. Hot isostatic pressure can create carbon-carbon composites that tolerate temperatures of 6000°F and resist thermal shock permitting rapid transition from –250 to 3000°F.

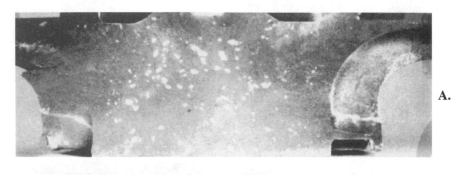

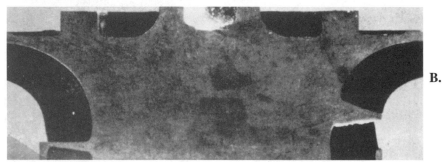

FIGURE 16.6 Densification of castings by HIP: A. Before HIP. B. After HIP. *(Industrial Materials Technology (IMT), Andover, MA)*

FIGURE 16.7 Large titanium investment casting. *(Industrial Materials Technology (IMT), Andover, MA)*

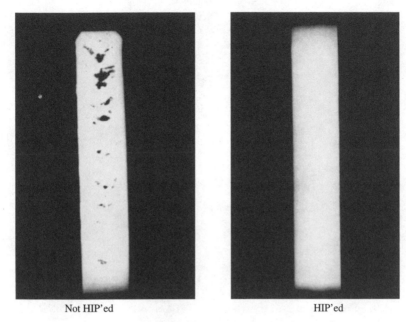

Not HIP'ed   HIP'ed

FIGURE 16.8 The beneficial effects of HIP. *(Industrial Materials Technology (IMT), Andover, MA)*

The HIP impregnation of carbon structures uses a temperature of 1800 degrees F and high pressure in an isolation chamber to impregnate woven graphite structures with pitch. Subsequent high-temperature treatments convert the pitch to a carbon matrix.

In a powder-metal HIP cycle, a pressure-tight metal container is used to transmit the gas pressure to the powder. Densities approaching 100% of theoretical can be achieved while retaining the fine structure inherent in the powder particles. The large HIP equipment at IMT's Andover, MA, facility can handle up to 45,000 pounds of powder metal billets in a single load. You can see the benefits that this company has obtained from HIPing to improve both Rene test slugs and a large void in another material, Figure 16.8.

Two applications for HIPing have been cemented carbides and powdered metal tool steels. The carbides are used as the basis for cutting tools and might end up being sputtered with a layer of titanium nitride. The powdered metals and ceramics are generally cold isostatic pressed before being HIPed. Both types of cutting tools maintain a sharp cutting edge much longer than they would without undergoing the HIP process. More about cold isostatic pressing later in the chapter.

As stated previously, HIP processing is the simultaneous application of heat and pressure created by heating inert gas in a closed, sealed container. This use of gas pressure provides uniform force all over the parts in the container and normally will not deform any parts. It is done without using expensive dies. One of the first developments for HIP was to join metal parts by diffusion bonding, and this bonding is still a vital attribute of the process. Currently, the process is used to create denser parts and materials. Castings

FIGURE 16.9 Loading large castings for HIP. *(Industrial Materials Technology (IMT), Andover, MA)*

and powder (metal and ceramic) parts all contain internal voids that often seriously affect mechanical properties. The HIP process will close these voids, then heal them by diffusion bonding. The mechanical property most improved by this process is fatigue strength. The voids in processed parts are similar to high-stress notches, which cause premature fatigue failures. IMT shows a large load of castings in Figure 16.9 going in just to offset this possibility.

In-plant HIP processing has become bill-of-material for virtually all titanium castings produced in the Howmet foundry. This processing reduces the frequency of weld repair, improves ductility and high-cycle fatigue properties, enhances uniformity in tensile properties, and, therefore, improves reliability in the titanium cast parts. Titanium castings are HIPed in the 1550- to 1850-degree F range at 15,000 psi for 2 to 4 hours.

Process

HIP equipment is basically an electric furnace within a high-pressure vessel. This equipment can perform functions that permit improvements in other manufacturing processes. As an example, a few years ago, parts customarily made by machining bar stock were changed to the powder-injection molding (PIM) process (Chapter 28) to save money. Unfortunately, porosity flaws in a stressed section caused early fatigue failures. After the PIM parts were HIPed, they performed successfully.

Industrial Materials Technology (IMT) Inc. of Andover, MA, is one of the largest companies in the world that provides HIP services. Figure 16.10 is a list of IMT facilities. It has been printed mainly to show typical work-zone sizes, and the maximum operating temperatures and pressures. The following is a typical sequence of events at their plants.

1. A fixture is loaded with as many castings as can fit within the container. The fixture, and/or support structure, must prevent distortion to the more-fragile castings.

2. The fully instrumented load is transferred into the pressure vessel and the insulation system is positioned.

COMPANY	LOCATION	COUNTRY	UNIT	WORK ZONE (IN)	MAX TEMP (F)	MAX TEMP (C)	MAX PRESS (ksi)	MAX PRE (MPa)
IMT,Inc.	Andover,MA	USA	6	6"x10"	2550	1400	30	207
			8	8"x26"	2900	1600	15	103
			16	16"X60"	2450	1340	15	103
			16	16"X60"	2450	1340	15	103
			16HP	16"X60"	2450	1340	45	310
			38	36"X96"	2350	1290	15	103
			38	36"X96"	2350	1290	15	103
			60	60"X120"	2250	1230	15	103
				65"X80"	2100	1150	15	103
IMT-Oregon	Portland,OR	USA	17	17"x57"	2250	1230	15	103
			24	24"x60"	2350	1290	30	207
			45	45"x96"	2175	1190	15	103
IMT-Ohio	London,OH	USA	4	4.5"x5.75"	2640	1450	30	207
				3"x5.5"	3990	2200	30	207
				3.5"x5"	2190	1200	30	207
			5	5.5"x12.5"	2640	1450	30	207
				4.5"x10"	3540	1950	30	207
			15	15.5"x35"	2640	1450	30	207
				15"x40"	3360	1850	30	207
			23	23"x58"	2640	1450	15	103
				18"x44"	3540	1950	15	103
			36(i)	35"x125"	2350	1290	15	103
IMT-Kentucky	Princeton, KY	USA	10	10.5"x29"	2640	1450	20	138
				9"x28"	3000	1650	20	138
			30	29" x 60"	2550	1400	15	103
H.I.P. Ltd	Chesterfield	England	1	11.5"x23.5"	2590	1420	20	140
			2	10"x30'	2590	1420	15	105
			3	16.5"x45"	2590	1420	20	140
			4	3"x4.5"	3630	2000	30	207
			5	45"x84"	2590	1420	15	103
			6	22.5"x50"	2590	1420	20	140
			7	8"x12"	3630	2000	29	203
H.I.P. Ltd	Hereford	England	1	20.5"x75"	2190	1200	15	105
			2	45"x98"	2560	1410	20	140
IMT-Europe	St. Niklaas	Belgium	1	2.5"x8"	3630	2000	29	200
			2	10"x30"	2640	1450	13	93
			3	34"x67"	2550	1400	29	200
			4(i)	45"x98"	2570	1410	20	140
IMT-Essen	Essen	Germany	1	27.5"x59"	2550	1400	15	105
Powdermet Sweden AB	Surahammar	Sweden	1	55"x138"	2280	1250	20	140
			2	20"x40"	3630	2000	30	207

(i) 9/95

FIGURE 16.10 Available HIP equipment. *(Industrial Materials Technology (IMT), Andover, MA)*

3. The cold system is closed, evacuated, flushed with argon, and evacuated again.

4. Normally, the container is pressurized to some initial pressure so that the desired operating temperature and pressure are reached at about the same time.

5. Dwell time at peak pressure and temperature will vary from 2 to 4 hours. When the power is turned off, the system cools by itself. The system is so thermally tight that this part of the cycle is the longest.

The choice of settings (dwell time, temperature, and pressure) is made experimentally. The first estimates are made on the basis of experience and engineering data. The control of temperature is more critical than the other two parameters. Gas purity is also critical. New gas for each run would be too expensive, so the gas is generally recycled. For this reason, on-line analytical tools are constantly monitored to detect contamination of the inert gas.

Commercial Castings. HIP is permitting changes to occur in the casting industry. Investment casting was a process much admired by the aeronautic industry and others because its products could be used with a minimum of machining. However, because the mechanical properties were somewhat inconsistent, the investment approach was not used as much as anticipated. With HIP, the fatigue characteristics were so much improved that investment castings are now very popular in the aircraft industry.

Castings have an inherent problem with porosity, and the industry has lived with it for decades. Since HIP became available as a tool to heal internal flaws, much attention is being paid to surface welding to avoid conventional encapsulation. Internal shrinkage and porosity are the major reasons for rejection of stainless-steel castings. The internal defects are detected by radiography while die-penetrant testing locates surface imperfections. Generally, it pays to weld the surface imperfections, then submit the repaired castings to HIP. This overlay welding repair is being routinely done to titanium, stainless steel, and high-alloy castings, which are expensive.

At this time, the majority of commercial castings do not require the mechanical property improvements offered by HIP to meet their specifications. Foundries have been satisfied with the strength of their castings and have been slow to try HIP because of the perceived high costs and slow turnaround. There are, however, other significant benefits to HIP that are beginning to turn the heads of these foundries by permitting them to improve their costs and expand markets.

The ratio of casting weight to the weight of casting gates used to run as high as 1:6. Now, when HIP is anticipated, this ratio is down to 1:3. When making large castings of titanium or a superalloy, the weight saved adds up to a large sum. Some of these gates cannot be remelted and reused. As a consequence, HIP is now written into many bills of material.

Now, casting designers actually anticipate the advantages of HIP, which allow them to use less gating and make a variety of other changes in the casting procedure. In this way, HIP will permit castings to replace the more expensive forgings in some places.

When high-temperature creep strength is a job requirement, casting is an excellent process to select. Although there are still problems, such as encapsulation requirements, to be considered, the combination of casting plus HIP can provide the required creep strength.

Today, a high percentage of castings are sold to an x-ray specification. Meeting these requirements can, at times, be both difficult and time-consuming—another phrase for costly. Some of these casting geometries are almost impossible to create, but if HIPing is utilized, the casting process is not only successful, but less expensive than forging or a weldment. Many foundries only try HIP when they're desperate, when they're experiencing 25 to 50% scrap, or when they're about to lose a contract. Then they learn that HIP is a cost-effective alternative.

Quite often a casting's subsurface porosity doesn't become apparent until machining exposes it, once again at great expense. Four examples are common:

1. *High-pressure pumps and valves*   Porosity in a side wall or on a seal surface can cause leakage.
2. *High-vacuum components*   Leakage caused by even a slight porosity on a machined surface.
3. *Sterile cleaning applications*   Equipment ranging from food processing to pharmaceuticals to surgical apparatus. The existence of surface porosity makes the equipment difficult, if not impossible, to sterilize.
4. *Highly polished surfaces*   The applications range from the decorative, where porosity detracts from appearance, to plated surfaces, where pits make it impossible to get an adherent coating.

Misconceptions About HIP.   There are three common misconceptions that keep foundries from considering HIP as an acceptable alternative. They think HIP is high priced, has a slow turnaround, and deals only with microporosity.

The cost of HIP services has decreased substantially over the past 25 years because of the influence of several factors:

1. *Size*   The first HIP units were small, experimental types. The size has increased dramatically and the price/volume has dropped.
2. *Efficiency*   Larger, more efficient (by sharing) support facilities (including liquid argon storage, gas reclaiming facilities, better vacuum pumps, gas analysis, computer control, and faster and more efficient loading cycles, including better tooling and rapid cooling).
3. *Generic HIP cycles*   The development of common HIP parameters has allowed commercial HIP facilities to load much more efficiently by sharing loads with multiple customers. These are called *Coach cycles*. These are of particular value for the large HIP units because sometimes the unit cannot be filled by products of one customer. Using the Coach cycle permits the vendor to load their HIP unit to full capacity. The HIP processing is then sold at a price/lb. or a flat price/piece and the HIP vendor assumes responsibility for filling the unit. Figure 16.10 is a compilation of equipment that IMT has across the world, together with a list of their temperatures and pressures.

Turnaround time has always been a concern of the foundry customer—especially now because of JIT (just in time) ordering. Three things have helped shrink turnaround times.

1. *Rapid cool*   Special cooling equipment has shortened the cycle by several hours.
2. *Coach HIP cycles*   These are regularly run on a schedule, once or twice per week (sometimes even three times). If a foundry plans ahead, it can schedule for a specific day and get a 2- to 3-day turnaround. The company, Industrial Materials Technology (IMT), which has furnished much of this information, has four regional facilities across the country, which helps speed up the service. Their main laboratory is in Andover, MA.
3. *Improved equipment*   The equipment now used in this process has been improved to the point where cycle time has been shortened.

HIP of Microporosity.   The capability of the HIP process to heal large voids is frequently not understood. A 1½" diameter bar was drilled at both ends. A ¼"-diameter hole was drilled at one end and a ⅛"-diameter hole at the other end. The holes were sealed by electron beam welding under vacuum. The bar was then HIP densified at 2050 degrees F for four hours at a pressure of 15,000 psi. Before and after x-rays show that the holes were eliminated with surprisingly little distortion of the outer profile of the bar because of the isostatic nature of the process; material feeds uniformly to the center of the pore/hole from all directions. Figure 16.4 and 16.8 indicate similar benefits of the HIP process.

The Chromalloy Corp. has performed a more striking experiment. They machined a 1" spherical void into two halves of a 3" cube of stainless steel. Again, the void was completely healed by HIP with a uniform, controlled shrinkage in the outer dimensions of the cube. The only limitation on the size of the pore that HIP can heal is the presence of a sufficiently thick surface skin to remain intact throughout the consolidation process. Once the skin tears, the densification process stops.

Encapsulation.   Encapsulation of parts is often a necessary first step. A part can be made by mechanically pressing powder or by cold isostatic pressing (CIP). It might be made by the powder-injection molding process. No matter which process is used to prepare the "green" part, internal voids are created, which might be connected to surface porosity. If this were the case, it would lead to internal contamination by gases during HIP pressurization. This cannot be tolerated.

Consequently, many parts must be securely sealed against gas penetration before pressurization. This includes parts that have already been sintered. Sheetmetal encapsulation is a very popular method of protection. Sometimes encapsulation is performed by heavy walled metal molds or ceramic molds. Glass is also commonly used to encapsulate parts to be HIPed.

The encapsulation technique selected depends on several features, not the least of which is cost. The utilization of workload space in the furnace is important. The more parts placed in the furnace in the same load, the less expensive the cost per part. The following describes three of the encapsulation methods.

The first method is to place the "green" body in an oversized capsule of silica or boron silicate glass. (*Note:* Care must be taken to avoid contamination. This glass can react with some materials.) The capsule is evacuated at an elevated temperature before being heat sealed. Then, the heated glass conforms to the shape within. During the HIP process, the heat and pressure compress the part within the glass without allowing gas to penetrate surface porosity of the part.

The second method is to immerse the porous parts in powdered glass, which has a low softening temperature. This method avoids the complications of the first method, but it has a complication of its own. Glass particles might penetrate the very pores you are trying to protect. Accordingly, a special material should be put over the entire surface of the parts being prepared for the HIP cycle. This step prevents glass from entering surface pores.

The third method is to carry the parts through a sintering step and then HIP them. Normally, parts that have been sintered have closed surfaces and do not require encapsulation prior to HIP. That permits you to load more parts per HIP cycle than is possible with the other encapsulation steps.

In permanent mold casting, the part's skin chills rapidly and should not be porous. This permits the HIP process to be used without requiring any encapsulation. Presently, encapsulation adds enough extra cost to the process to price it out of competition in too many cases. Research is being done to lower its cost. Sources are working on highly viscous glasses and ceramics for encapsulation.

PROCESS DIFFERENCES

The following is a flow process for silicon-nitride parts going through various HIP processes. The pressure and heat of hot gases during the HIP process is what provides the uniform consolidation of porosities; even those appearing in weld joints. These pressures and temperatures can be adjusted during the HIP cycle. The cycle can begin with a heating step to initiate the encapsulation. Then, it continues in a prepared sequence, which makes the cycle ideal for computer control.

As stated previously, the inert gas used in this process is most often argon. Each different material processed requires an adjustment of pressure, temperature, and time to achieve optimum results. That is also true for the same material in different form. That means a powdered material will require settings different from the same material in wrought or cast form. Each part and type of material preparation process requires its own specific adjustments. For instance, a stainless steel casting has a different set of adjustments than the same material in powder form. The HIP settings for a sintered part are different than the settings of the same part, dry pressed or cold isostatic pressed.

Another HIP process, called *high-pressure reaction sintering*, uses high-pressure nitrogen gas, instead of argon, which is the most commonly used inert gas. The nitrogen sometimes provides better results than argon. However, when processing nitride ceramics, nitrogen gas could react with carbon materials and form a poisonous gas. Suitable precautions must be taken to prevent such an occurrence.

The cold isostatic pressure (CIP) process, mentioned several times previously, uses high pressure, through hydraulics, to densify powders in rubber molds, to approximately 92 to 95% of theoretical density. Then, the parts can be either sintered or put through the HIP process. If mechanical properties are significant, the HIP process should be used to raise density to nearly 100%. See Figure 16.11 for a laboratory cold isostatic pressing.

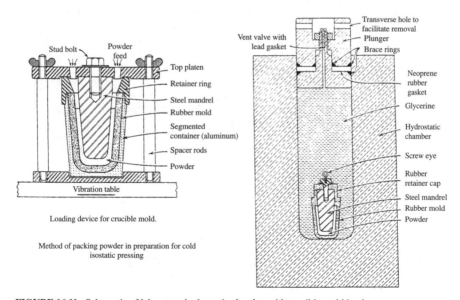

Loading device for crucible mold.

Method of packing powder in preparation for cold
isostatic pressing

FIGURE 16.11 Schematic of laboratory hydrostatic chamber with crucible mold in place.

In the CIP process, no heat is used, but the rubber mold is submerged in water so that the pressure is uniformly applied all over the mold, just like the pressure in HIP. Some vendors supply both CIP and HIP services. They should be consulted about the sequence of steps your product should have before the part drawings are finalized.

THE ROLE OF HIP IN INDUSTRY

HIP has become exceedingly significant to modern industrial technology. It is basically a batch process, but it has increased its output per batch cycle from pounds to tons. Its required use is noted on drawings and purchase specifications. This increased popularity is, of course, because of its unusual results (increased mechanical properties), which, in turn, are caused in large measure to innovations in the design and construction of HIP equipment.

A wide range of applications for HIP have become apparent because of extensive experimentation by a few dedicated vendors and scientific communities, such as Battelle. The most important applications of HIP in current use are:

1. Densification of internal flaws in many materials: castings, powdered metal, ceramic parts, and weldments.

2. The diffusion bonding of similar and dissimilar materials and the healing of densified, internal porosity.

3. Powder compaction production of both metal and nonmetallic parts.

Hot isostatic pressing can be used to create high-strength bonds between similar and dissimilar materials through the use of diffusion bonding. IMT, and probably other HIP vendors, have experience in the application of powder-to-powder, powder-to-solid, and solid-to-solid bonding. The resultant HIP-clad surfaces can be engineered to be more wear and/or corrosion resistant. This technology makes it possible to add a highly resistant coating to a less costly substrate to increase the operational life of the part, thus achieving substantial overall cost savings. See Figure 16.12 for a couple of these applications.

An example of this coating is in valves of different sizes. The valve body is made of an inexpensive, alloy steel. The face of the valve requires wear resistance. HIP is used to bond a fracture-tough microstructure that contains wear-resistant particles to the valve body. The processing can is designed to permit application of the cladding just to the working surface of the part. The composite cladding is 100% dense and metallurgically bonded to the alloy steel. The rejuvenation of service-run parts is especially attractive to the aircraft industry. The cost of materials and labor has increased so rapidly that HIP processing could provide a welcome alternative.

CASES

The HIP process was at first used for space and aeronautic components. Then, it started to be used for automotive parts, such as pistons, rods, crankshafts, and engine blocks. Next, it had applications for medical implants, turbine blades, pump impellers, and boat propellers. One company making stainless steel impellers that were extremely highly stressed,

B.

A.

C.

FIGURE 16.12 HIP clad wear surfaces: A. HIP tooling for bonding a wear-resistant surface to a steel valve body. B. The microstructure of a wear-resistant steel matrix. C. These components are HIP bonded together to produce a highly wear-resistant tungsten carbide-surfaced valve lifter for use in a diesel engine. *(Industrial Materials Technology (IMT), Andover, MA)*

went from 90% defective parts to almost 0. Now HIP is being used by many companies trying to improve the reliability of their castings.

A helicopter manufacturer tried HIPing a cast titanium hub for a helicopter rotor. It was processed at 1680 degrees F and 15,000 psi for two hours. The casting's properties were similar to the forging previously used, but cost much less.

Pratt & Whitney reported that HIPing increased the tensile strength of Inconel 718 to 80% that of the wrought material, compared to 60 to 70% for as-cast. At the same time, fatigue strength increased from 55 to 85% of the wrought metal.

In many cases, HIPing can double the life of cemented carbide tool inserts by eliminating internal flaws. The pressure used to create cemented carbides often causes internal tears. HIPed carbides are used for products that require extremely smooth surfaces, such as extrusion punches, mandrels, dies, and strip mill rolls. A tiny defect in a strip mill roll could result in the scrapping of thousands of feet of sheet metal, costing far more than the added expense of HIPing.

Rene 120 high-pressure turbine blades had been plagued with poor castability. Shrink porosity in the dovetail was often exposed during dovetail grinding. The result was a high reject rate and a much higher cost than had been anticipated. HIP reduced the reject rate from 28% to only 4% at the dovetail grind.

The cost of manufacturing an aircraft fitting was compared using three different methods. These methods were machining from bar stock, a titanium forging, and using titanium-powder metal technology. The fitting was finally made from P/M by cold isostatic pressing, then vacuum sintering to a density exceeding 95%. Material utilization was 90% by P/M, 30% as a forging, and 5% machined from bar stock.

QUALITY CONTROL

In the beginning, HIP was a loosely controlled research process. But over the years, problem areas have been identified and proper quality procedures incorporated. There are five main areas of HIP quality control:

1. Gas purity has historically been one of the most significant variables in the process. Contaminants (such as nitrogen, oxygen, methane, carbon monoxide, and hydrogen), even in trace (ppm) concentrations, can cause undesirable surface reactions. Experience with these contaminants has led to the establishment of limits (in ppm) that control impurities. Howmet, for instance, permits total impurities of no more than 100 ppm. These purity levels are monitored by on-line gas chromatographs.

2. A temperature tolerance of ±25°F is critical in the HIP process. Over temperature can cause incipient melting and insufficient temperature can result in incomplete void closure. As a further check of these quality procedures, small test cubes cast in a superalloy with a thermally sensitive microstructure are included in every run at the thermocouple locations. These are examined metallographically to check thermal response and to confirm thermocouple readings.

3. Pressure is kept within the desired tolerance by a setpoint controller, which references the output of a pressure transducer.

4. Metallography is checked by including in every run at least two "closure bars" cast in the alloy of the parts being HIP processed and placed at the top and bottom of the load. These bars are also used to monitor the surface condition and the absence of incipient melting.

5. Mechanical testing is employed to establish optimum production cycles before process approvals are granted. In some cases, testing is done on a regular basis to satisfy customer requirements.

COST

You have to expect a certain amount of porosity in any casting. Industry has lived with it for many years. Since 1975, attention has been paid to HIP as a means of eliminating that porosity and providing the casting with properties that are normally associated with a forging. Despite the obstacles that HIP has caused to schedules and costs, industry must ask itself some key questions about the expense of not HIPing their castings. The manager of the Howmet Foundry claims that 95% of the castings that Howmet produces are HIPed.

Castings of expensive superalloys are made to near-net-shape, a fact that has stimulated industry's interest in HIPing. These castings have very little extra metal to be machined off. For example, the radial compressor rotor for the F-107 turbine engine required 32 pounds of material to fabricate from a forged billet, but only a little over 5 pounds with HIPing. The final weight of the rotor is 3.6 pounds. Material savings are very important when dealing with high-performance metals.

The cost per HIP cycle depends on the size of the container. A large container might cost $4000 per cycle. A medium-sized container might cost $2000 per cycle and a small one could cost $1000. Naturally, you should compute the quantity of parts that could be HIPed in each size of container, then proceed with the least expensive.

Industry has to question the cost of not HIPing. "What is the rejection rate on castings? What is the cost of radiographic inspections? How much is spent on scrap review, rework, and reinspection? How many man-hours are wasted waiting for repairs? How does all this affect deliveries? How much of a premium would the customer pay to receive only good castings?" The true, actual costs of rejected parts surprises most people who are not familiar with the sequence of events.

It is hoped that research and development will decrease the costs of HIPing. The IMT Corporation is equipping its larger units with a rapid-cooling system that will allow reduction of the cooling period to ½ the present floor-to-floor time. Several companies are working on surface-connected internal flaws, searching for a step that would preclude the expense of the current encapsulation process. This would permit HIP to be used on the many small, inexpensive castings that now cannot afford the process. The difference between acceptability and rejection of a casting is generally internal flaws. Currently, only the expensive castings are salvaged. But if costs decline, more borderline castings would use HIP consistently or at least for repair.

In the period of 1982 to 1992, the change in pricing of the IMT 38" HIPing unit dropped 50%. This was computed using the Consumer Price Index for real-dollar adjustment. This example reflects the cost reductions caused by improvements in the efficiency of HIP facilities. Changes that have contributed to these savings are: more efficient facility-wide support systems (liquid-argon storage, gas-reclaiming equipment, gas analysis, vacuum pumps, computer controls, etc.), and faster, more efficient cooling and loading. The result is lower costs and quicker throughput of both military and commercial products.

Coach cycles, which permit more economical densification of ferrous, nickel, superalloy, and aluminum castings run on a fixed, periodic schedule, once or twice per week. Space in each cycle is available at one fixed price per pound. These cycles have a dependable, rapid turnaround. Foundries and other customers can plan ahead and schedule work for a specific day and get a 2- to 3-day turnaround.

Some vendors are developing special low-cost HIP cycles. One example is a proprietary cycle called Densal, which IMT has available for specific aluminum castings. This cycle is faster and uses less gas than a traditional HIP cycle. It was developed around 1990 by an English HIP company; in the U.S., it is licensed to IMT. Studies are in progress to attempt to extend this concept to other alloys.

The development of common HIP parameters has led to more efficient load cycles that combine the work of multiple customers in one HIP load. Most vendors now offer these coach cycles (also called *shared* or *piggy-back cycles*). This is of particular value for the large HIP units that have a low price per unit volume, but frequently cannot be filled by one customer. In coach processing, the HIP vendor assumes responsibility for filling the unit. The net result is a cost advantage for small-lot customers if they can use the common (generic) HIP cycles.

In 1970, the cost of HIPing ran about $10 per pound. Today, in spite of rising expenses for labor and materials, the cost for the coach cycles runs between $1 to $3 per pound, depending on the casting material, size, shape, and quantity. Cycle times now typically range from 5 to 8 hours, with a good portion of this period spent waiting for parts inside the pressure vessel to cool.

The purpose of this chapter is not to offer HIP as a panacea for all casting problems. The process is certainly not a substitute for good casting design and good foundry techniques. It is of no help whatsoever for ceramic or other inclusions. However, HIP can be a cost-effective tool that the conscientious casting engineer should include among the tools used to solve problems and develop new applications.

CHAPTER 17
LATHE TURNING

INTRODUCTION

Fundamentally, lathe turning generates cylinders with a single-point cutting tool. Most of the time, the tool is stationary and the workpiece is rotating. In many respects, this is the most straight-forward machining method in common use throughout the world. Consequently, much development has occurred in cutting tools for lathe turning. Then, the results were carried over to milling cutters and drills. It should become evident that the evolution of the most modern cutting tools has been continuing for decades.

Four basic turning operations are depicted in Figure 17.1. Figure 17.1A is typical longitudinal turning. Figure 17.1B is typical facing and Figures 17.1C and D are profiling and angle cutting. Turning is a combination of two movements: rotation of the workpiece and feed movement of the cutting tool. Turning is best described as the removal of material from the surface diameter of a rotating workpiece with a single-point tool.

The workpiece rotates in a lathe with a certain spindle speed at a definite number of revolutions per minute (rpm). At the same time, the tool is cutting in a horizontal direction in an adjusted speed of 0.010" per revolution (or 0.020", 0.030", 0.060", 0.125" or more). This speed is adjustable and the tool could be machining threads instead of just cutting. If the tool was ground to an angle of 60 degrees, it would cut a standard U.S. thread (more about threads later).

It is infinitely better for the tool to continuously cut in one long pass than to cut intermittently, as in machining a casting (Figure 17.2). The stresses imposed by intermittent cuts, as in the first cut of a forging or a casting, can break (or at least dull) your cutting tool. One precaution you can take is to turn the tool to take the pressure on a longer surface (Figure 17.2). This spreads the force over a larger cutting edge, thereby being advantageous to tool life.

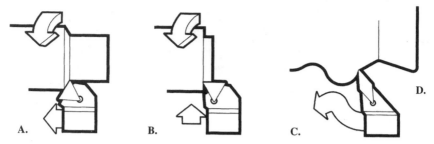

FIGURE 17.1 Lathe tools turning, facing, contouring, and angling. *(Sandvik Coromant Technical Editorial Dept.)*

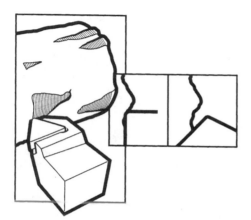

FIGURE 17.2 Intermittent machining. *(Sandvik Coromant Technical Editorial Dept.)*

MATERIALS TESTING

As a rule, entering angles of 60 to 80 degrees should be selected for general turning. At least that is a sensible starting point, which will generally lead to longer tool life. The hardness of the workpiece material will determine what lead angle you use. The hardness is measured by the Rockwell, Brinell, or Vickers test. The harder the material, the more lead you should give or the smaller the entering angle.

The Charpy or Izod machine tests the materials for shock or impact resistance. These machines have a pendulum that swings down and fractures a notched specimen. By measuring the weight of and the swing height of the pendulum, the energy required to break the specimen is determined (Figure 17.3). A high Charpy or Izod number has the same effect as a high Rockwell number.

TOOL SHAPE

The tool's nose radius has a significant effect both on the roughing and finishing of a turned workpiece. If the point is left sharp, the finish appears like a thread; if ground and stoned, the rounded shape gives a smooth finish. It should be evident that the surface finish is influenced by the size of the nose radius (Figure 17.4).

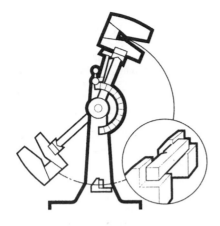

test piece placed in a test machine having a pendulum. These machines are usually one of either the Charpy or Izod type. By measuring the weight and swing height of the pendulum, the required energy is established.

The materials tested will break differently depending upon how ductile, or brittle, they are. The ability to withstand cracking is also measured due to the use of notched test pieces, providing useful information about the need to eliminate stress indications, such as machining marks and sudden form changes. Impact resistance of materials undergoes a marked deterioration at indications and sudden transitions to another thickness.

The impact resistance is the measured energy needed to break a test piece. Energy is absorbed and toughness is established through a notched

FIGURE 17.3 Charpy and Izod testing. *(Sandvik Coromant Technical Editorial Dept.)*

FIGURE 17.4 The nose radius affects surface finish. *(Sandvik Coromant Technical Editorial Dept.)*

Top rake is the angle at the top of the tool. It is generally positive; sharp, flowing back and down, but it can be negative. In fact, it is normally negative when a heavy cut is being taken, or sometimes it is used when an intermittent cut is required. The nose radius and the top rake have much to do with the type of chip produced. A small alteration in either can change the chip from one that is long and curling to one that breaks off into more easily handled pieces. Experience will show what size of chip should be attained in any situation. For instance, too light a chip might cause the tool to be pushed away from the cutting surface. In fact, a modest cut might provide a better finish. Yet, some materials can be finished with a very large nose radius on the tool's nose.

A guiding rule is to select the largest toolholder size that will fit in the machine because you want to reduce the tool overhang and provide the most rigid base for the tool.

INSERT SHAPE

Plenty of high-speed steel tools are used, generally from 2 to 4" long and ¼ to ¾" square. However, this section is confined to tools that are square, triangular, or round inserts (which are held in toolholders with a screw). The effective cutting length should be established because you don't want to change tools unless it is unavoidable. When selecting the insert size, a larger, thicker insert will provide edge security, although at a higher cost. The assortment of inserts in Figure 17.5 are marketed by Sandvik Coromant Company, but there are plenty of competitors.

Generally, for roughing, a large nose radius should be selected for strength, but it should be checked for tendencies to cause vibration. When establishing the feed rate for

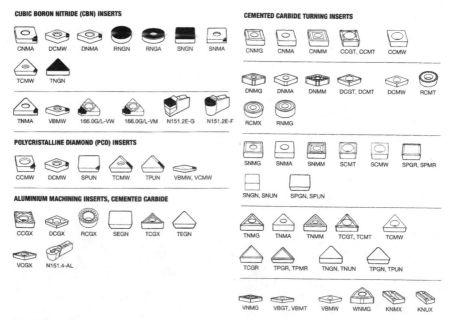

FIGURE 17.5   Lathe-turning cutting-tool inserts. (*Sandvik Coromant Technical Editorial Dept.*)

FIGURE 17.6 A copying situation where one tool is used for the complete cut. *(Sandvik Coromant Technical Editorial Dept.)*

the roughing operation, it is essential that the feed rate does not exceed the tool radius. In fact, it should be kept at half the radius or less.

Coated, cemented carbides (GC) dominate modern machining applications and provide the best alternative for most turning operations in steel, stainless steel, and cast iron. These should be considered for most turning operations. The uncoated, cemented carbides (C) are excellent for certain grades of cutting.

Cermets (CT), which are titanium-based cemented carbides, are suitable for light roughing and finishing. Cubic boron nitride (CB) is a very hard material that is suitable for turning hardened steel, chilled cast iron, and nickel- or cobalt-based alloys. Polycrystalline diamonds (CD) have a completely different area of application. It is unsuited for materials containing carbon. It should be used only for nonferrous metals.

The correct selection of turning inserts should be picked out of a reference handbook that you can get from any insert distributor. The cutting speed should also be looked up in the same catalog. The feed rate will be related to the material hardness. All of this cutting data is available in the catalog. The overview of correct insert geometries, showing the nose radius and cutting edge angles with typical chip-breaker design, should be looked up until you remember them.

There are similarities—areas where more than one tool design will be satisfactory. Originally, copying or duplicating by reproducing the shape of a template or prototype has given way to a metal-removing operation, where the direction of feed changes during the machining process. These traditional inserts have angles of 35- to 55-degree points. This should enable the workpiece to be completed without changing tools. See Figure 17.6 for a view of a copying situation.

Generally, the round-shaped insert should be used as often as possible because it has the strongest edge and produces a small entering angle. The recommended cutting depth can be up to ¼" insert diameter.

PRIMARY GUIDELINES FOR TURNING WITH CERAMICS

1. Begin by chamfering a start to the cut with a cemented-carbide tool.

2. Use a round insert especially for a roughing operation.

3. Plan tool path carefully to avoid shock loads.

4. Use a cemented-carbide insert for initial roughing if the first cut is irregular, giving the appearance of an intermittent cut.

5. Have dry machining as first preference—especially if the cut is intermittent. If machining wet, ensure that the cutting area is flooded.

6. Stability is the essential factor throughout. Use the correct modular tooling system.

These instructions are for any type of turning (horizontal or vertical) lathe; old-fashioned, plain-machine lathe; or new machining center.

PARTING AND GROOVING

In parting operations, the workpiece rotates while the tool moves inward, 90 degrees toward the center of the work, as with face turning. Normally, the saddle is locked to the table so that the only direction the tool can take is straight in toward the center (Figure 17.7). As the machining progresses, the cutting action is slower and slower. When the cutting tool approaches dead center, it generally breaks off a pip at the center (Figure 17.8). The size of the pip can be reduced by supporting the workpiece. In fact, the pip can be directed toward the cutoff section or the main piece by biasing the cutting edge slightly to

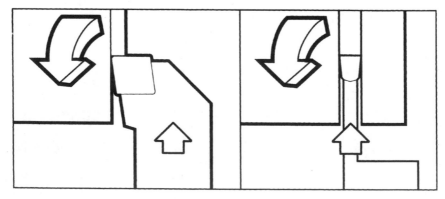

FIGURE 17.7 Facing and parting involve radial feed, toward the center. *(Sandvik Coromant Technical Editorial Dept.)*

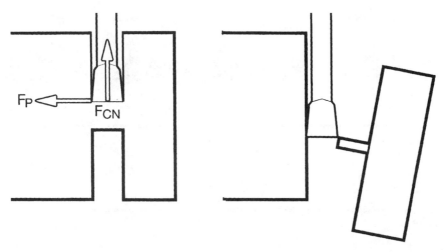

FIGURE 17.8 Pip-formation at the center at parting operations. *(Sandvik Coromant Technical Editorial Dept.)*

one side or the other. Another way to reduce pip size is to reduce the feed by 75% when the tool nears the center.

As the depth of the cut gets deeper and deeper, the cutting tool has to be longer and stability decreases. Only a small surface is present to draw off the heat, which makes the lubrication very important. Grooving operations are like an aborted parting operation. The tool cuts only part way down, but the operation is similar.

This simple operation, parting, is one of the most difficult lathe operations. Because the tool is cutting on both sides, the chip must be narrower than the groove, and they must be formed in such a way that they can be evacuated from the groove and not become a long unwieldy chip. This can be achieved by grinding a chip breaker in the tool.

In addition to all the precautions you must take, the tool must strike the center of the workpiece within ±0.002". If it strikes higher, it isn't cutting, it's pushing. If it strikes too low, it will dig into the work and probably break. Vibration can also arise as a result of the deflection of the workpiece. The closer the parting position is to the chuck, the less vibration and tendency to deflect. Consequently, the machining should be done as close to the chuck as possible, as in Figure 17.9.

THREADING

Development of threading tools has come a long way since the days of high-speed tool bits ground to shape and slowly fed by the lead screw. Most of today's threading is performed by indexable insert tools as part of a rapid CNC process. Now threading is routinely handled by correctly shaped tools and the CNC mechanism. During the feed passes, the tool is traversed longitudinally along the workpiece, then withdrawn and moved back to the starting position for the next pass along the same thread groove.

The feed rate is the important key that must coincide with the thread pitch. This coordination is obtained by various means, depending on the type of machine being used. The

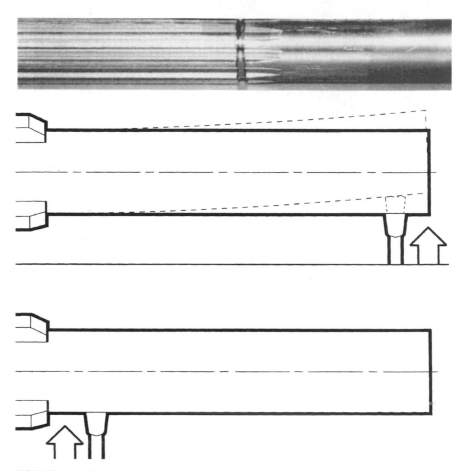

FIGURE 17.9 Eliminating vibrations by machining closer to the chuck. *(Sandvik Coromant Technical Editorial Dept.)*

national standard for threads in the U.S. is 60 degrees for the included angle. There was a time when engaging the lathe's lead screw was the only way to cut threads. The job required 6 to 16 passes with a shaped tool reducing the depth of cut with each pass.

Internal thread cutting is a little more difficult than external threads because of the chip-evacuation problems and tendency to instability. A thread gauge is used to set the threading tool square to the workpiece before threading begins. This is most important for both external and internal threads. The thread profile of 60 degrees is truncated by having the sharp points cut off and left flat or given rounded crests and roots.

The Unified thread is designated by the diameter in inches, the threads per inch, followed by UNC for coarse threads (¼-20 UNC), and UNF for fine threads (¼-24 UNF). There are left- and right-handed threads, but most are right-handed. Right-handed threads are always cut toward the chuck and left-handed threads are always toward the tailstock.

Thread-cutting speed is generally 25% slower than ordinary turning. When threading with carbides, a high enough speed must be used to prevent the formation of a built-up edge. Threading to a shoulder can cause problems. It is difficult to withdraw the tool at the same place each time. The only solution is to lower the cutting speed.

With each pass, an increasingly larger portion of the tooth is cut so that the load on the tool increases. If the cut is kept constant, it will lead to a rough, undesirable finish. Consequently, the last two passes are very light to eliminate deflection, caused by plastic deformation. When threading to a fit, taking a 0.001" cut can make a profound difference. You are cutting three surfaces simultaneously, which makes quite a difference. Just cleaning out the last cut without any infeed can make a tight fit just right.

No matter what rules and suggestions are given, you will form your own biases because there is no substitute for experience. For instance, some machinists will grind their tools a little shy of 60 degrees with perhaps a slight side rake. Each machinist will have his trick for finishing the last tooth forward of the shoulder. A good machinist will have a method of figuring deflection in internal threading. This is sometimes a problem in vibration. And no machinist worthy of the name fails to round off points on thread roots and crests.

MULTIPLE-START THREADS

Threads can have two or more parallel threads, which means two or more thread starts. In a single thread, the lead is equal to the pitch, the distance that the thread advances in one revolution. The lead of a thread with two starts is double a single. Four starts means that one turn would advance the screw four times a single pitch. These are used where fast movement is desired. The old-fashioned fountain pen would close with one twist.

This is a good spot for a true story. During World War II, a British destroyer came into the Boston Naval Shipyard and a young apprentice was given the job of replacing the forward gun's breech block, which had cracked. He wasn't able to board the ship, but a bright engineering officer gave him a sketch with all the required information—he thought. The young apprentice machined a beautiful quadruple thread, but it wouldn't even start in the female thread. One look told the story. The apprentice had machined an American standard UNC thread of 60 degrees, but the breech block was a Whitworth standard of 55 degrees.

CHAPTER 18

DRILLING

INTRODUCTION

The methods of making short or deep cylindrical holes, whether through holes or to a shoulder, is called *drilling*. Unless it was done with the use of a drill jig, it means center drilling and pilot drilling. The subject of drilling also includes subsequent machining: reaming, counterboring, and ball burnishing. Formerly, drilling was carried out via conventional, vertical machines and it was such a slow process that it became a bottleneck in production. Nowadays, this process is performed on many different machines (such as lathes, machining centers, and milling machines), and it is a fast operation.

Drilling is the most common machining operation performed, and the great majority of holes are under 1" in diameter. In all machining operations, the evacuation of chips is important, but in drilling, the removal of chips is of paramount importance.

In conventional drills, the workpiece is stationary with the drill both turning and advancing as a feed. In lathes, the workpiece (not the drill) rotates, although the drill does advance in feed. Normally, most drilling is done with a common helical drill. Trepanning (Chapter 8) is taking its share of drilling jobs (Figure 18.1). A very large share of drilling is being done by drills fitted with inserts; a lesson learned from lathe turning.

DRILLING DEFINED

The evolution of lathe cutting tools occurred first, then the fruits of that labor spilled over and benefited drills and milling machines. The basic definition for drilling is the same whether you drill with a helical drill, a spade drill, or a drill with inserts. The material-removal rate can be computed easily. Simply multiply the speed times the feed.

When inserts are being used, you must be concerned about rake angle and chip breakers, just the same as though you were lathe turning. Drilling with modern cemented-carbide

FIGURE 18.1 Large-diameter holes are generally trepanned. *(Sandvik Coromant Technical Editorial Dept.)*

drills enable high material removal rates to be achieved. Large volumes of chips are created, which are flushed away with high-pressure cutting fluid.

PRECAUTIONS FOR DRILLING

Manually feeding the drill is a cardinal sin. Most of the time, the feed is insufficient and the drill rubs and loses its edge. Every machine has a feed; engage it. Not supplying enough coolant is another cardinal sin. If drills with coolant holes are available, use them. If unavailable, flood the scene externally with coolant.

The most common cause of short tool life and damage to the drill is faulty centering. In applications where the drill is turning, centering is achieved by predrilling a center hole or using a drill jig. For applications where the workpiece is rotating and the drill is stationary (except for feed), it is important to ensure that the centerline of the drill coincides with that of the workpiece. Otherwise, the cutting edges of the drill are exposed to uneven loading and the hole will be oversized.

Much trouble can be averted by ensuring that the drill is perpendicular to the workpiece. This will save unnecessary wear on the drill and have a tendency to correctly size the hole.

If you have many ⅛" diameter holes to drill and don't have a reamer handy to accurately finish them, try drilling the holes undersized—perhaps ³⁄₃₂" diameter, but in the correct positions. Then, you can use the ⅛" drill to correctly size the holes. They will be fine, just as though they had been reamed.

When drilling holes that cross the axis of another hole, the drill exits a concave surface, then is forced into another one. There is the possibility of deflection occurring. The safest way to approach this situation is to drill the hole from two directions. If, however, it is decided to cross drill the hole in one operation, great emphasis should be placed on the stability of the drill. Slow down the drill when you are breaking into the second concave surface.

The tolerance for straightness is most critical when drilling deep holes. Hole straightness is best obtained when both the workpiece and the drill rotate. If this is not possible, opt for a turning workpiece. The worst conditions for straightness occur when you have a rotating drill and a nonrotating workpiece. The choice of tool, feed, and workpiece material affect the stability, the power required, and the choice of machine.

Generally, surface finish and tolerance are the criteria for discarding a drill or changing inserts. Another method is to use 80% of the estimated tool life. Increasing the cutting speed moves the build-up area toward the center of the drill. The increased temperature reduces the ability of the cutting edge to resist wear. With too low a feed, the long chips that form are liable to break inserts and clog the workpiece. The longest tool life is obtained by using cutting data that wears the tool from center to periphery evenly.

When trepanning, whether horizontal or vertical, make some provision to prevent the core from damaging the drill when the operation is complete. One suggestion is to place spring-loaded pads against the core to keep it off the trepanning tool. It is best to trepan in the vertical position and dispense with the problem entirely.

Common to all drilling tools are the problems that arise from the cutting speed varying from zero at the center to a maximum at the periphery. The drilling tool is defined as a rotating tool with one or more end-cutting edges and one or two helical or straight flutes. The chips that are formed in the hole must get out without damaging the hole's surface. This means that the chips must go up the helix or down the straight flute. That's why the lubrication is so important.

APPLICATION OF DRILLING TOOLS

The tool geometry for indexable inserts will permit drilling short holes up to ¾" diameter at a stock-removal rate about three times faster than twist drills. These indexable inserts can be changed 30 to 40 times during the service life of the drill. However, they aren't available in all of the common drill sizes, so twist drills have to be used for many of the small holes. The wise approach is to get as many sizes of the insert drills as possible and keep them on hand.

When close tolerances must be met in a short hole, several steps are possible.

1. A deep-hole drill will be accurate, but it is necessary to have the special equipment needed for deep-hole drilling.
2. A short-hole drill can be used; then finish boring to size or ream (Figure 18.2).
3. A twist drill can be used; then finish boring to size or ream.

Whenever it is possible to machine at high spindle speeds, cemented carbides should be used to increase production. Every drill manufacturer has handbooks that are available at no cost. They should be studied before buying drills. The Sandvik Coromant handbook says that in the medium-sized hole diameters, there is an overlap between Coromant U-drills and Coromant Delta drills.

Close tolerances or a hole depth that restricts the use of indexable insert drills shows you that Coromant Delta is the only choice. However, if the initial penetration surface is not flat, if the hole is predrilled, or if cross-drilling is necessary, then Coromant U-drill is the only option. U-drills have inserts that can be changed, which is a cost advantage (Figure 18.3).

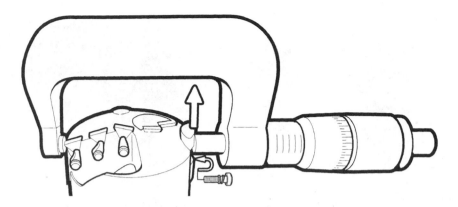

When setting the diameter of the drill, the clamping screw is freed (1). Then the diameter is adjusted using two adjustable screws in the insert seating unit of the peripheral insert (2). The adjustable screws are first set to a diameter which is smaller than that required and then screwed to the right diameter. The last hundredths of a millimetre can be set in with the clamping screw tightened.

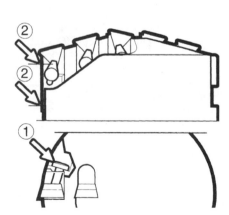

STS

FIGURE 18.2 Setting the inserts on a drill to cut accurately. *(Sandvik Coromant Technical Editorial Dept.)*

FIGURE 18.3 Modern insert and brazed carbide make short work of drilling. *(Sandvik Coromant Technical Editorial Dept.)*

When drilling large-diameter holes, consider short-hole drills; when machine power is limited, trepanning is the way to go. Short-hole drills with indexable inserts are not as accurate as twist drills, but they offer high productivity and are used principally for the larger hole diameters. Some of these indexable insert drills have asymmetrically placed inserts that overlap. This design permits the inserts to be changed so that different grades and geometries can be used at the center from the type used at the periphery. The insert can be triangular, round, or square. It is held in place on a three-point support by a screw.

Indexable insert drills can be used on surfaces that are a little out of flat. It doesn't matter if the surface being penetrated is concave, convex, or just irregular. Slow down the feed until you make the initial penetration. Although the twist drill is the most common short-hole drill, there is a considerable difference between HSS drills, cemented-carbide drills, and those with modern drill geometries. Modern geometries keep tight tolerances and haven't the trouble with pushing the center as twist drills. In fact, their center has a rake and can penetrate steel at high speed.

The choice of insert geometry is important when many holes are to be drilled. This is where a distributor's handbook should be studied. The handbook will have recommended inserts and various geometries for low, medium, and high cutting speeds. It will also show various flute designs for the most efficient evacuation of chips at high penetration rates for various materials.

The hard-cut drill is a cemented-carbide drill with straight flutes and is designed for drilling difficult-to-machine materials, such as chilled cast-iron and stellite. Broken taps and hardware can be removed with a hard-cut drill. A center reference must first be put in the head of the tap with a larger drill, which centers the tap-removal drill. After the head of the tap is removed, very often the remainder of the tap can be urged out with two pointed hand tools.

DEEP-HOLE DRILLING

The techniques described from here on apply to deep-hole drilling. A deep hole is normally a depth ranging from 5 times diameter to 100 times diameter. In fact, they have managed 200 times diameter by using more than one steady rest. This type of machining has a high accuracy of diameters, straightness of holes, and a high removal rate of material. This type of drilling places high demands on the machine, the cutting tools, and all of the associated equipment.

Workpieces rejected in the wide range of industries (such as chemical, nuclear, steel, oil, and aerospace) can have huge economic effects (if caught) and catastrophic consequences (if not caught). Reliability gets high priority here. This type of machining demands good chip breaking and excellent lubrication. The chip evacuation must be the very best.

Deep-hole drilling machines present a choice of what is rotating. The workpiece, the drill, or both, can be turning. Of course, the feed is always to the drill. There are three methods of deep-hole drilling. The small-diameter holes, the most common operation, does the drilling in one step. The large-diameter holes are also trepanned in one step. If the trepanning tool is removed to change inserts, the weight of the hanging core can make it difficult to guide the trepanning tool back into the hole again. The third method is to drill a smaller hole and go back and finish it to size.

Two guiding principles are followed in deep-hole drilling. The cutting fluid is supplied internally through the tool and the chips are carried away externally through a groove in the shaft of the tool. This is the well-known gun-drilling system. Then, there is the system that supplies the fluid externally and the chips are removed internally through the tool. Two different systems have been developed with this principle: STS (single-tube system) and the Ejector system, which is a twin-tube system (Figure 18.4).

The cutting geometry of indexable insert drills is asymmetrical, meaning that in deep-hole drilling, pads must be used to support and guide the drill. Without this assistance, the extra stress would damage the drill and score the walls of the hole. A heavy flow of lubricating oil also assists in this protection. One additional requirement of deep-hole drilling is a drill bushing to start the job accurately (Figure 18.5).

Many mixtures of fluids are used in machining: some are synthetic, some semi-synthetic. Some are soft, some are hard, and many are in between. Some are neat (not mixed with any water) and some are water-soluble. There are dozens of reputable cutting oil distributors and their advice should be sought. The choice of cutting fluid depends on the machining operation, the workpiece material, the tool material, and the cutting data.

Generally, better lubrication should be sought with low-speed operations, difficult operations, difficult-to-machine materials, and a necessity for superior surface finish. The table in Figure 18.6 shows various machining operations listed in order of metal-cutting demands, from relatively light operations (such as grinding in Figure 18.6(1)) to demanding operations (such as thread turning in Figure 18-6(9) and thread tapping in Figure 18-6(10)).

REGRINDING DRILLS

It is important to follow instructions when attempting to sharpen a modern drill. A wheel with a sharp edge is required. If it is an HSS drill, you must retain the original point geometry. The chisel point has grown in size and the shine on the edge indicates dullness. That chisel point must be thinned until the shine disappears. If the drill was coated with TiN, you will have to reduce speed when you use that drill again. The special machines sold for drill sharpening are reliable, although much of the sharpening can be done by hand on a suitable grinding wheel (Figure 18.7).

For regrinding hard-cut drills, a metal-bonded diamond or silicon-carbide wheel is recommended, plus a simple fixture that is specially produced for the hard-cut drill. The drill is placed in the fixture so that the angles of the drill line up with the corresponding angles on the fixture. A setting block is used to set the drill properly in the fixture. The fixture and the setting block are positioned on a level surface and the drill is lined up using the setting

A.

B.

FIGURE 18.4 Deep hole drilling: A. STS system. B. The ejector system. *(Sandvik Coromant Technical Editorial Dept.)*

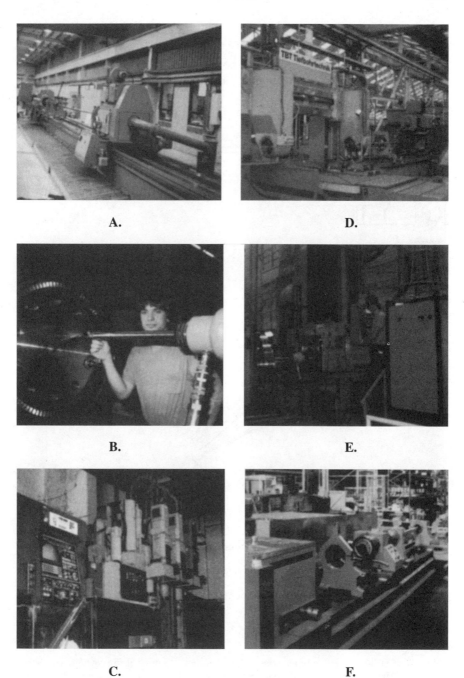

FIGURE 18.5 Deep hole drilling: A. Ejector and STS. B. Ejector. C. STS. D. STS. E. Ejector. F. STS. *(Sandvik Coromant Technical Editorial Dept.)*

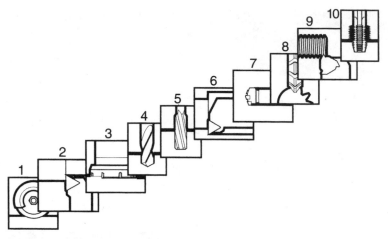

FIGURE 18.6 Need of thorough lubrication from light (1) to heavy (10) *(Sandvik Coromant Technical Editorial Dept.)*

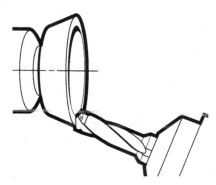

FIGURE 18.7 Regrinding modern drills. *(Sandvik Coromant Technical Editorial Dept.)*

block. The two main cutting edges are ground first with the fixture turned 30 degrees. Then, both surfaces are ground at a 60-degree angle (Figure 18.8). Instructions for the regrind of all Coromant drills are in the Sandvik handbook. All drill distributors have similar handbooks for their clients.

CHOICE OF DRILLING SYSTEM

With very small-diameter holes, the only option: the gun drilling system. For very large-diameter holes, trepanning with STS (single-tube system) is the wisest solution. Gundrilling is used for small holes, and with extra steady rests (a stand placed on the ways of a lathe to support workpiece and keep it turning about its axis), it is possible to drill 200 times the hole diameter if the pump can take the added load. This system requires a drill bushing and a support for the workpiece.

The Ejector is the only system that is adaptable to most machines. The Ejector system requires good chip breaking because the cutting fluid pressure and volume are only 50% of an STS system. The Ejector system and the STS and trepanning tools also can use brazed drills or indexable insert drills.

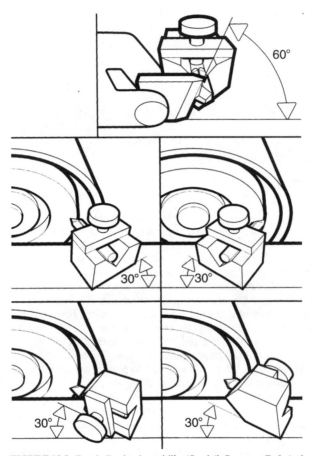

FIGURE 18.8 Regrinding hard-cut drills. *(Sandvik Coromant Technical Editorial Dept.)*

CHAPTER 19
MILLING MACHINES

INTRODUCTION

Milling performs metal cutting by a coordinated movement between a cutting tool and the feed of a workpiece. Today, that means moving the tool against the workpiece in any direction. The milling cutter has several cutting edges, each of which removes a certain amount of stock. Milling in a modern machine is very efficient, providing excellent surface finishes, high accuracy, and tremendous flexibility in geometry and flexibility. CNC is furnishing this new ability in contour milling.

Machining centers is the latest in milling machines. But milling has been performed since the turn of the century. At first, it was called *planing*. There was a single stationary tool and a flat, moving table that held the workpiece. Then, the milling cutter was invented, and it rotated as the workpiece was fed into it to form the first milling machine. In World War II, we used exceptionally large planers that were big enough to hold the tremendous low-speed turbines of naval ships. Generally, few planers are around now. For the sake of production, they have given way to fast-machining milling machines, some of them twin-cutting machines.

In high-production shops, you can see these machines doing two pieces at a time, pulling blue chips off, so hot that a thick plastic or metal shield must be placed to catch the chips. The insert-type blades, held on the cutter by a screw, are remarkably efficient. It pays to have some machinists go to a milling cutter school for a week and learn, not only the best inserts to use for different materials, but also the feeds and speed to use for their machining. They learn the rules of basic milling, how to compute cutting speed for fast stock removal. Some companies send a lead mechanic to such schools and use them to set up each job on the milling machines.

Milling has a great variation in the jobs it can do. In spite of the sophisticated, new CNC multi-axis machining centers, these older, twin millers can excel at certain jobs. The choice of milling machine is the first thing to do. Will the machine be a vertical, horizontal, universal, gantry, CNC, or machining center? Factors such as stability, accuracy, and finish

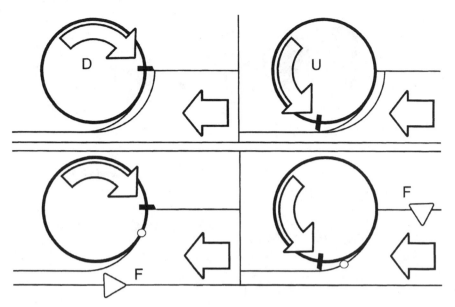

FIGURE 19.1 The two methods of milling: Down milling/climb milling and up milling/conventional milling. *(Sandvik Coromant Technical Editorial Dept.)*

should then be assessed. No matter how good the machine and tooling are, instability is the greatest threat to performance.

The two main types of milling are: up and down milling. In up milling (also called *conventional milling*), the feed direction of the workpiece is opposite to the cutter rotation at the area of cutting. In down milling (also called *climb milling*), the workpiece feed direction is the same as the cutter at the area of cut.

In up milling, the chip thickness starts at zero and increases to the end of the cut. There is always the danger that the inserts will rub the workpiece or make contact with a work-hardened surface, caused by the preceding insert. Forces (*F*) tend to lift the work off the table, so the work must be securely fastened to the table. In down milling, the insert starts its cut with a large chip, which avoids the burnishing effect of up milling, and the workpiece is pulled into the cutter holding the insert in the cut (Figure 19.1).

Of the two, down milling is preferred to up. However, there can be no play in the lead-screw or looseness anyplace that could cause instability. Backlash anywhere cannot be tolerated because it would cause insert breakage, at the very least. If you don't know the equipment too well, up milling should be selected. To obtain the right performance, it is critical to monitor chip formation.

SURFACE TEXTURE

The surface texture in metal cutting is the resulting irregularities caused by the plastic flow of metal during a machining operation. This will vary slightly as the machining type, condition

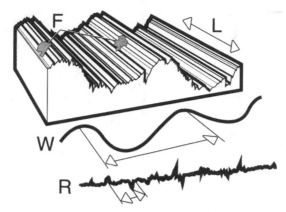

FIGURE 19.2 Surface texture defined. *(Sandvik Coromant Technical Editorial Dept.)*

of tooling, workpiece material, and general stability changes. The following elements define surface texture (Figure 19.2):

1. *Roughness* is the finely spaced geometric deviations on a short length of the workpiece.
2. *Waviness* is the larger irregularities on the next level up of sampling length. The spacing of peaks and valleys is larger than roughness.
3. The next level up has the deviations from actual component form, such as straightness and roundness. These deviations are not normally classified under surface texture, like roughness and waviness.
4. Surface lay is the orientation of surface pattern that describes the direction of the dominating pattern of the machining method.
5. *Flaws* are faults not included in the measurement of the surface. They include scratches, cracks, and other unintentional deformation of the surface. For roughness, which might be within 10 microns in height, a very short evaluation distance is sufficient. For waviness, which might be 30 microns in height, a longer distance is needed.

Generally, the only thing in the surface you have to be concerned about is the microinch finish.

MILLING CUTTERS

A milling cutter has a variable number of teeth, and its selection is very important. At least two cutting edges must be engaged in the cut simultaneously. At least a chip depth of 0.100" should be taken. The pitch of a milling cutter is the distance between a point on one tooth to the same point on the next tooth. Cutters are classified into coarse, close, and extra-close pitches.

The coarse pitch means fewer teeth and large-chip pockets. Close pitch means more teeth and moderate chip pockets. *Extra-close* means small-chip pockets and very high table feeds. This type of cutter is especially useful when the cutting speed must be kept low, as in machining titanium.

Vibrations can cause trouble in milling. Sometimes the frequency of the cutter inserts will coincide with the natural frequency of the machine or possibly the workpiece. A method of overcoming this problem is to use a differential pitch, which is nothing more than unequal spacing of the teeth. Controlling the natural frequency of the workpiece is more likely to be the problem than the machine. See Figure 19.3 to view the separation of teeth in a differential cutter, as opposed to even distribution in ordinary cutters.

It is emphasized that in facemilling, the cutter should be larger than the surface it is cutting by 50 percent. The chip thickness at the entry and exit of the cut is affected by the cutter diameter. This means that you could have a similar disadvantage as you would with insufficient feed per tooth.

The number one rule in successful machining is maintaining stability. Surface texture will suffer when vibrations are excessive. You must have sufficient chip thickness to maintain satisfactory cutting action. When a thin chip is started, especially one that starts at zero, the burnishing action begins immediately. Power is consumed to try and separate tool from workpiece by a slight plastic deformation. A thick chip is more power efficient than a thin chip. When the exit of the cut occurs, Figure 19.4 shows how the chip tends to change from compressive to tensile. This should be avoided by the proper selection of the cutting tool.

EFFECT OF TILTED SPINDLE

Sometimes milling machines should have their spindles tilted slightly in the direction of the feed. This is to ensure that the cutter doesn't lie flat to the workpiece. The tilt, but a fraction of a degree, is to prevent the cutter from back cutting, which can mark up a surface. Even with the proper spindle tilt, back-cutting problems can arise because of workpiece or spindle deflection. This can be combated through:

1. Improved workpiece support.
2. Reduce cutting force by changing to a positive cutter.

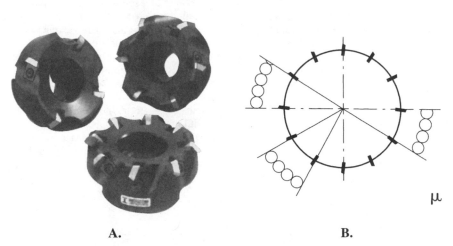

A.   **B.**

FIGURE 19.3 Various pitches: A. Same milling cutter with various pitches. B. Differential pitch. *(Sandvik Coromant Technical Editorial Dept.)*

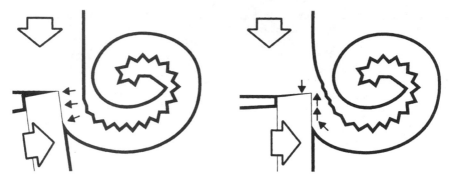

FIGURE 19.4 At the exit of the cut release of force should be maintained compressive. *(Sandvik Coromant Technical Editorial Dept.)*

3. Check cutter mounting for burrs, dirt, or misalignment

4. Reduce forces by reducing feed, depth or width of cut, or by increasing the cutting speed.

5. Reduce spindle overhang to a minimum.

EFFECTS OF VIBRATION

Nothing will deteriorate a machined product as fast as a situation leading to vibration. Machine condition and rigidity will affect the workpiece's surface quality. Excessive wear of the spindle bearings feed mechanism will cause vibrations. Without proper maintenance, any machine will develop vibrations and its attendant poor finish. Vibrations can also be caused by spindle overhang, which can be at least partially eliminated by tuning the toolholder. This can be suspected if the overhang is more than four times the diameter. Avoid cuts on an unsupported overhanging section of a workpiece. The proper tool should be selected. That is the tool's diameter to the workpiece-width ratio should be correct. Where possible, use a positive geometry cutter—one with not too many teeth involved in the cut at one time. Feed per tooth is always important. Usually, it is better to add a tooth than to subtract one.

The type of workpiece material should count heavily in choice of cutting tool. Some all-round facemills will machine steel, stainless steel, cast iron, titanium, and to some extent, aluminum and copper. This would be alright for a mixed schedule of work. But for real production, it is always best to choose the special cutter for your exact material. The volume of chips that can come off workpieces is amazing when first noticed.

A thicker insert can be required to give extra cutting strength. Chipbreaking is not an issue in milling because the length of cut is limited, but chip formation is important because it has to do with chip congestion and positive/negative geometry in the cutter. Positive milling has gained a lot of ground partly influenced by more widespread facemilling and less-powerful milling machines.

END MILLS

The end mill has a working range that is defined by its diameter and length. Some end mills can be seen as small-diameter square shoulder facemills. They can also make a groove or do side milling up to the point of deflection. It's important to be sure that the chips are evacuated—especially at high machining rates. Down milling is preferred for peripheral-type milling because the feed per tooth can be varied in accordance with the range from full slotting to lighter edging.

Solid-carbide and brazed-carbide end mills are available in larger diameter sizes, up to 2" with long cutting-edge capability. These can be used in heavy milling production and can also serve as drills. The overhang of end mills should be always be kept as short as possible to prevent deflection.

The end mill is designed to machine axially, which makes it more susceptible to deflection and vibration—especially at high machining rates. Down milling is preferred for this operation because the feed per tooth can be varied from full slotting to light edging. When long and heavy cuts have to be made, the long-edge milling cutter is the one to use (Figure 19.5). This is like a large helical end mill, where the cutting edge is composed of indexable inserts.

FIGURE 19.5   Edging with long-edge cutter. *(Sandvik Coromant Technical Editorial Dept.)*

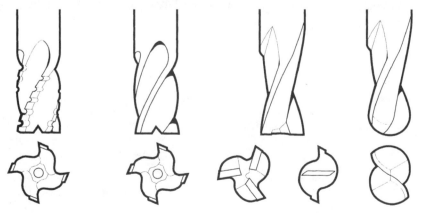

FIGURE 19.6   Two-, three-, and four-helix end mills. *(Sandvik Coromant Technical Editorial Dept.)*

FIGURE 19.7 Drilling endmills. *(Sandvik Coromant Technical Editorial Dept.)*

FIGURE 19.8 Gang saws working, bathed in lubricant. *(Sandvik Coromant Technical Editorial Dept.)*

A change between down and up milling can sometimes improve surface quality. When machining soft materials, the correct coolant would also improve the finish. Two-, three-, and four-helix end mills are shown in Figure 19.6. In all end-mill operations, it is important to hold the end mill very securely with the shortest overhang possible and have plenty of coolant or air remove swarf from slots. Drilling end mills are shown in Figure 19.7.

SLITTING SAWS

There is the whole gamut of slitting saws to consider. Figure 19.8 is an example of using gangs of slitting saws together on an arbor. This is one place where the cutters should engage the workpiece all together, simultaneously. There are cases where this is impossible and there is a danger of vibration. A flywheel is just about the only way to eliminate the problem forever. It is important to keep the distance between the cutter and the flywheel as small as possible.

A flywheel can be built up from a number of carbon steel discs—each having a center hole and a keyway to fit the arbor. For a given flywheel weight, the moment of inertia increases as the diameter increases, which means that if the arrangement permits a large diameter, the weight of the flywheel can be reduced. One other suggestion might eliminate the need for a flywheel.

Make two keyways on cutters larger than 4". This will permit the cutters to be displaced half the pitch in relation to each other when used in gangs. This would even the load variations.

Side and face milling on horizontal machines can cause torsional vibration with detrimental machining results. A flywheel can often solve the problem and lead to improved productivity. The best position of the flywheel is close to the main frame. It is important to keep the distance between the cutters and flywheel as small as possible.

For a given flywheel weight, the moment of inertia increases as the diameter increases. If conditions permit a large-diameter flywheel, the weight can be reduced. The following suggestion might preclude the necessity of a flywheel. Double keyways on side and face mills 4" and larger to permit the cutters to be moved half the pitch from each other to even out the load variation and eliminate need for a flywheel.

THINGS TO CHECK ON

There are three main things to check on, not just for vibration, but whenever milling:

1	2	3
Workpiece	Machine	Operation
Material	Type	Working allowance
Hardness	Power	Surface texture
Quality	Capability	Intermittent cut
Stability	Limitations	Entry
Cutting width	Condition	Exit
Cutting depth	Stability	Number of passes

FACEMILLING

Facemilling is a combined action of the cutting edges, the ones on the periphery and those on the face of the tool. They are available in various shapes for a multitude of different cuts. There is a facemill for cutting soft material, such as aluminum. A square shoulder facemill is used to finish short chipping material (such as cast iron) or it could be a square shoulder facemill for long chipping material (such as steel). There is the general-purpose strong-edge facemill. They have both negative and positive geometry. Negative geometry is very strong and is used to machine hard alloys. Positive geometry cuts easier, with less power. A variety of cutters are shown in Figure 19.9. Wiper finishing inserts can be used on some facemills to improve the surface texture of the finished piece. They are set about 0.002" below the other inserts. The facemilling operation is shown in Figure 19.10.

The insert size for a facemill depends on the application. The cutting depth should be less than ⅔ of the cutting-edge length. Variations in capability are great, depending on the entering angle insert size, geometry, and the facemill design. A thicker insert will increase the cutting-edge strength. Chipbreaking is not an issue in milling because the length of each cut is limited and produces only a short chip. Chip formation, on the other hand, is an issue—especially with chip congestion and the design of pockets and clamping.

A dozen various inserts are shown in Figure 19.11. The catalog will describe each one in detail and every distributor will have their own identification letters and numbers. In

A.

B.

C.

FIGURE 19.9   Facemills: A. Square inserts in 90-degree facemill. B. Triangular inserts in 45-degree facemill. C. Facemills for various applications. *(Sandvik Coromant Technical Editorial Dept.)*

A.

B.

FIGURE 19.10 Facemilling: A. Vertical facemilling of large steel-plate component. B Large-component facemilling in a horizontal mill. *(Sandvik Coromant Technical Editorial Dept.)*

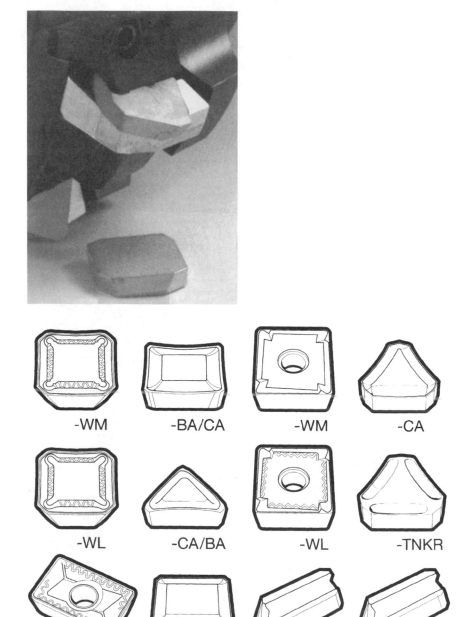

-WM   -BA/CA   -WM   -CA

-WL   -CA/BA   -WL   -TNKR

-WL   -11   -31   -32

Selection of modern, positive milling inserts

FIGURE 19.11   Inserts for facemilling. *(Sandvik Coromant Technical Editorial Dept.)*

facemilling, the width of the surface to be machined is the main factor. When there are shoulders and grooves to machine, end mills are to be considered.

SPECIALTY MACHINING

Chamfering, weld preparations, and deburring are performed by rotary burrs of cementite with chip-breaking facilities along cutting edges. These are very effective as small-milling cutters.

I've included the many angles of milling because modern milling is very versatile. The 90-degree square shoulder facemill and the 45-degree cast-iron cutters, and the 75- and 60-degree entering angle cutters have been included. But the round insert facemill has been ignored until now. The chip thickness changes with the feed per tooth. The round cutting edge is the strongest and possesses many advantages. These can be used to machine most materials, such as high-strength steel, heat-resistant alloys, titanium alloys, and stainless steel. Flood coolants should be used (Figure 19.12).

The use of machining centers and less-powerful machines have required more-efficient machines. Rake angles have suddenly become very significant. There has been a decided swing to positive rake angles because they cut with less power. In fact, positive geometry inserts have become the first choice for facemilling and end-milling performance. But many other tools have been used successfully, including HSS end mills and helical end mills. The router is mainly a pocketing end mill that is designed for the high-production machining of aluminum. The router has one or two edges and can operate at speeds up to 10,000 RPM.

Cermet inserts can provide a mirror finish and can be an alternative to grinding. New types of milling are being discovered and used in specialty cases. Some of these new inserts actually run better without a coolant, but if one is used to get rid of the swarf, it must be used copiously. When you investigate new types of inserts, always see if they need a coolant. If your operation can avoid coolants, it will save about 15% of the cost of machining components.

FIGURE 19.12 Rough-milling of die with round-insert cutter. *(Sandvik Coromant Technical Editorial Dept.)*

CHAPTER 20
MACHINING CENTERS

INTRODUCTION

Generally, machining centers consist of stationary workpieces and rotating cutting tools. A pallet carries workpieces to the center and a tool changer gets the tool it needs to perform the next operation from a magazine of tools set up in a turret or from a lineup of tools kept out of the way of coolants and any large workpiece fastened to the table. A multitude of tools are necessary to make this machining center work.

Dozens of companies sell the adapters needed to make the cutters fit into the machine. With these basic holders also come front-end adapters, which permit extension down into the work area of the spindle. Figure 20.1 shows the Varilock modular tooling system with the basic holder and two different extenders. As mentioned, there are many distributors of this type of necessary hardware.

All machining centers have some kind of modular, quick-change tooling system. When more than one machine is in the center, there is the possibility of a reduction in tooling inventories. Sometimes other more-conventional machines are put there to save part-handling costs. Modular tooling is the key element in these centers. A common center height is established and maintained for uniformity. In-process measuring probes are economically viable in this environment. Figure 20.2 shows manually inserted modular tooling. Every possible method of speeding up production is being devised. There are modular quick-change tools that will fit on a lathe, milling machine, boring mill, and drill. It can form the true universal tool system for the machine shop of the future. Figure 20.3 shows such a setup for a universal milling machine.

Manual machining centers were introduced during the 1960s and automatic tool changing and CNC were introduced during the 1970s. In this period, standards were developed for the tool/spindle interface. The International Standards Organization's (ISO) taper standard's were used to design modular tooling. Today, 75% of all machining centers are ISO 40 or 50 interfaces. The cutting tools have been standardized also and now tooling sets are being prepared in the tool rooms to be placed in the magazines of the machining centers. This is an example of standardization advantages to be gained outside of the machine itself. See Figure 20.4 for a view of modular tools for machining centers.

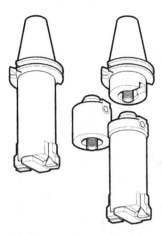

FIGURE 20-1 The Varilock modular tool locking system. *(Sandvik Coromant Technical Editorial Dept.)*

FIGURE 20-2 Semi-automatic modular tooling. *(Sandvik Coromant Technical Editorial Dept.)*

Careful foresight should be used when establishing inventories of holding tools, adaptors, extensions, tooling sets, and cutting tools. These tooling systems are designed to withstand the high cutting forces imposed by today's production demands. They hold the cutting edge at the centerline where they are set, no matter how many times they are used. For the maximum rigidity of the modular toolholder, the very reliable centerbolt type should be used. The front-clamped type should be used when quick change is the priority. Both types are necessary. Always get the type with an integral hole for coolant.

The Varilock (one reliable make) front-clamping mechanism operates with a differential screw. Opposite sets of serrated clamping jaws grasp and pull the tool back into the coupling with a matching drawbolt (Figure 20.5). Contact is always made on the backside of the serrations to generate strong axial clamping force. Whenever you look for a modular holding system, look for the features mentioned here.

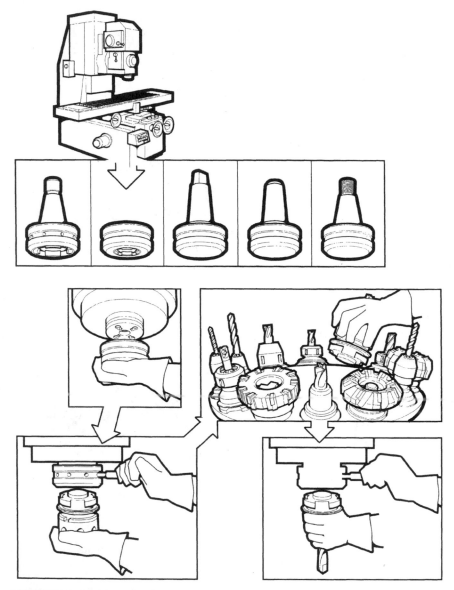

FIGURE 20-3 Modular tooling for a universal millng machine. *(Sandvik Coromant Technical Editorial Dept.)*

BLOCK TOOL SYSTEM

The block tool system (BTS) has changed tooling for machining centers, CNC lathes vertical, and multispindle machines. Standard cutting units use ISO inserts and spare parts,

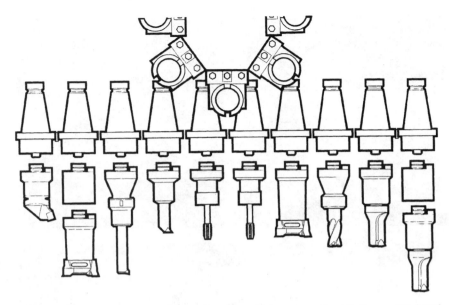

FIGURE 20.4 The principle of modular tooling for machining centers. *(Sandvik Coromant Technical Editorial Dept.)*

which form a simple, cost-effective system that can be integrated into almost any turning machine. Various cutting tools are shown in Figure 20.6.

BTS coupling permits no play in any direction when in the clamped position. The center position of the drawbar ensures that the clamping pressure directly opposes the cutting forces. The force on this drawbar makes the block tools as rigid as a solid tool. The accuracy of the coupling provides excellent repeatability when the unit is clamped and unclamped.

Semi-automatic installations are manually handled with press-button operation to release or grab the unit. The design is based on mechanical clamping and hydraulic release. The unit is compact and operates through a series of Belleville washers. Pressing a button activates a hydraulic piston, which compresses the Bellevilles and releases the unit. The Belleville washer is just a concave washer, not flat. When you press it over center so that the concavity goes to the other side, it moves the release just the same way it arms a projectile in a bomb-arming device (fuse).

The advantage of a machining center is that you can do all the work required at one machine. You can turn, mill, drill, ream, bore, tap, and thread. Part-handling time is saved, which means more machine time. Tool-storage magazines permit a continuous supply of tools to be available in the machine. They can be integrated as complete assemblies into most machines. Disc magazines can hold up to 24 different tools. Drum magazines are necessary if a large number of backup tools are required. Chain magazines can hold up to 60 different tools.

All tools and accessories should be supplied with identification tags. These memory tags can carry their own data and communicate with machine PCs, CNCs, and most data-processing systems. Tool preparation, inspection, and machining data can accurately be performed without intermediate manual errors.

On average, conventional tool changes require 9 minutes per hour. Modular tool changes take 1 minute, saving more time for machining. Modular tools have several added

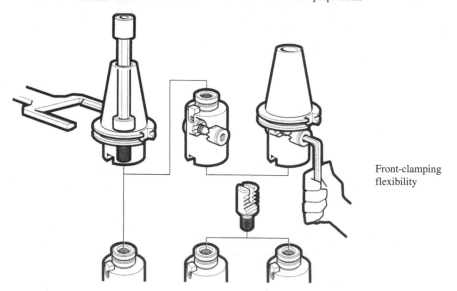

Modular quick change tools mean short down-tiems and efficient tool preparation

Front-clamping
flexibility

FIGURE 20.5   Modular quick-change tools. *(Sandvik Coromant Technical Editorial Dept.)*

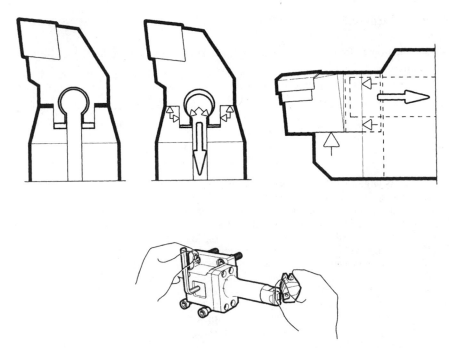

FIGURE 20.6 Block tool system (BTS) turning tools. *(Sandvik Coromant Technical Editorial Dept.)*

advantages. Tool management can be established at any level, whether you're managing a shop full of CNC machines or just one machining center. A computerized plan of administration will do the following for your organization:

1. Run the complete manufacturing program.
2. Tool identification tracks all tools and their data.
3. The tool-storage system facilitates mechanical tool handling.
4. Prepares a path toward automation.
5. Tool monitoring monitors the machining process.
6. The tool probes permit in-process measuring and inspection.
7. Minimizes the tool inventory.
8. Because you are in control of manufacturing, it allows additional work to start quicker.

CHECKLIST BEFORE SELECTING MODULAR TOOLS

1. Ensure that the contemplated tools will be suitable for all current machines and for foreseeable machines in your company's future.
2. Ensure that the contemplated tools will be usable without the burden of adapters that slow down the tool change.

3. Ensure that the stability, repeatability, and accuracy aren't compromised by operational demands.

4. Establish values for built-in safety margins—especially for heavy-duty applications.

5. Test for workpiece quality.

6. Establish priority for quick-changing ability.

7. Select from a dependable supplier who can provide technical support in a broad level.

CASE

A moderately large company producing hardware for the transportation and fuel industries had a situation develop in their largest machine shop, which almost caused them to purchase an unnecessary machine. They had three machining centers that were busy in three shifts. In fact, these machines had a backlog of almost three month's work ahead of them.

The work performed by these centers was of the highest quality and everyone was trying to get work scheduled there because the machined components looked so good. The question facing the scheduler was simply, "Should the work be scheduled to begin in three months in the centers or should it be scheduled to begin in one week in a conventional lathe and milling machine?" Management decided to buy a fourth machining center and was making inquiries when another machinist was hired for the third shift.

Before the purchase was made, the new machinist presented management with another problem. His pile of finished parts was as high as the work done on the two other shifts. The foreman watched the new machinist; he was running his machine faster than anyone else and taking prodigious chips. He had attended classes at the machining center's factory school and had learned what cutters to use for stainless steel. To make a long story short, he was used as a setup man. In three weeks, the scheduler was only one month behind and the gap was closing fast.

P · A · R · T · 4

FASTENING AND JOINING PROCESSES

CHAPTER 21
INDUSTRIAL ADHESIVES

WHAT ARE THEY?

Around 1965, few reliable adhesives were available, and dozens of companies and the military were anxious for the appearance of dependable products. In 1961, I, while attending a lecture on the subject of plastics, heard William McSheehy, Manager of Plastic Fabrications at Raytheon, make the statement that the country's first qualified plastics engineer had not yet graduated from college. He claimed that most of his engineering staff were mechanical, electronic, or chemical engineers. Those personnel learned plastics by reading advertisement literature and practicing. We certainly have come a long way in the last 40 years.

Currently, many companies are involved in the manufacture of reliable adhesives; the products available for use include epoxies filled with a multitude of various materials, cyanoacrylates, urethanes, putties of various materials, stick adhesives, underwater adhesives, and multipurpose adhesives. Some adhesives are loaded with carbides, ceramics, quartz, zinc, and there are room-temperature vulcanization (RTV) silicones. Various types of silicones are used as rubber cements or gasketing materials. There is an adhesive for any fabrication or repair application conceivable.

No doubt, it is best to contact the adhesive producer and obtain advice concerning your application. Yet experience has demonstrated that you will generally be successful with repairs you make using only what is available. That is simply because most industrial adhesives are truly "user-friendly." A large leak in an automobile gas tank was sealed with a sheetmetal screw (to slow down the leak) and a dab of quick-setting epoxy. It was still holding when the car was turned in two years later. A helper, rebuilding an engine, omitted the small gasket under the gooseneck leading to the radiator hose. Naturally, the gooseneck cracked when tightened to the cylinder head and another couldn't be obtained for at least four days. The crack was permanently sealed with an epoxy adhesive.

Rust spots threatened to eat through the base of a dishwashing machine. Because the spots were within an inch of a red-hot resistance water heater, the first epoxy used for the patch was one advertised to withstand a high operating temperature. The patches dissolved the next time that the machine was used. Then, an underwater epoxy was used for the next patch (should have known the constant water presence would keep the patch cool) and is still maintaining integrity one year later. The point is that many epoxies are good for any emergency repair. But when you have a production decision facing you, it is prudent to obtain the very best adhesive available for the specific application. The recommended adhesive might surprise you.

At least two dozen reliable adhesive suppliers are just a telephone call away and every one would work with you to solve production problems or advise what type of adhesive would be best for your application. Six adhesive producers have supplied information for this chapter concerning the latest material they currently have on the market.

They all agree that the most important single step in using any adhesive is the proper surface preparation. The following is a general procedure for repairs and should be reviewed when writing a procedure for new work assemblies.

GENERAL SURFACE PREPARATION

1. Unless using a wet-surface adhesive, the surface must be dry.
2. First, stop all leaks or seepage. This can be done many ways.
   a. Shut off the flow or pressure.
   b. Fit the wooden peg or sheetmetal screw into the hole to stop the flow.
   c. Stuff hole with wax, cork, Mortite, or even a cloth.
3. If the leak is caused by corrosion, the side wall might be weak. Open the orifice until good metal is exposed and the wall is thick enough to be plugged.
4. Remove surface condensation, "sweating," and dampness by using a hot-air device or at least a dry cloth.

Degreasing

The degreasing of surfaces that are immersed in oil is usually the task most difficult to do. There is always the possibility that oil absorbed in the metal surface will cause a problem after the adhesive cures. Oily pump shafts or bearing housings are good examples of such a part.

1. Apply a good cleaner to the surface. Devcon sells Cleaner Blend 300 for this purpose, but each adhesive producer will suggest their own product.
2. Absorbed oil on surfaces that have been constantly immersed in oil should be removed by heat.
3. Allow the surfaces to cool, then degrease them again with the cleaner from step 1.

Surface Roughening

Always abrade a smooth surface to increase the adhesion of the epoxy (or other material) to the substrate. Many field applications are redone because someone forgot to get a good surface profile before applying the patch.

1. Surface cleaning by abrasive blasting or rough grinding should be enough to proceed with the coating.
2. Otherwise, roughen the surface with coarse emery paper or a hand file.
3. Oxidation of aluminum and copper surfaces will reduce the adhesion of an epoxy to the surface. This oxide must be removed before coating and the repair must be done immediately because aluminum (especially) starts oxidizing as soon as it is exposed.

Generally, these steps will provide a suitable surface for adhesive bonding. However, when these steps do not provide satisfaction, the surfaces might require chemical etching. Each material needs its own, specific chemical solution. There is a completely different etchant for aluminum, copper and copper alloys, stainless steel, magnesium and its alloys, and rubber. The parts to be bonded have to be immersed for a period (such as 5 to 10 minutes), then rinsed in water and dried.

Underwater Repairs

Underwater surfaces cannot receive the same surface preparation previously mentioned in this surface-preparation guide, but they do need the following attention:

1. Remove all dirt, barnacles, flaking paint, and algae/seaweed from the substrate.

2. Obviously, degreasing underwater cannot be done, but at least wipe it with a clean cloth.

3. Abrade the surface any way possible. The use of a file or emery paper would be satisfactory.

4. Oxidation must be removed by any mechanical or chemical means.

METAL REBUILDING

A shaft usually has two types of repair areas: the shaft itself and the keyways machined into the shaft. Shafts are worn by vibration, rubbing, fretting corrosion, and abrasive contaminants. Shaft areas that are repaired easily are those that see a seal, bushing, sleeve, or bearing. A shaft worn by mechanical packing is sometimes not practical to repair because the frictional heat generated by fast movement within a packing will soften most epoxies.

Figure 21.1 shows a common repair to a shaft. The worn area (Figure 21.1A) needs to be degreased first, and then machined in a lathe (Figure 21.1B). This will require undercutting the shaft diameter. A general guideline for machining depth is to undercut $\frac{1}{16}$" for shafts $\frac{1}{2}$" to 1" in diameter and to undercut $\frac{1}{8}$" for shafts 1" to 3" in diameter. Dovetailing (Figure 21.1C) would lock the repair into place.

Finish the undercutting by machining a coarse thread like a phonograph profile over the repair area (as indicated in Figure 21.1D). Because these photographs are from the Devcon handbook, we then apply a thin layer of Devcon's Metal-Filled Epoxy over the repair area while the lathe is turning at a very slow speed. Apply the epoxy with a tool stiff enough to force the epoxy into the machined grooves. Continue to apply the product until the threaded area is completely filled and there is an excess of epoxy above the shaft diameter (Figure 21.1E). Permit the product to cure for at least two hours before machining. If you need to accelerate the process, apply a heat source (such as a hot air gun or a heat lamp) to the repaired area. Machine at the lowest speed possible with a sharp tool. As a precaution, you can test the heat of the turning operation by touching your hand to the area and ascertaining that it is not too uncomfortable to hold.

This precaution is necessary. Many years ago, we solved a fixturing problem in a rather unique fashion. We took a small part, shaped like an automatic pistol trigger, and used it to fire a power tool. First, we had to drill a small hole in it. Unfortunately, the "trigger" was very curved and it seemed impossible to hold it for the drilling. We experimented. We drilled a small hole in a steel block and filled the hole with Devcon. Gradually, the "trigger" was sunk deeper and deeper into the epoxy. In about two hours, the "trigger" was held securely. After the epoxy hardened, we were able to insert a "trigger," hold it with a tiny De-sta-co clamp, and drill it. The operation proceeded for a few hundred parts smoothly,

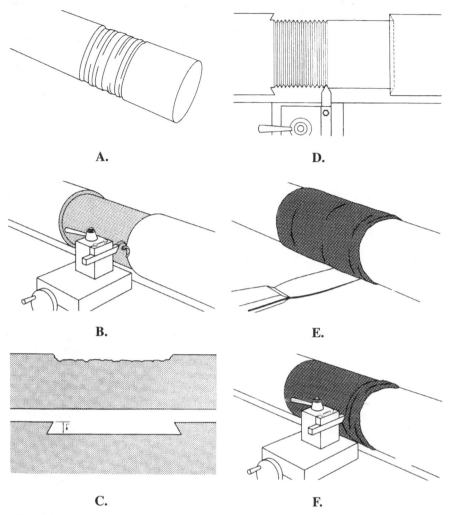

A.

D.

B.

E.

C.

F.

FIGURE 21.1 Shaft repair. *(ITW Devcon, Danvers, MA)*

until the drill dulled, heated the epoxy, and destroyed the fixture. The resultant damage demonstrated the value of using a sharp cutting tool and keeping things cool.

The other common shaft repair concerns the keyway. When a keyway becomes worn through the constant pressure of starting and stopping the shaft's rotation, it is no longer effective, but it can be repaired by the following procedure. Prepare the keyway as you would any other metal surface, by cleaning and degreasing. Then, remachine the keyway by removing a small amount of material on both sides and from the bottom (Figure 21.2A). This can be accomplished by milling or grinding. Filing could remove the necessary material, but that would be an onerous task because about 0.010" of material should come off.

Degrease the area again. Next, apply a coating of Devcon Release Agent #19600 to a new key and to the keyway area where you don't want the epoxy to stick. Every adhesive supplier will have their own release agent (Figure 21.2B). Also, coat the keyway area of the hub with the same release agent.

Apply the epoxy to the keyway area (Figure 21.2C) using a spatula. Build up thicker layers on the side walls, rather than the bottom, to ensure that the key will not be raised too high. Insert the key and remove excess epoxy from all surfaces where it doesn't belong. Immediately assemble the hub to the shaft and leave the assembly rest until the cure occurs. The hub will be removable because of the coatings of the release agent.

METAL SURFACE REPAIRS

Many metal cracks can be permanently repaired. Those that can't be permanent can carry on until new parts can be obtained. Examples of repairable parts are: pump casings, valve bodies, oil pans and their covers, bearing housings, and gear boxes. The following directions should be followed:

1. Clean and degrease the crack area.
2. Drill holes ⅛" larger than the crack at both ends of the crack.

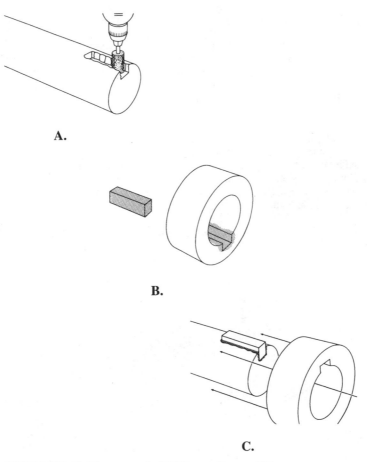

A.

B.

C.

FIGURE 21.2 Shaft keyway repair. *(ITW Devcon, Danvers, MA)*

3. If the crack is long, drill additional holes to relieve the stress within the crack.

4. Use the "V" point of a drill (or grind) on the outer edge of the crack to "V" out the entire length of the crack. This will provide more surface area for the epoxy to take hold.

5. Figure 21.1 shows a common repair to a shaft. The worn area (Figure 21.1A) needs to be degreased first, then machined in a lathe (Figure 21.1B). This will require undercutting the shaft diameter. A general guideline for machining depth is undercut ¹⁄₁₆" for shafts ½ to 1" in diameter and undercut ⅛" for shafts 1 to 3" in diameter.

6. Dovetailing (Figure 21.1C) would lock the repair into place.

7. Use a spatula to force the material deep into the crack and overlap the crack up to 1" at both ends.

8. Where great strength is required, a piece of reinforcing material (fiberglass, nylon, or wire screening) could be used over the patch. Push it into the patch and cover it with additional epoxy (Figure 21.3).

9. Because these photographs are from the Devcon Handbook, a thin layer of Devcon's Metal-Filled Epoxy is applied over the repair area while the lathe is turning at very slow speed. Apply the epoxy with a tool stiff enough to force the epoxy into the machined grooves. Continue to apply the product until the threaded area is completely filled and there is an excess of epoxy above the shaft diameter (Figure 21.1E).

Permit the product to cure at least two hours before machining (Figure 21.1F). If you need to accelerate the process, apply a heat source (such as a hot air gun or heat lamp) to the repaired area. Machine at the lowest speed possible with a sharp tool. As a precaution, you can test the heat of the turning operation by touching your hand to the area and ascertaining that it is not uncomfortable to hold.

This precaution is necessary. Many years ago, I solved the problem of machining a flat on a small part that looked like an automatic pistol trigger. I continuously and gradually sunk the small curved part deeper and deeper into a hole in a steel block that had been filled with Devcon. After the epoxy hardened, we were able to insert the sample part into the formed hole and secure it with a tiny De-sta-co clamp. We machined the sample trigger. Then, we replaced the sample trigger with many more until the end mill became dull and heated up the epoxy. The damage to the fixture indicated the importance of ascertaining that the fixture remained cool at all times.

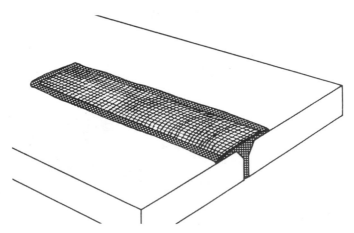

FIGURE 21.3 A high-strength crack repair. *(ITW Devcon, Danvers, MA)*

These preparation instructions are so important that they are being repeated.

1. Clean and degrease the area prior to filling crack with epoxy.
2. Use a spatula to force the material deep into the crack and overlap the crack up to 1" on both sides.
3. Where great strength is needed, a piece of reinforcing material (fiberglass, nylon, or wire screening) could be used over the patch. Push it into the patch and cover it with additional epoxy (Figure 21.3).

Adhesive producers supply charts that show chemical resistance and typical physical properties of their products. They also have application guides to indicate the kind of repairs that each product is designed for. You must have a product list from each company to appreciate the spectrum of products available. There are adhesives for fastening every category of material, such as wood, most metals, most plastics, most fabrics, and just about all materials used in modern manufacturing. This is true for repairs of those materials, as well.

Devcon markets such specialty products as epoxy putty, underwater putty, carbide putty, floor patch, epoxy concrete sealer, cold galvanizer, urethanes, methacrylates, and cyanoacrylates. One product is used to anchor nuts and bolts to holes in granite (layout tables). There is a material for securing hardware against loosening. Obtain handbooks from all suppliers to learn what each can do for you.

Devcon's Flexane is an ideal solution for rebuilding and repairing rubber rollers and conveyor belts. It does an excellent job protecting metals, such as pump impellers, from impact and wear by forming a medium-hard or hard rubber covering.

Most epoxy producers package their products in many convenient ways. One epoxy hardens in five minutes. A two-ton epoxy hardens with that strength. With the stick epoxy, you can break off a piece and knead it to shape. The epoxy putty works like the epoxy stick. The patch epoxy is an adhesive-impregnated patching system that hardens at room temperature for a field, waterproof repair for pipes, tanks, containers, and other equipment.

Multipurpose epoxies can bond many different substrates. The underwater repair putty, the pouch epoxy, is sold in small, plastic bags with a stopper of some kind separating catalyst from epoxy until the repair is about to be made. Methacrylate adhesive can bond dissimilar substrates, in addition to hard-to-bond plastics.

Conductive epoxies, which include high-performance materials for coating, bonding, and sealing electrical components, have electrical and thermal conductivities that are similar to metals. This makes them particularly useful where hot soldering is impractical or where a conductive epoxy is desirable. Conductive epoxies are available in both two-part and one-part systems. Some are pastes, liquids, and some have even been put into spray cans. A brush-on surface coat can be silk screened onto PC boards.

Most of these adhesives are filled with either silver or gold, but others use aluminum, steel, beryllium, or carbon, when superior conductivity is not required. Service temperatures vary from 220 to 500 degrees F. At least one coating can withstand a much higher temperature after it is fused to a substrate of ceramic or glass.

Anaerobics are made by a dozen companies to lock hardware. Some prevent vibration loosening, but permit easy adjustment and disassembly. Others provide positive locking, sealing, and superior environmental and chemical resistance. They are available in various grades for specific purposes.

Cyanoacrylate products, some of which are slightly thixotropic and can be used in spaces as wide as 0.010", while others run like water. Of all adhesives, these have the most manufacturers. Certain precautions must be taken when using cyanoacrylates. These are available in dropper bottles because generally one drop is all that is needed. That is the problem. It flows so quickly that it catches the unwary. This material manages to sneak out and gets between fingers that are instantaneously bonded. That is why it is imperative to wear rubber gloves when handling cyanocrylates, or any epoxy.

Before starting any adhesive job, you should read all directions and advice—especially how to clean your hands when the job is done. If your fingers get stuck together accidentally with any cyanoacrylate, try twirling the tapered point of a pencil between them. A small taper provides a great separating force that can sometimes save your skin.

CIBA-GEIGY PRODUCTS

Ciba's Araldite 2000 epoxy and polyurethane adhesives are formulated for easy handling and reliable bonding in repair, maintenance, and assembly operations. This comprehensive family of products includes materials designed for use on aluminum, steel, rigid plastics, engineering thermoplastics, rubber, ceramics, and a variety of substrate combinations. These adhesives feature a range of work lives from four minutes to two hours, along with high shear strength. Some systems provide outstanding chemical and heat resistance.

In addition to versatile performance capabilities, the Araldite 2000 line of adhesives is supplied in different packages: 50- and 200-ml dual-barrel dispensing cartridges and easy to mix, preweighed quart kits. Ciba furnishes charts that are helpful in determining which adhesive is appropriate for a particular project. The charts list performance that can be expected for bonding various materials. For each product, the following is listed: work life, time-to-handling strength, cure time, lap shear strength, chemical resistance, sag resistance, impact resistance, and heat resistance. Some charts list viscosity at mix, color, gel time, service temperature range, and mix ratios of two-part systems.

MASTER BOND ADHESIVES

Master Bond adhesive systems are another group of reliable products. That company's chart lists 70 adhesive systems. The system that first attracted my attention was the one for bonding to polyethylene and polypropylene. These two materials are often used on worktables to prevent adhesives from stacking to the table. This adhesive will actually adhere to them.

This company makes casting, bonding, and sealing epoxy systems. Some have remarkable resistance to thermal cycling and chemicals, including water, oil, and most organic solvents over a temperature range of -60 degrees F to more than 250 degrees F. Some do not contain solvents or volatiles, and the viscosities range from extremely low to thixotropic, which can be applied overhead. Shrinkage after cure can be as low as 0.0003" per inch. They can be used to bond dissimilar materials (such as wood, ceramics, glass, rubbers, and plastics). Master Bond makes an epoxy adhesive that sets up in five minutes and one that has an unusual peel strength.

PERMATEX PRODUCTS

This company's products run the entire gamut from epoxy, cyanoacrylates, and silicone adhesives to pipe sealants, gasket materials, and their (Loctite) brand of thread sealers. Like most of the manufacturers of these products, the Permatex company has been around for a long time. In this case, more than 50 years.

Their "Spray-A-Gasket" is used to hold gaskets in position during assembly. Their "Weatherstrip" fast-drying, neoprene-based adhesive dries tack-free in four minutes while it bonds rubber, insulation, weatherstripping, and other porous materials. Permatex sells handy, easy-to-use "Poxy-Pak" epoxies, cans of silicone "Form-A-Gasket spray (Figure 21.4), high-temperature silicone, and pipe sealant with Teflon.

This pipe sealant with Teflon is amazing. Pipefitters handling large lines like 4 to 10" diameters, very often have the problem of lining up pipes and fittings going in odd directions, like to different floors. They used to manhandle them, changing sealing tape, trying to make a joint tight, while at the same time aligning the pipes accurately. Now they apply a coating of a Teflon sealer and the joint might be left slightly loose, but without leaks.

They make a "Quick Gel" nonrun, instant adhesive that stays where you put it on leather, rubber, metal, many plastics, cork, and wood. They manufacture the same type of products that all the major producers do. Their "Super Glue" bonds close-fitting, nonporous surfaces in seconds, with a full cure in 1 to 2 hours and no clamping is necessary. If that sounds like something you read earlier in this chapter, it certainly should because there are many competitive products. The following adhesive manufacturers should be sought for a reliable solution to your bonding problems because they are among the best: Lord; Tra-Con; ITW Adhesives; Click Bond, Inc.; Devcon; Armstrong; Ciba; Dexter; Crest Products; Permatex; Deco-Coat; Emerson-Cuming (a division of W. R. Grace Co.), and Fel-Pro.

In 1939, I began a four-year machine shop apprenticeship at the Boston Navy Yard. One of my first assignments was to the department where brass bushings about 10" in diameter were installed on destroyer rudder shafts. They were roughly six feet long and a very large torch was positioned at each end of the bushing to heat it because the inside diameter of the bushing was about 0.010" smaller than the rudder shaft's outside diameter. After much heating, noise, and diameter checking with a long-handled gauge, an overhead

FIGURE 21.4 Applying a gasket. *(Permatex Industrial, Rocky Hill, CT)*

crane would pick up the enlarged bushing and rush it to the shaft, where it was quickly slid down to a shoulder. That was quite a sight. I never missed that assembly because if the bushing didn't make the full distance to the shoulder, it would have to be machined off in a huge lathe.

Around 1970, I visited the Navy yard for sentimental reasons and I noticed a change in the installation of the rudder shaft bushings. The ID of the brass bushings were now machined with a small clearance over the OD of the rudder shaft, and some kind of watery, super glue was used to fit the two together much easier than ever before. When the escorting admiral was questioned about the reliability of that bond, he just smiled and said, "Do you really think the Navy would take any chances with that assembly?"

So, don't be concerned about these adhesives. Choose any reliable manufacturer, follow their advice and directions, and don't worry about safety or reliability. Something else that you should remember is that most of these preparations are user friendly; they will generally be stronger than advertised.

Easy-to-use, environmentally safe, hand-held dispensing equipment for both two part and one part adhesives is resulting in higher productivity and large cost savings for many companies. These systems reduce material used, clean up time and materials, and operator fatigue. The old method of buying adhesives in bulk, transferring the contents to smaller packages and then hand-mixing the two part systems, has been an environmental nightmare. The material wasted as well as the disposal problem have created much concern for managers.

The ConProTec Company is one of several that furnish hand-held dispensing systems for two part epoxies, urethanes, acrylics, silicones, and polyesters. Their "Mixpac" cartridges simplify dispensing whether done manually or pneumatically. See Figure 21.5.

The EPD Company of Elmstead NY, produces a remarkable, versatile sealant called "Astroseal." It seals through oil, under water, and in high temperatures. It can be applied in the rain or under water and its service temperature range is −50 degrees F to 220 degrees F. It has been used successfully at much higher temperatures as a gasket compound. The

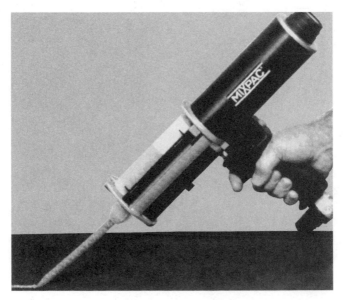

FIGURE 21.5   A Mixpac manual dispenser. *(Conprotec, Inc., Salem, NH)*

FIGURE 21.6   A versatile sealant. *(EPD, Inc., Elmsford, NY)*

material is a high molecular weight, elastomeric copolymer, which is nontoxic when fully cured.

It will seal problem materials such as: TFE, Hypalon, stainless steel, ferrous and non-ferrous metals, aluminum, and galvanized surfaces. Its chemical resistance is greatly affected by temperature, cure time, and the presence of moisture. Consultation with the producer is strongly advised. When the job is finished, the Astroseal should be removed from the gun with solvents. Otherwise, it will have to be removed with a sharp tool.

In its cured state, it will rejoin to itself if the bead is broken. Just add material to the broken area. It is unnecessary to remove the old sealant. After a 24-hour wait it can be painted. This material stretches and remains flexible. It will not become brittle. Before using Astroseal, all surfaces it will touch must be cleaned. Some cleaning chemicals require a waiting period to ensure that no residue remains to spoil the seal.

It is important to recognize that this is a sealant and not a cement. If used under pressure, a strengthening agent (such as fiberglass or steel plate) should be used in conjunction with Astroseal (Figure 21.6).

Twice, I have experienced severe production problems on account of inconsistency of adhesive application. Had I known of the EFD Corp. of Providence, RI, much frustration could have been eliminated. This company manufactures a world-class adhesive-dispensing system. What you want is to apply assembly fluids as inexpensively as possible. You need fast, accurate deposits without waste, mess, mistakes, or operator fatigue. This is unobtainable from manual applicators or squeeze tubes.

Dispensing equipment will easily double production when compared to hand syringes. Oozing stops, drip rags are no longer necessary, and the equipment pays for itself in a short time. EFD advertising literature describes dozens of case histories, worldwide, where costs of assembly have been drastically lowered by the three types of dispensing equipment distributed by EFD: hand operated, air powered, and timed air pulse.

Unlike hypodermic syringes with tight-fitting neoprene pistons, the EFD hand model, the first basic dispenser, uses a molded piston that has negligible friction and requires very little effort. In fact, there is an internal mechanical advantage of 10 to 1. This gives the operator better control—especially when thick fluids are used. When the barrel has been emptied, remove the reusable plunger and discard the barrel and piston. No cleanup is necessary.

Instead of worrying about how hard to squeeze a tube or press a syringe, air power, the second basic dispenser, permits the user to concentrate on getting the fluid exactly where it's needed. EFD has several models of this equipment (Figure 21.7). This has a foot control so that the operator has both hands free. There is no drip or ooze. The air is adjustable from 0 to 100 psi.

Combining the advantages of adjustable air pressure with an automatic, timed pulse of air, is the third type of basic dispenser, timed pulse. The operator makes consistent, repeatable deposits without guesswork. This type of equipment reduces rejects and saves a lot of adhesive material. If zero defects is your company goal, this equipment is indispensable (Figure 21.8).

Tips and syringes are made of different materials for specific uses. Auxiliary equipment is available to facilitate operations. For instance, there are barrel loaders to make that job easier, barrel stands, finger switches, vacuum tweezers, and production kits holding sample components.

The viscosity of the various adhesives covers a large range. Consequently, the syringes must be able to handle them all. EFD syringes have no seals to create friction. Instead, a close-fitting piston in the syringe permits easy control of all fluids. See Figure 21.9 for an explanation. Figure 21.9A demonstrates a situation when using a thick liquid and air tends to get trapped. This type of syringe permits the air to escape. Figure 21.9B shows how thin fluids can leak or drip between dispense cycles and outgas fumes. This type of syringe prevents that. Figure 21.9C shows how easily this type of syringe can be filled. Air escapes past the piston and leaves a contained system without drip or ooze.

Figure 21.10 shows the type of dispensing valve that is behind this very successful system. In many dispensing applications, such as the manufacture of loudspeakers, the quality of the finished product depends on precise control of the adhesive. Dripping and leakage can affect the adhesive deposit size. Changes in operating speed and ambient temperature can also affect production. EFD's controller is compact enough to be placed next to the valve so that the operator can purge the system and refine the program, if necessary.

Cyanoacrylates excel at forming quick, strong bonds between dissimilar materials. They polymerize by exposing a thin film of adhesive to moisture ions on substrate surfaces. If this moisture gets into automated dispensing systems, it would begin the curing process prematurely. There is always the risk that cyanoacrylates will bond the wetted parts on the valves that are used to apply them.

When investigating valves for this application, look for four criteria:

1. For trouble-free operation, seals should not be used.
2. For fast cycling with minimal wear, the open-close stroke should be very short.
3. The stroke should be adjustable to provide for precise control of deposit size.
4. The valve should be compact and lightweight to permit easy positioning. See Figure 21.11 for a view of three such valves in use.

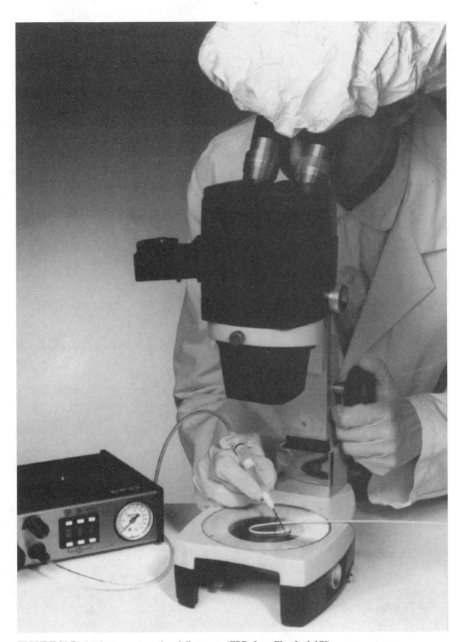

FIGURE 21.7 A microprocessor-timed dispenser. *(EPD, Inc., Elmsford, NY)*

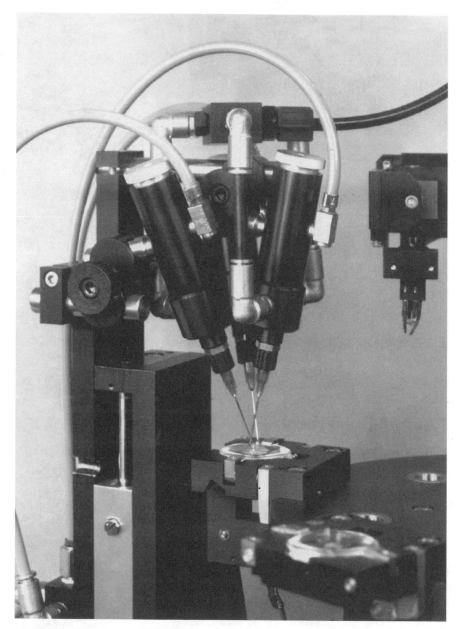

FIGURE 21.8 The operator bonds platinum marker bands to a heart catheter with cyanoacryate, using a Model 1500XL-15 dispenser. *(Spectra-Physics Scanning Systems and EFD, Inc., East Providence, RI)*

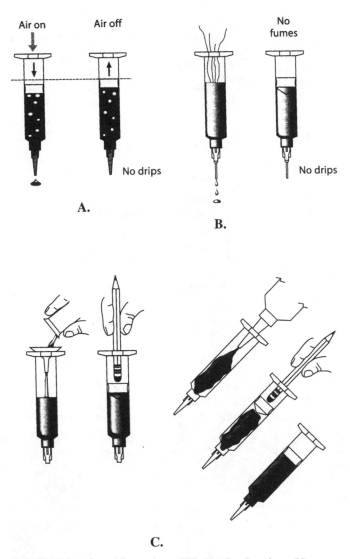

FIGURE 21.9 A Smoothflow syringe. *(EFD, Inc., East Providence, RI)*

Adhesive bonding has been a reliable, cost-effective process for more than a decade. Worthwhile information concerning this assembly method is available from the producers. Any type of adhesive or adhesive-dispensing equipment can be obtained directly from the manufacturer or one of the many distributors located in most large cities.

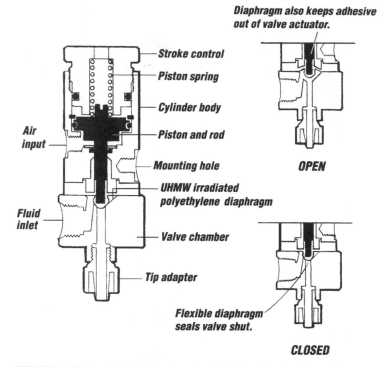

Diaphragm also keeps adhesive out of valve actuator.

Stroke control

Piston spring

Cylinder body

Air input

Piston and rod

Mounting hole

OPEN

UHMW irradiated polyethylene diaphragm

Fluid inlet

Valve chamber

Tip adapter

Flexible diaphragm seals valve shut.

CLOSED

FIGURE 21.10 An adhesive-dispensing valve. *(EFD, Inc., East Providence, RI)*

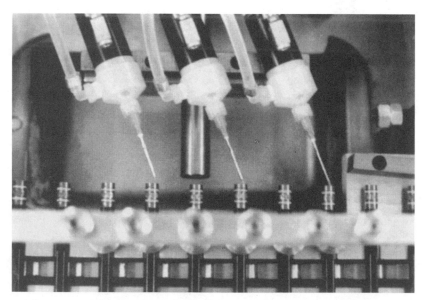

FIGURE 21.11 An adhesive application system. *(Harvey Machine Co. and EFD, Inc., East Providence, RI)*

CHAPTER 22
MAGNEFORMING

HOW MAGNEFORM WORKS

The basic magnetic pulse principle has been in use in metal-forming equipment since about 1970. It is the same principle as that which activates an electric motor. When an electric current generates a pulsed magnetic field close to a metal conductor, a controllable force is created that can be used to shape metal without actual contact (Figure 22.1).

The basic components of the Magneform machine are: energy storage capacitors, a work coil, and switching devices. High-voltage capacitors are charged, then discharged through a coil, inducing an intense magnetic field. This field, in turn, induces current in the conducting workpiece, setting up an opposing magnetic field. The net magnetic force does the forming.

During forming, pressures as high as 50,000 psi move the workpiece at velocities as great as 900 feet per second. The strength of this force can be closely controlled, which is important for versatility. These magnetic forces can produce up to 180 parts per minute (Figure 22.2).

Magneform is a proven, widely accepted method of forming metal parts. Controlled forces are placed where they are wanted. The machine cycle can be easily synchronized with conveyors and other feed mechanisms. This is what permits such fast production cycles. The actual forming can be done in 100 millionths of a second.

The following is a sample of parts previously made:

Applications	Production Rate/hour	Mode of operation
Ball joint seals	2400	Automated
Automotive fuel pumps	900	Automated
Timer rotors	720	Automated
Baseball bats	450	Semi-automated
Projectile rotating bands	400	Manual

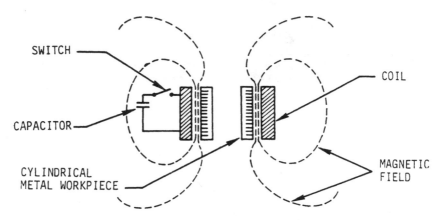

FIGURE 22.1 In the basic Magneform circuit, high-voltage capacitors are discharged through a coil, inducing an extremely intense magnetic field. This field, in turn, induces current in the conducting workpiece, setting up an opposing magnetic field. The net magnetic force does the forming. *(Maxwell Laboratories, Inc., San Diego, CA)*

BENEFITS OF MAGNEFORMING OVER CONVENTIONAL FORMING

Magneform enjoys the following advantages:

1. It is precisely controllable and allows forming of metal parts over plastics, composites, glass, and other metals.

2. Because Magneform makes no contact with the workpiece, no lubricant is required. Parts have no tool marks, heat deformation, or surface contamination.

3. Magneform parts are uniform in appearance and there is no tool wear.

4. Magneform generally makes joints equal to or stronger than the material of the workpiece.

5. Magneform machine operators are easily trained.

6. The machines are energy efficient and easily installed.

No conventional equipment can duplicate Magneform's ability to concentrate a uniform force on a selected area on the periphery of a workpiece, without mechanical contact. Two parts can be joined concentrically and without excessive heat. Even if one of the parts has eccentric protuberances, the two parts can be joined concentrically. Figure 22.3 shows how an outer shell will fill in serrations. Although the parts in these sketches are concentric, if the inner parts had eccentric protuberances, the outer parts would still deform completely around the inner.

Figure 22.4 shows five joints that would be better made by Magneforming.

1. A spun joint is a type often used when heat cannot be applied. Spinning requires a skilled operator to ensure minimum eccentricity. This is another place to use Magneforming that would do the work faster and simpler. A low-skill operator could

Magneform is used to subassemble the nosecone and join it to the warhead of the shoulder-fired viper missile.

The outer aluminum tube section of this missile canister is swaged onto a grooved internal spacer ring at a rate of 350 per hour using a 12 kJ Magneform system. The resulting joint is strong and smooth.

An operator demonstrated the use of a semiautomatic 12 kJ Magneform machine to join bandings to small caliber projectiles. The 20 mm projectiles are banded at a rate of 1000 units per hour.

Using this four-station rotary index table, an aluminum skirt is swaged to a steel fragmentation cup. The joint must withstand a 600 lb push-out test and a 15 psi, four-second leak test. Epoxy is applied to the joint area before the Magneform operation to insure sealing. Two assemblies are formed in each machine cycle. The production rate exceeds 1200 per hour.

This military standard electrical connector is produced by compressing a threaded metallic shell over the non-metallic receptacle. A 6 kJ Magneform machine produces 400 of these each hour.

Right angle ducting intersection and contoured tubing part.

Contoured tubing parts for right angle intersections of various diameters.

FIGURE 22.2 A Magneform machine and some assemblies made by it. *(Maxwell Laboratories, Inc., San Diego, CA)*

handle the work and even if the parts were out of round, Magneforming would join them with little difficulty.

2. There is a pinned or dowelled assembly. It requires drilling, then assembling pins, rivets, or dowels. Such an assembly creates localized stress, which is avoided by Magneforming.

3. Bolting has disadvantages similar to pinning.

4. Welded joints would be neater and cheaper if they could be Magneformed instead. Problems that occur with welded joints are: changes in physical properties in metal surrounding the weld; spatter requiring removal, special care removing flux, and a rejection rate—especially when seals are made.

5. Brazing has disadvantages similar to welding. In addition, brazing aluminum presents certain difficulties.

Magneforming can simplify and improve production procedures in several ways. The inherent advantages of this method preclude certain problems that are common to the conventional joining methods.

The assembly of axially loaded structural members by the Magneform® method exhibits certain advantages over other presently used methods, especially in applications such as aircraft structure or actuator rods where weight is a prime consideration.

Figure 22.3A illustrates a simple joint to be used for lightly loaded members. The end of the tubular component is formed into a single groove in the end fitting.

A more sophisticated joint configuration such as Fig. 22-3C is recommended, where the assembly will be highly stressed. Three generously radiused grooves are used to avoid stress concentration. The grooves become progressively deeper toward the outboard end of the fitting. Tests indicate a better distribution of stress to all grooves under axial load due to the variation in groove depth. Tests also indicate a significantly lessened tendency for the end fitting to move relative to the tube

The radiused groove type of magnetically formed joint has shown outstanding quality characteristics in vibration and fatigue tests. Figure 22-3D shows results of a test comparing fatigue life of magnetically formed and riveted assemblies. Twenty-one magnetically formed and twelve riveted assemblies were tested.

On this type of magnetic assembly where the ultimate strength of the tube must be achieved, it is preferable that the fitting (inside part) be of higher yield strength or higher modulus of elasticity than the tube. The fitting must be designed with proper cross-section.

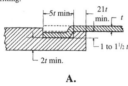

A.

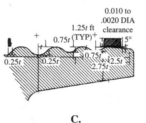

C.

This is preferred to forming a groove bridged by the tube as shown in Fig. 22-3B. Considerably less energy is required in the first example. Both types of joints may be made easily and inexpensively by the Magneform® method. Close tolerances and close fit on the parts are not required.

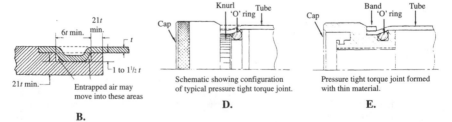

B.

Schematic showing configuration of typical pressure tight torque joint.

D.

Pressure tight torque joint formed with thin material.

E.

FIGURE 22.3 Schematics of the magneform process: A. Grooved joint configuration for lightly loaded members. B. Joint configuration showing tube bridging groove. C. Joint configuration for high stress and repetitive load reversal. D. Schematic showing the configuration of a typical pressure-tight torque joint. E. A pressure-tight torque joint formed with thin material. *(Maxwell Laboratories, Inc., San Diego, CA)*

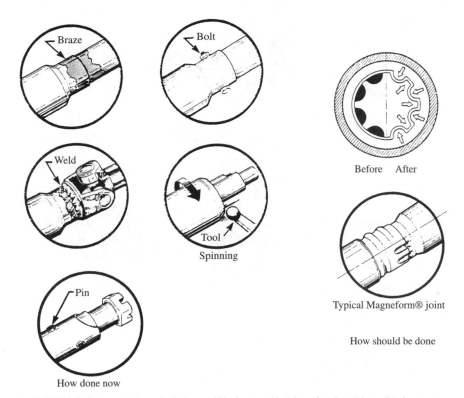

Braze

Bolt

Before   After

Weld

Tool

Spinning

Typical Magneform® joint

How should be done

Pin

How done now

FIGURE 22.4 Five assembly methods that could be improved by Magneforming. *(Maxwell Laboratories, Inc., San Diego, CA)*

MAGNEFORM APPLICATIONS

Magneform applications include forming and assembly operations in the automotive, appliance, aerospace, electrical, nuclear, and ordnance fields. Magneform, generally replaces swaging, spin rolling, soldering, pinning or welding operations. Some jobs cannot be done at all by any other method (Figure 22.5).

Electromagnetic metal forming works best with light- to medium-gauge materials that have high electrical conductivity, such as copper, aluminum, steel, and brass. Stainless steel and other materials with poor conductivity can be formed by using a high conductivity driver, such as an aluminum or copper sleeve. The workpiece must provide a continuous electrical path.

The magnetic work coil can be located remote from the energy storage and control unit to accommodate any production requirement. The work station can be designed to include automatic workpiece orientation and processing.

Some of the more common applications of Magneforming are:

1. Replacing nuts and washers that secure bearings and other components.

2. The process is useful for push-pull devices, such as handles on shafts or control rods.

3. To secure gear hubs on shafts thereby avoiding press fits.

Underwater flares are assembled with this simplified Magneform work station. It uses a two-piece magnetic field shaper and a standard compression coil. Parts are loaded and positioned manually.

Metal parts of this light antitank missile assembly are joined with a 6 kJ Magneform machine. Parts are loaded and unloaded manually at a rate of 350 units per hour.

Steel torque tube .070" wall with steel test fittings. Magnetically formed joint was torque tested to approx. 32,000 inch pounds, at which point the tubing twisted as shown, with no slippage of the joint.

High voltage fuse test assembly. Cap at left magnetically swaged into groove on fiberglass fuse body. Cap at right, magnetically swaged into smooth end surface of fiberglass fuse body

Example of torque joint made by compressing outer component on knurled surface of inner component.

Automotive drive shaft

FIGURE 22.5 Examples of products using the Magneform process. *(Maxwell Laboratories, Inc., San Diego, CA)*

4. To replace swaging of tube attachments in many situations, such as coaxial high-voltage cables and operating cables for aircraft.

5. To assemble end caps to components, such as the fiberglass or phenolic fuse bodies.

6. To create flanges on tube sections for sheetmetal pipe assemblies.

7. To magnetically compress bands that lock end bells onto frames of small motors. Concentricity of assembled parts is maintained by a fixture. For a motor to be assembled in this highly competitive environment, magnetic pulse forming offers the possibility of extensive cost reduction while maintaining quality.

FORMING PRESSURE-TIGHT TORQUE JOINTS WITH MAGNEFORM

Pressure-tight joints, capable of withstanding high wrenching torque, can be quickly assembled by Magneform. Considerable money can be saved by using this method, rather than welding or brazing. Also, the appearance of the joint will be improved.

Typical high-quality assemblies can be made with a single magnetic impulse. The following parts were assembled for testing by Magneform: The tube was 3003-H 14 aluminum with an OD of 4.8". The cap was 6061-T6 aluminum. The requirements are: The joint must be leak tight to 400 psi and withstand a wrenching torque of 140 ft.-lbs. and a tensile load of 7000 lbs.

This design uses an O-ring for a seal and a coarse straight knurl for torque resistance. Test results indicated the assembly exceeded all requirements. When considering the use of magnetic impulse forming in this type of assembly, the groove configuration must be tailored to the function. This will vary, depending on the material and the operating environment, which might prohibit the use of a rubber O-ring. The addition of a knurl increases the joint strength in torque. When thin wall tubing material is to be used, a band can be formed around the assembly very easily (see Figure 22.3 again).

To eliminate the contamination caused during the manufacture of high-performance fluid-system components, Michigan Dynamics, Inc., devised an unusual method for producing ultra-clean 10-micron wire cloth-filter assemblies for the Sidewinder missile launcher. The filter components are sealed in plastic bags in a clean room after first being decontaminated and semi-assembled (Figure 22.6).

These filter components are easily cleaned in the clean room prior to assembly. Still in the clean room, the parts are placed and held in the proper relationship by the snug fit between components. Next, the loosely assembled parts are sealed in a polyethylene bag. Then, the packaged parts are taken out of the clean room to a Magneform machine in the manufacturing area without the possibility of contamination. The bagged parts are placed in a magnetic field shaper, which concentrates a magnetic field around the area to be swaged.

CONTOURED TUBING TRANSITIONS

The round-to-square transition is being formed with a single magnetic impulse from a Magneform expansion coil. The unformed tubing was placed in a split die with the

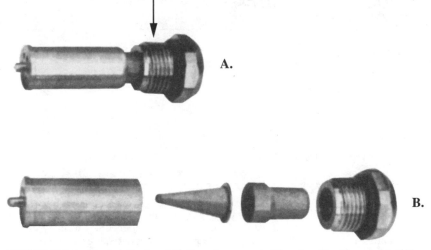

FIGURE 22.6 A unique application of the Magneform process: Ultra-clean 10-micron wire cloth filter assemblies. A. The assembled filter (arrow indicates the area magnetically swaged). B. Unassembled components. *(Maxwell Laboratories, Inc., San Diego, CA)*

Magneform coil inside the tubing in a position where the maximum field intensity was exerted in the area where the transition was to be formed. Transitions of this type can be readily formed at the ends of any lengths of tubing in diameters from approximately 1.5" and up and in wall thicknesses of from approximately 0.020 to 0.080".

PRODUCTION COSTS OF DUCTING INTERSECTIONS

Tubing parts used for ducting intersections are used in a wide variety of diameters, contours, and intersecting angles. Most of these parts were formerly produced as welded joints. Using Magneform expansion coils and female dies, the parts can now be formed in a single operation with significant cost savings. So many different diameter contours are required that thousands of dollars in production cost savings are possible.

MAGNEFORMED JOINTS COMPARED TO OTHERS

Magneform swaging is entirely a machine operation. Once the parameters of the operation are established, and the joint is positioned and aligned by fixtures, the only step remaining to take is the release of energy by actuating a switch.

By contrast, welding requires highly specialized operator skill. The strength of the joint is variable and depends on the skill of the craftsman. The output of the Magneform process is more reliable and consistent. Sometimes high strength and cost are secondary to reliability. Consistency of strength is the greatest asset of the Magneform process. Because this process doesn't use heat; expansion, shrinkage, and bowing problems are avoided.

COMPARISON OF MAGNEFORM AND WELDED JOINTS

Conditions	Welded Joints	Magneform Joints
Load at failure	5680 lbs.	5377 lbs.
Ultimate tensile stress	12,000 psi	11,400 psi
Variation in samples %	17.8	13.97
Corrosion properties	Good	Good
Resistance to vibration	Good	Good
Strength uniformity	Fair	Excellent
Inspection	Radiography and dye check	Visual

No conventional equipment can duplicate Magneform's ability to quickly join two parts concentrically, without heat or mechanical distortion. Sometimes welding distorts and embrittles. Brazing can distort metal, and brazing aluminum requires a fair amount of skill. Spinning joints requires a skilled operator. Bolting concentrates the load in a joint.

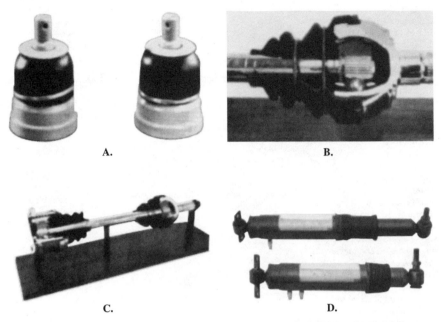

A. B.

C. D.

FIGURE 22.7 Special uses for the Magneform process: A. Automotive ball joints. B. Automotive constant-velocity universal joint. The cut-away shows magnetically swaged band penetration. C. Automotive independently suspended drive axle, showing clamping bands magnetically swaged on the c.v. joints and axle. D. Automotive airlift shock absorbers showing the magnetically swaged outer clamping band (the inner band cannot be seen). *(Maxwell Laboratories, Inc., San Diego, CA)*

On the other hand, Magneforming doesn't embrittle or distort metal, nor does it spatter. It speeds up assemblies and should be the selected method for this type of work. Unique Magneform applications are shown in Figures 22.7 through 22.9.

The assembly of certain filters such as that shown in Figs. A and B is readily accomplished with the Magneform® machine. In this filter, a fine mesh, stainless steel screen is sealed into stainless steel housings using aluminum drivers. As illustrated in Fig. A, the aluminum driver forces the lip on the stainless steel housing to capture and seal the stainless steel filter mesh. It is possible to arrange the aluminum driver in such a way that it removes itself automatically during the swaging operation. Figure B shows an assembled filter, a dis-assembled end piece with the drivers shown alongside. The cleanliness, speed, uniform circumferential pressure and absence of excessive heating are of particular advantage over conventional assembly techniques.

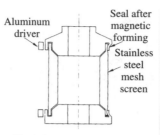

Fig A. Stainless steel filter mesh positioned in recesses of end fittings. Left, before magnetic pulse forming. Right, after magnetic pulse forming. Note position of aluminum driver.

Fig B. Typical stainless steel filter assembly top view shows assembly before forming with aluminum drivers at right. Bottom shows assembled filter.

Fig C. Flanged sections of stainless steel tubing formed with ring die, Magneform® expansion coil and aluminum driver.

Magneform® machine with a 2.5" inch O.D. expansion coil provided a fast and economical means of flanging stainless steel tubing. Flanged sections of stainless steel tubing are shown in Figure A. The flanged sections were formed from stainless steel tubing 2.5" O.D. × 0.020" thick. Since stainless steel is a poor electrical conductor, it does not respond to the amount of energy available from the expansion coil. Consequently, to form the stainless

steel material, it must be pushed or driven by a disposable driver. In the case of parts illustrated, two turns of 0.015" aluminum strip wrapped around the expansion coil (with the stainless steel workpiece slipped over the foil) drove the end of the stainless steel tube against the ring die and formed the flange. The magnetic pulse method does away with a slow spin roll operation and eliminates costly die required by conventional method.

FIGURE 22.8   Two special jobs for Magneforming. *(Maxwell Laboratories, Inc., San Diego, CA)*

FIGURE 22.9 Notice the two machines (upper center and left) feeding two coils. *(Maxwell Magneform Corp., San Diego, CA)*

CHAPTER 23
ULTRASONIC TECHNOLOGY FOR PRODUCTION

Ultrasonic energy can be used for production in a number of different ways. The most common are machining, welding, hardware insertion, staking, and spot welding.

ULTRASONIC MACHINING

Ultrasonic machining (impact grinding) provides the capability of machining hard, brittle materials, where it is not feasible to complete the jobs with more conventional means. The process is nonchemical, nonthermal, and nonelectrical, so the parts being processed are not affected metallurgically. Slots, irregular shapes, blind cavities, and through cavities all can be machined by this method.

The actual method involves the use of an abrasive slurry, such as silicon or boron carbide, which flows between the workpiece and a tool. The tool moves vertically only a few thousands of an inch while vibrating about 20,000 times per second. The flow of abrasives under and around the tool permits particles to strike the workpiece, chipping off microscopic flakes of material until a counterpart of the tool is imprinted in the workpiece (Figure 23.1).

Bullen Ultrasonics, one of the leaders in ultrasonic machining, has developed skills in providing innovative solutions to unique application problems (Figure 23.2).

Common materials machined ultrasonically are glass, quartz, fiberoptics, zirconia, germanium, ferrite, alumina, and other ceramics and hardened steels. It is not difficult to drill holes in these hard and brittle materials and hold 0.005" tolerance on diameters (Figure 23.3). When the ultrasonic machine is configured to appear and operate like a Bridgeport milling machine, it can maintain close tolerances on hole location and edge machining (Figure 23.4).

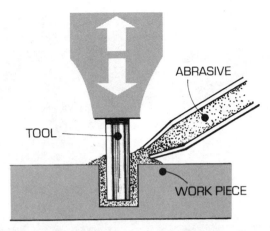

FIGURE 23.1 Machining steel with ultrasonics. *(Dukane Corp., St. Charles, IL)*

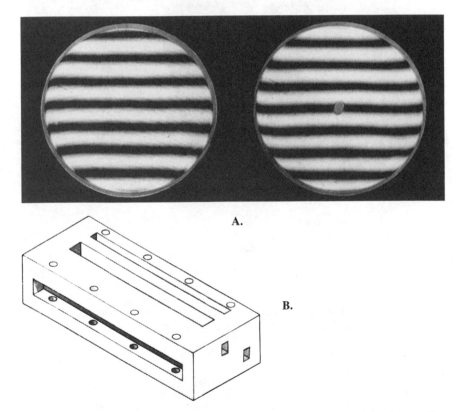

FIGURE 23.2 Difficult machining operations with ultrasonics: A. Ultrasonic machining does not affect the optical surface. There must be no chipping on the front surface, where the hole passes through or any distortion of the optical flatness. B. More than 40% of the mass of this fused silica block was removed without stressing or disturbing the optical surfaces. *Bullen Ultrasonics, Inc., Eaton, OH)*

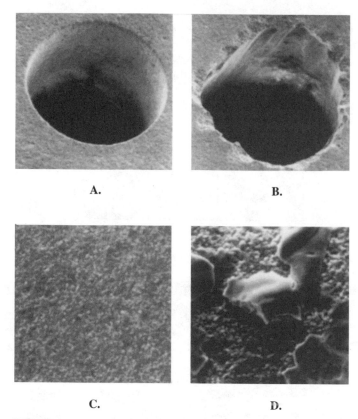

A.   B.

C.   D.

FIGURE 23.3 Ultrasonic vs. laser machining (SEM photos of 0.025" holes in 0.025 alumina substrate): A. Ultrasonically machined (100X). B. Laser drilled (100X). C. Ultrasonically machined (500X). D. Laser drilled (500X). *(Bullen Ultrasonics, Inc., Eaton, OH)*

WELDING

Ultrasonic welding is used to join plastic parts. When two plastic parts are mechanically vibrated together at the ultrasonic frequency, friction generates localized heat that causes the materials to melt, flow, and fuse. The equipment required for this process consists of a power supply (generator), transducer, booster, horn, and handgun (Figure 23.5). This equipment can be integrated by sophisticated automation into a system of repeatable production.

An ultrasonic generator accepts 60-cycle current and delivers 20 to 40 kHz to a piezoelectric transducer. Piezoceramic wafers convert the electrical energy to mechanical motion. This motion is then imparted to acoustically resonant tools called *horns*. When the horn is placed on the workpiece, it passes vibratory energy to the joint, producing a controlled melt at the point of least resistance. The weld cycle lasts about one second; generally, some type of shear joint should be used for maximum strength.

Shear Joints

Although the common butt weld, the step weld, the tongue-and-groove weld, and variations of them might make up the majority of plastic welds; they do have drawbacks that
are covered later in the chapter. The basic shear joint is shown in Figure 23.6, before, during, and after welding. The DuPont company recommends that type in their book on plastics. Figure 23.7 shows three variations of the basic shear joint.

In forming the shear joint, initial contact is limited to a small area, which could be a recess or a step in one of the parts. First, the contact surfaces melt. Then, as the parts tele-
scope together, they continue to melt along the vertical walls. The smearing action of the

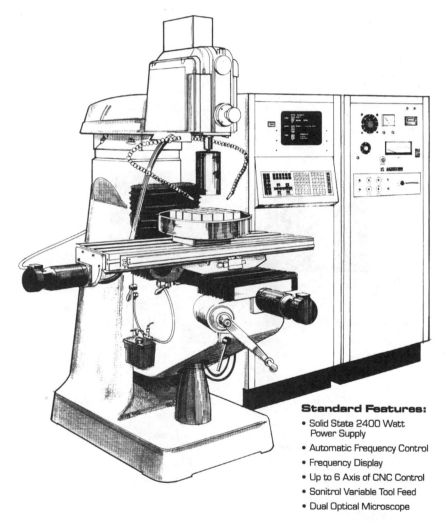

Standard Features:

- Solid State 2400 Watt
  Power Supply

- Automatic Frequency Control

- Frequency Display

- Up to 6 Axis of CNC Control

- Sonitrol Variable Tool Feed

- Dual Optical Microscope

FIGURE 23.4   A Bullen ultrasonic CNC 2400-watt impact grinder. *(Bullen Ultrasonics, Inc., Eaton, OH)*

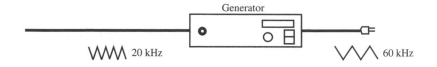

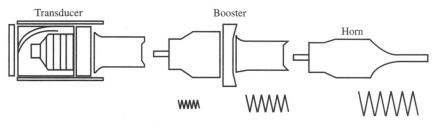

FIGURE 23.5 How it works. *(Dukane Corp., St. Charles, IL)*

Shear joint

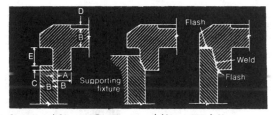

Before welding   During welding   Welding completed

Dimension A:
.016 inches. This dimension is constant in all cases.
Dimension B:
This is the general wall thickness.
Dimension C:
.016 - .024 inches. This recess is to ensure precise location of the lid.
Dimension D:
This recess is optional and is generally recommended for ensuring good contact with the welding horn.
Dimension E:
Equal to or greater than dimension B.

FIGURE 23.6 Shear joint. *(Dukane Corp., St. Charles, IL)*

Variations of basic joint

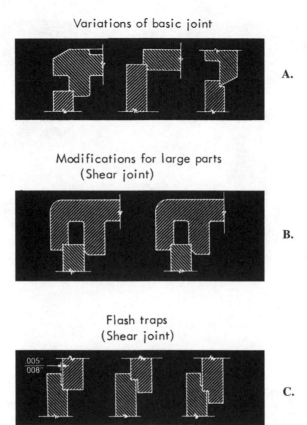

Modifications for large parts
(Shear joint)

Flash traps
(Shear joint)

FIGURE 23.7 Variations of the basic shear joint. *(Dukane Corp., St. Charles, IL)*

two melt surfaces precludes leaks and voids, making this the best joint for leak-free seals. Like all plastic welds, there is a strong, molecular bond.

The fastest joint to make, and one that requires very low energy, is the shear joint. Heat generated at the joint is retained until ultrasonic vibrations cease, which is good because during telescoping and smearing, the melted plastic isn't exposed to air, which would cool it.

Weld strength is determined by the depth of the telescoped section, which is a function of weld time and part design. Joints can be made stronger than the adjacent walls by designing the depth of telescoping at 1.25 to 1.5 times the wall thickness. The tensile strength of the welded material approaches the strength of the as-molded plastic.

Several significant features to this type of design should be considered. The top part should be as shallow as possible; in effect, just a lid. The walls of the bottom section must be supported at the joint by a holding fixture in order to avoid expansion during welding. Inferior welds will result if the upper part slips off the lower part, or if the stepped area is

FIGURE 23.8   Welded plastic assemblies. *(Dukane Corp., St. Charles, IL)*

too small. The fixture should prevent the upper part from slipping out of position, and if the contact area is too small, review the design to consider how to enlarge it. The design should make provision for flash when it cannot be tolerated. Figure 23.7C shows a design where displaced molten material could be directed.

The statement that the weld is stronger than the rest of the part can be true, provided these instructions are followed. Welding provides safer working conditions by eliminating the need for toxic solvents and messy adhesives. Welding creates consistent, strong, integral bonds and lowers rejection rates. It saves part and labor costs by eliminating hardware (Figure 23.8).

Butt Welds

The common butt joint, which includes a weld concentrator, is very popular. Upon the application of ultrasonic energy, material from the concentrator provides a solid weld as it disperses throughout the weld area. The base of the concentrator should be about 20% of the overall width, but not over twice the height; the height should be between 0.010 and 0.025" high (Figure 23.9).

The DuPont Company has issued a notice referring to Delrin (acetal), Zytel (nylon), and Lucite (acrylic). They suggest avoiding the butt weld for these materials. Apparently, the "V"-shaped concentrator flows away from the weld and crystallizes before sufficient heat is generated to weld the full width of the joint, so only spotty welding occurs. Because of this type of open joint, the melt is exposed to air during welding, which can cause oxidative degradation.

Step Welds

Using the step design will add strength to the weld. The melted plastic from the concentrator moves into the slip fit area, adding strength to the joint. This weld generally provides a clean appearance (Figure 23.10).

Tongue-and-groove Welds

Tongue-and-groove welds are a combination of both butt and step welds. This presents a good appearance with excellent strength (Figure 23.11).

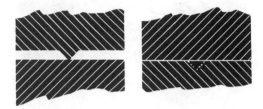

FIGURE 23.9   A butt weld. *(Dukane Corp., St. Charles, IL)*

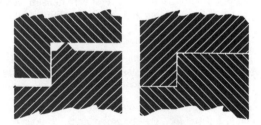

FIGURE 23.10   A step weld. *(Dukane Corp., St. Charles, IL)*

FIGURE 23.11   A tongue-and-groove joint. *(Dukane Corp., St. Charles, IL)*

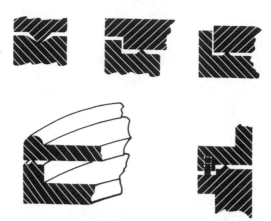

FIGURE 23.12 Design variations. *(Dukane Corp., St. Charles, IL)*

Weld Variations

You can adapt many variations of the butt, step, and tongue-and-groove welds for successful plastic welding. Modifications are sometimes necessary on special applications to obtain a required weld. See Figure 23.12 for several variations in welds, then see Figure 23.13 for a chart of ultrasonic weldability and the compatibility of thermoplastics for ultrasonic welding.

ULTRASONIC HARDWARE INSERTION

Very often, it is necessary to get metal hardware into plastic parts. Usually, small metal parts, such as threaded inserts or pins, must be embedded. It is done by premolding a hole in the proper position in the plastic piece, which is slightly smaller than the component to be embedded.

This hole guides the metal insert and creates an interference fit between the insert and the plastic piece. As soon as the melt occurs at the interface, the melted plastic flows into the knurls and undercuts of the metal insert. This locks in the encapsulated metal part (Figure 23.14).

There are several advantages to this method:

1. It is a fast assembly, doing one or more inserts in a fraction of one second.
2. It takes very high-pullout and high-torque loads.
3. The inserts locate and align themselves.
4. It eliminates most problems of threads that are molded in plastic.
5. Very little stress is created in the surrounding area, thus eliminating cracking while maintaining dimensions.
6. The process parameters can be consistently monitored and controlled.

ULTRASONIC WELDABILITY – COMPATIBILITY CHART FOR THERMOPLASTICS

	SPOT WELDING	DEGATING	SWAGING	STAKING	INSERTING	WELDING		COMPATIBILITY
ABS	G	G	G	G	G	G		A
ABS/P.C. (CYCOLOY)	G	G	G	G	G	G		A
ABS/PVC (CYCOVIN)	G	F	F	F	F	G		A
ACETAL (CELCON, DELRIN)	F	G	G	G	G	C		C
ACRYLICS	G	G	G	G	G	G		A
ACRYLIC/PVC (KYDEX)	G	G	G	G	G	G		A
ACRYLIC/PVC (KYDEX)	G	G	G	G	G	G		A
ASA	G	G	G	G	G	F		A
CELLULOSICS (CA, CAB, CAP)	G	G	F	G	G	G		F
P.P.S. (RYTON)	G	G	G	G	G	G		C
P.P.O. (NORYL)	G	G	G	G	G	G		A
NYLON	G	F	G	G	G	C	H	C
POLYCARBONATE	G	G	G	G	G	G	H	A
P.C./POLYESTER (XENOY)	G	G	G	G	G	G	H	C
POLYETHYLENE (P/E)	P	P	G	G	G	C		C
POLYPROPYLENE (P/P)	G	G	G	G	G	G		C
POLYSTYRENE	G	P	G	G	G	G		C
POLYSULFONE	G	G	G	G	G	G	H	A
P.V.C. (RIGID)	G	G	G	G	G	G		A
SAN/NAS	G	F	F	G	G	G		A
POLYESTER (CELENEX, VALOX)	G	G	G	G	G	G		C
STRUCTURAL FOAMS	F	P	P	F	G	C		C
POLYIMIDE	G	G	G	G	G	C		C

WELDABILITY – READ ACROSS

COMPATIBILITY – READ ACROSS AND UP

X = COMPATIBLE AT TIMES BASED ON MATERIAL COMPOSITION

■ = GOOD COMPATIBILITY

A = AMORPHOUS RESINS   C = CRYSTALLINE RESIN

H = HYGROSCOPIC, SHOULD BE DRY BEFORE WELDING

WELD CHARACTERISTICS: G = GOOD   F = FAIR   P = POOR

FIGURE 22.13 Ultrasonic weldability: a compatibility chart for thermoplastics. *(Dukane Corp., St. Charles, IL)*

287

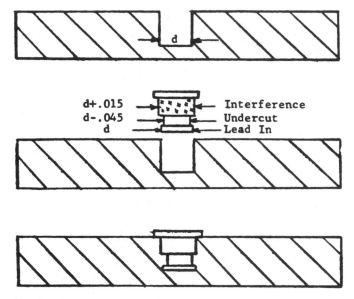

FIGURE 23.14   Ultrasonic insertion. *(Dukane Corp., St. Charles, IL)*

Specialized equipment is available:

1. Handgun units for inserting up to a size of ¼–20.
2. There are small, standard presses for both single and multiple insertions.
3. Hardened steel horns can be used for greater wear resistance.
4. Slow speed control to provide better melt flow with large inserts and multiple insertions.
5. Micrometer adjustment for setting precise insertion depth. The technique to use is low amplitude (0.001 to 0.002"), high pressure, and short weld time (usually less than one second). It is also a good idea to use hardened steel fixtures when inserting into glass-filled parts.

STAKING

Staking is an assembly procedure used to join dissimilar materials: usually metal to plastic or two or more plastics. Generally, the plastic part will have protrusions or posts that extend up through holes in the metal piece. A specially contoured horn contacts the post and melts it so that it reforms to establish a locking head over the metal part. This is also the scenario if the two parts are plastic or if a plastic rivet is used (Figure 23.15) (sketches 1 through 10).

In any process involving localized heating by ultrasonics, the designer must control where and how fast the temperature rise will occur. Geometry plays a vital role in determining the location of high strain which creates a desirable heating zone. Therefore, an energy

A.

Examples: Rosette style (see figures D, E); Hollow stud (see figure H).

Advantages: The rosette style is used primarily in press operations. This style gives better results with sharp transitional materials (e.g., nylon). However, hollow studs can also be used with hand-held units, depending upon material and size. The hollow stud can easily be removed for repair, which leaves a pilot hole for a screw.

B.

Examples: Simple profile (see figure C); Flat head profile (see figure F); Pan head profile (see figure G).

Advantages: This is an excellent design for handgun operations involving staking. The pointed stud design yields excellent results with broad transitional materials (e.g., ABS). it provides for easier alignment of parts (looser alignment tolerances) and can be used with a wide variety of head configurations to meet special design requirements. The pointed stud design also reduces horn wear problems with glass-filled plastics.

Staking configurations

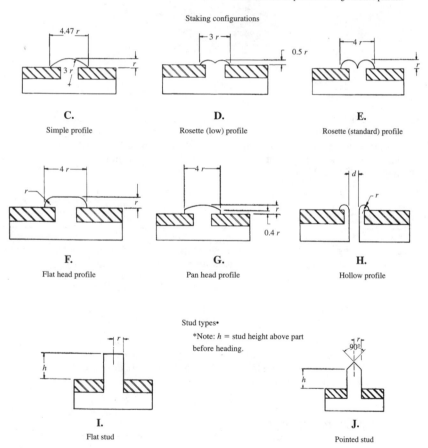

C.

Simple profile

D.

Rosette (low) profile

E.

Rosette (standard) profile

F.

Flat head profile

G.

Pan head profile

H.

Hollow profile

Stud types•

*Note: *h* = stud height above part before heading.

I.

Flat stud

J.

Pointed stud

FIGURE 23.15   Staking configurations. *(Dukane Corp., St. Charles, IL)*

Stud Type	Head Form	Stud Diameter	Head Diameter	Head Height	Stud Height
Flat Stud	Rosette Low Profile	2r	3r	.5r	1.20r
Flat Stud	Rosette Stand. Profile	2r	4r	r	3.14r
Flat Stud	Hollow Profile	2r + d	4r + d	r	$\frac{3.14r\,(2r+d)}{2\,(r+d)}$
Pointed Stud	Simple Profile	2r	4.47r	r	2.33r
Pointed Stud	Flat Head Profile	2r	4r	r	2.90r
Pointed Stud	Pan Head Profile	2r	4r	r	2.50r

FIGURE 23.16   Formulas for deriving stud height. *(Dukane Corp., St. Charles, IL)*

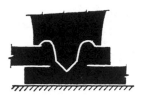

FIGURE 23.17   A spot weld. *(Dukane Corp., St. Charles, IL)*

director should be employed when using the ultrasonic staking technique. That is, the cross sectional area/height ratio of the material at the location where initial dissipation is to occur is drastically reduced as compared to the adjacent segments; in this case the body of the horn and the piece part containing the stud. There are formulas for deriving stud height. See chart Figure 23.16.

ULTRASONIC SPOT WELDING

This type of welding is recommended when two thermoplastic parts are to be joined at localized places eliminating the need for more complex joints (Figure 23.17).

MISCELLANEOUS

A good number of highly qualified ultrasonic machine manufacturers are in this country making equipment that will produce this work. You should have no trouble obtaining competent advice. I have noticed quite a few different manufacturers during my research; any one of these could supply the various types of ultrasonic equipment you require.

ANCILLARY EQUIPMENT

A number of small ancillary units are marketed by various companies producing ultrasonic equipment. The Dukane Company of St. Charles, IL, markets several units.

A general-purpose probe system is designed to be used for small, fragile parts. Its power source is 120 volts, 50/60 Hz. The output power is 350 or 700 watts with an output frequency of 40 Hz. It can be adapted to continuous seam welds (Figure 23.18).

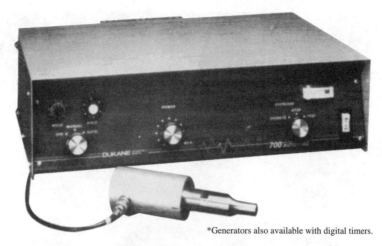

*Generators also available with digital timers.

FIGURE 23.18   A general-purpose probe system. *(Dukane Corp., St. Charles, IL)*

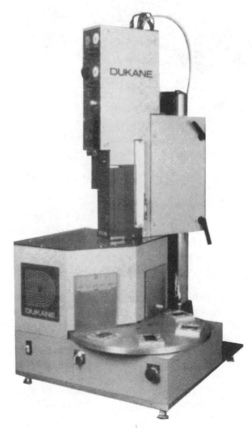

FIGURE 23.19   A rotary parts system. *(Dukane Corp., St. Charles, IL)*

FIGURE 23.20   Ultra-com microcomputer control and monitoring system. *(Dukane Corp., St. Charles, IL)*

Dukane makes a rotary parts-handling system, which is designed for economical manual or semi-automatic operation with free-standing base, welder, six-station rotary table, dial plate, and safety guard. This system can perform spot welding and any other common welding, staking, inserting, and swaging for medium- or heavy-duty service (Figure 23.19).

The Ultra Cutter is designed to cut the new aerospace composite materials, such as graphite fibers, aluminum honeycombs, and kevlar. A 1" blade vibrates at 40 kHz, which cuts the materials with minimum force. At the same time, the cut edges of fabric materials are fused to prevent unraveling.

The microcomputer helps automate the use of ultrasonics. It has a modular design that permits simple field installation of interface features. It is a real-time, multitask operating system, simultaneously controlling and monitoring multiple-process parameters. It provides input to a screen display and to a printer for permanent documentation (Figure 23.20).

The horn analyzer can be very useful in the manufacture of trial and production horns of both 20- and 40-kHz frequencies. It also serves as a valuable troubleshooting tool for existing ultrasonic equipment.

ULTRASONIC LEAK DETECTORS

Thousands of dollars are wasted annually by large users of compressed air; even small users lose their share of money. A common problem seems to afflict so many who use compressed air: leaks. They exist at steam traps, pipe flanges, fittings, joints, and tools. Ultrasound can locate these leaks. The Predictive Support Service Co. Division of U.E. Systems of Elmsford, NY, manufactures a hand-held system for locating and recording these sites.

Operators and mechanics can hear large leaks, and sometimes they will go to the trouble of securing them. But it's the small ones that never get caught and end up costing thousands of dollars. The system sold by this New York concern functions as follows: The operator carries a pistol-shaped object that is pointed at every part of the compressed-air system in series: fittings, pipes, tubes, everything. This gun is an airborne ultrasonic detector and it receives sound through the air at 20 to 100 kHz.

Pressure and vacuum leaks of any kind of gas create sounds in this range because of the turbulence caused by escaping gas. The high-frequency sound component is heard by a passive receiver and passed to a piezoelectric transducer element in the gun. Circuitry amplifies the signal, then it is heterodyned into a lower frequency, which can be heard by the earphoned operator, who is carrying an audio recorder or some data-logging instrument. The leak can't be fixed until it is found.

Superior Signal Co. of Spotswood, NJ, is another source of ultrasonic detectors that are utilized for the same purpose. Using dynamic noise discrimination (DND), their handheld tubular assembly detects a broad band of ultrasonic waves and automatically selects the proper frequency band of the fault. The detector then transmits the otherwise undetectable sound to a headset for analysis. A touch-probe attachment will also detect mechanical problems in bearings and gears.

ULTRASONICS IN QUALITY CONTROL

In Chapter 46, you will notice that inspecting the assemblies of resins and reinforcing elements is very difficult. However, ultrasonics is useful in inspecting the skin and its bond to the substrate and is effective in locating flaws in the part. Currently, liquid-coupled ultrasonics is the most widely used method, but dry-coupled ultrasonics is the state-of-the-art.

Ultrasonic Sensors

Ultrasonic sensors are providing sensing and control solutions for applications, ranging from simple, high-speed counting and level detection to very sophisticated closed-loop process control. They service the marketplace by supplying a cost-effective sensing method with unique properties. By using a wide variety of ultrasonic transducers and several different frequency ranges, an ultrasonic sensor can be designed to solve many application problems that would be cost-prohibitive or simply could not be solved with other sensors.

In industrial sensing, more applications are now requiring detection over distance. Some of these ultrasonic units can detect up to 20 feet, but limit switches and inductive sensors cannot. Some photoelectric sensors can detect over long distances, but they can't detect over a wide area without using a large number of sensors. Ultrasonic sensors cover both wide and narrow areas. Only the selection of the proper transducer (Figure 23.21) is required.

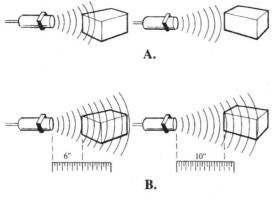

Q: What is the difference between proximity detection and ranging measurement?

A: Proximity detection: An object passing anywhere within the preset range will be detected and generate an output signal. The detect point is independent of target size, material or degree of reflectivity.

Ranging measurement: Precise distance(s) of an object moving to and from the sensor are measured via time intervals between transmitted and reflected bursts of ultrasonic sound. The example shows a target detected at six inches from the sensor and moving to ten inches. The distance of change is continuously calculated and outputed.

FIGURE 23.21 Two ultrasonic sensor types: A. Proximity detection. B. Ranging measurement. *(Dukane Corp., St. Charles, IL)*

Ultrasonic sensors are impervious to target material composition. The material can be clear, solid, liquid, porous, soft, hard, wood, metal, and any color. Sound is timed from when it leaves the transducer to when it returns. This system makes distance measurement easy and accurate to 0.05% of range, which amounts to ±0.002" at a distance of 4". These sensors are hardened to survive shop use.

ADVANTAGES OF ULTRASONICS

The following are unique advantages that ultrasonic sensors have over conventional sensors:

1. Measures and detects distances to moving objects.
2. Impervious to target materials and colors.
3. Solid-state units have virtually unlimited, maintenance-free lifespan.
4. Able to detect small or large objects over long distances.
5. Resistant to external disturbances, such as vibration, infrared radiation, ambient noise, and EMI radiation.
6. Unaffected by dust, dirt, or high-moisture environments.
7. A full line of more than 150 ultrasonic sensors are available that can: loop control, provide robotic sensing, sort and transfer products by height, count and measure, provide level detection of liquids and solids, measure distances, provide part distinction, provide web brake detection, measure roll diameter, provide tension control and winding, provide unwinding regulation, provide stacking height control, provide proximity and ranging detection for automatic-guided vehicles, provide people detection for counting or security, measure thickness, and much more for OEM applications (Figure 23.22).

It should be apparent that ultrasonics is another field in the arena of sophisticated processes that is destined for increased usage in the near future.

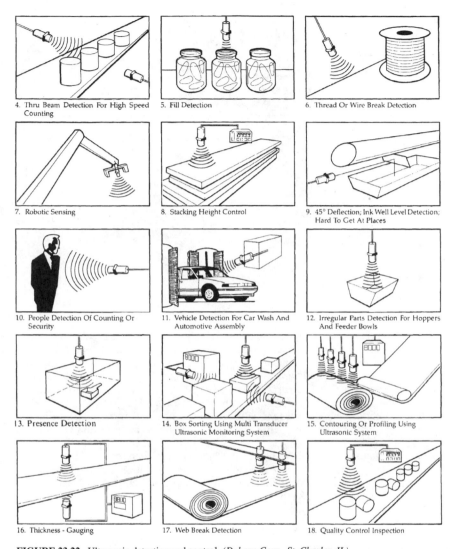

4. Thru Beam Detection For High Speed Counting

5. Fill Detection

6. Thread Or Wire Break Detection

7. Robotic Sensing

8. Stacking Height Control

9. 45° Deflection; Ink Well Level Detection; Hard To Get At Places

10. People Detection Of Counting Or Security

11. Vehicle Detection For Car Wash And Automotive Assembly

12. Irregular Parts Detection For Hoppers And Feeder Bowls

13. Presence Detection

14. Box Sorting Using Multi Transducer Ultrasonic Monitoring System

15. Contouring Or Profiling Using Ultrasonic System

16. Thickness - Gauging

17. Web Break Detection

18. Quality Control Inspection

FIGURE 23.22   Ultrasonic detection and control. *(Dukane Corp., St. Charles, IL)*

CHAPTER 24

INTRODUCTION TO ALUMINUM BRAZING

Each year, millions of aluminum parts are joined by brazing. They are used in automobiles, trucks, airplanes, spaceships, liquefaction plants, missiles, TVs, and other products.

Aluminum assemblies, ranging in thickness from thin sheet to heavy plate and also including castings, are routinely brazed in shops throughout the world. When desired, tolerances are held to better than ±0.002" and distortion is kept close to zero. Temper in heat-treatable alloys can be restored by post-brazing thermal treatment.

Brazed joints are strong, vacuum tight, and neat. The fillets formed by brazing have good fatigue resistance. Properly dip-brazed aluminum units can withstand vibration and shock to 125 Gs.

Brazing is no longer an art, but an established science. Therefore, it warrants serious consideration by all those who have need to join aluminum to itself and to other metals.

The preceding is by The Aluminum Association, Inc. of Washington, DC, as is much of this chapter on their favorite subject.

GENERAL INFORMATION

If you bring two pieces of metal within four Angstrom units ($A = 10^{-10}$ meter) of each other, interatomic attraction will bind them together permanently. This phenomenon is the basis of brazing and soldering and is done by "wetting" the metals to be joined with molten metal, which forms the joint when it cools.

The American Welding Society has a simple definition for brazing and soldering. Above 800°F, the process is called *brazing* and uses brazing filler metal; below 800°F, the process is called *soldering* and the metal filler is called *solder*. Welding differs from these two processes because the base metals to be joined are themselves molten at the moment of joining.

There are many advantages to brazing. Strong, uniform, leakproof joints can be made inexpensively and rapidly by modern techniques. Inaccessible joints and parts that might not be joinable by any other method can often be joined by brazing.

Complicated assemblies with thick and thin sections, odd shapes and differing wrought and cast aluminum can be made into one integral component by a single trip through a brazing furnace or a dip pot. Metal as thin as 0.006" and as thick as 6" can be brazed.

Brazed joint strength is very high. The nature of the inter-atomic bond is such that even a simple joint, when properly designed, will have equal or greater strength than the base metal.

Heat treatable alloys can be solution heat treated by quenching immediately after brazing. Thus, they can be strengthened by aging alone.

Brazed aluminum assemblies have excellent corrosion resistance when properly cleaned of residual flux. These aluminum joints generally resist corrosion as well as welded aluminum joints.

Brazed aluminum assemblies conduct heat and electricity uniformly. Brazed aluminum heat exchangers, evaporators, and similar complex fabrications are long lasting and highly efficient. The meniscus surface, formed by the filler metal as it curves around corners, is ideally shaped to resist fatigue.

Complex shapes with greatly varied sections are brazed with little distortion. Aluminum's excellent thermal conductivity assists in providing even distribution of the moderate temperature required for brazing.

Brazing makes precise joining comparatively simple. Unlike welding, in which the application of intense heat to small areas tends to move the parts out of alignment, parts joined by furnace and salt-pot techniques are heated fairly evenly. Brazing makes part alignment easier. Brazed joints with tolerances of ±0.002" are commonly used in microwave component production.

Properly brazed joints are leak tight. A vessel was sealed by brazing and evacuated to 2×10 torr for 100 hours. After that time, leakage increased internal pressure to only 1.6×10 torr, which is excellent for any metal joint.

Finishing costs are negligible. The capillary action that draws filler metal into a joint also forms smooth concave surfaces. Little mechanical finishing, if any, is required. When using a flux in brazing, removal of residual flux is required. The color match between parent metal and filler is generally good.

Personnel training is minimal. Production brazing equipment has been refined to where semi-skilled and unskilled people are sufficient for most operations. When torch brazing is required, mechanically adept personnel can be trained in a few hours to do the job.

PROCESS DESCRIPTION

The basic techniques used to braze and solder aluminum are similar to those used to join other metals. In fact, the very same equipment used for brazing and soldering aluminum can be used to join other metals in this fashion. This is frequently done commercially.

Aluminum and its alloys have a number of physical properties that differ markedly from those of other metals that are commonly brazed and soldered. Aluminum's thermal conductivity is very high; it oxidizes rapidly; its coefficient of thermal expansion is greater than that of many other common metals; and aluminum doesn't change its color as its temperature changes.

The oxide that forms on aluminum as soon as the bare metal is exposed to air has a very high melting point, 3622°F. Aluminum oxide is neither melted nor reduced by

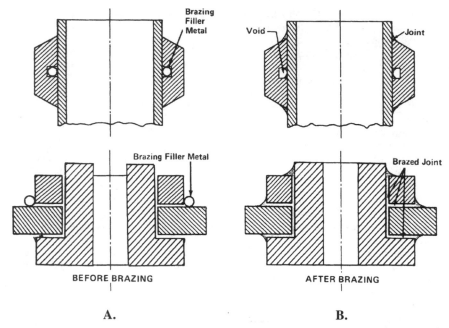

FIGURE 24.1 Brazed joints: A. Before. B. After. *(Aluminum Association, Washington, DC)*

temperatures that melt the metal itself. When aluminum is brazed or soldered, a flux is used to break up the oxide, float it away, and protect the bare metal base from further oxidation. Also, various mechanical means can be used to break up the oxide and expose the bare aluminum to the molten brazing filler metal.

These differing physical properties by no means limit aluminum's response to brazing, soldering, and welding techniques. They merely require a bit of mental adjustment on the part of the designer and engineer, who comes to aluminum fully trained and experienced in working with other metals.

The general procedure for atmospheric brazing aluminum is as follows. The surfaces to be joined are cleaned and spaced a few thousandths of an inch apart. A piece of brazing filler is positioned in or near the joint to be formed and the joint is coated with a suitable flux. Heat is applied. The flux reacts, displaces the oxide on the surface of the base and filler metal and shields the bare metal from contact with the air (Figure 24.1).

The brazing metal filler melts and is drawn into the joint by capillary attraction. As the filler flows, it displaces the flux and wets the hot base metal, adapting to submicroscopic irregularities, and dissolving the small high points it encounters. This process occurs in a few seconds. After cooling and cleaning, the joint is ready for use.

Because the brazing fillers must melt at temperatures lower than the metals they join, their chemistries must be different. This results in a diffusion at the bond line, which produces alloying in both the filler and base metal. Because diffusion always accompanies brazing, the filler metal always loses a portion of its silicon to the base metal. This raises the melting point of the completed joint. Therefore, a second joint can be brazed next to an already-cooled joint with far less danger of softening and damaging the entire assembly than the listed temperatures indicate.

Aluminum alloys are brazed with filler metals that are similar to the base metals they are joining. The brazing filler metals have liquidus temperatures close to the solidus temperature of the parent metals. It is, therefore, very important to maintain close temperature control when brazing.

Liquidus and *solidus* are terms that define the melting zone of alloys. Whereas pure metals (and eutectics) melt and flow at the same temperature, alloys begin melting at one temperature, called a *solidus*, and are completely molten at a higher temperature, called the *liquidus*. Below the solidus temperature, the alloy is completely solid. The alloy is completely liquid above the liquidus temperature. In between, the alloy is mushy (partially molten).

ALUMINUM ALLOYS THAT CAN BE BRAZED

Most of the nonheat-treatable aluminum alloys and many of the heat-treatable aluminum alloys can be brazed. The heat-treatable alloys most frequently brazed are 6061, 6063, and 6951. The nonheat-treatable alloys that respond best to brazing are 1100, 3003, 3004, and 5005.

Unless casting quality is very poor, castings can be brazed just as easily as wrought material. Casting alloys that are brazable are 356.0, A356.0, 357.0, 359.0, 443.0, 710.0, 711.0, and 712.0.

ALUMINUM ALLOYS THAT CANNOT BE BRAZED

Alloys 2011, 2014, 2017, 2024, and 7075 are not brazed with existing fillers. The melting points of these alloys are too low for the fillers developed thus far. Alloys with a magnesium content of 2.0% and greater are difficult to braze because present fluxes do not effectively remove the tenacious oxides that form on these alloys. High-magnesium alloys of the 5000 series can, however, be brazed by vacuum techniques.

BRAZING ALUMINUM TO OTHER METALS

Aluminum can be brazed to many other metals. A partial list includes the ferrous alloys and nickel, titanium, beryllium, kovar, monel, and inconel. Aluminum cannot be brazed directly to magnesium because an alloy forms that is so brittle that it fails under very little stress.

Aluminum can be brazed to copper and brass, but the brittle compound that forms limits the joint's application. Aluminum can, however, be joined to copper and brass by means of a transition joint. Aluminum is first brazed to steel, which is then brazed to the copper.

BRAZING METHODS

The numerous brazing methods all include the same basic steps. The main differences between the various methods lie in the way that the parts are heated and in the way that flux and filler metal are applied. The only exceptions are vacuum and vibration brazing, which require no flux.

JOINT AND FIXTURE DESIGN

Simple, strong, satisfactorily brazed joints can be made by lapping one clean, flux-covered piece of aluminum with another. Brazing filler metal is positioned between or at the edge of the lap. The parts are heated until the filler metal melts and is drawn into the joint by capillary force. The joint is cooled, cleaned, inspected, and placed in service. Thousands of these simple joints are produced every day in the course of manufacture, field construction, and repair. However, when superior-quality, tight-tolerance, distortion-free, complex aluminum assemblies are to be brazed, careful designing is required.

BASIC DESIGN CONSIDERATIONS

Proper design normally begins with a careful study of the relationship between the brazed joints, the parts they are joining, and the dimensional criteria of the completed unit. During this study, the designer should bear the following parameters in mind:

1. The distance between faying surfaces (joint gap clearance) is important. It might be estimated on the basis of past experience, but actual tests might show it to be different.
2. The coefficient of expansion of aluminum is almost twice that of the metals commonly used for the brazing fixture.
3. Aluminum is soft at brazing temperature and barely self-supporting. Thin sections tend to droop if unsupported.
4. Some distortion can be expected if a complex assembly is severely quenched after brazing.

   Design anticipation can reduce these considerations to negligible factors, for example:

1. Necessary joint gap changes can be accommodated by using a lap joint. A change in joint clearance from optimum generally results in lowered joint quality and is to be avoided. Lap width, however, can be varied considerably with no penalty (Figure 24.2).

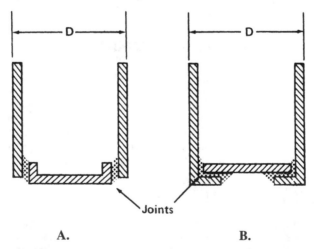

A.   **B.**

FIGURE 24.2 Lap joints: A. Dimension *D* can only be varied by changing joint clearance. B. Dimension *D* can be changed by varying the joint lap width. *(Aluminum Association, Washington, DC)*

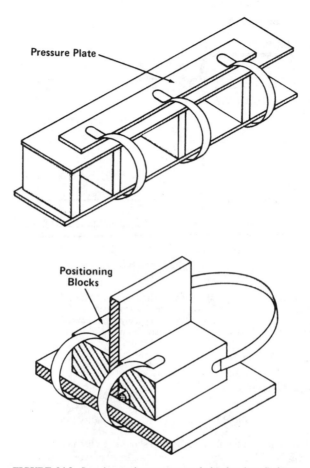

FIGURE 24.3 Securing against movement during brazing. C-clamp springs can be used to hold parts lightly, but dependably, during brazing. *(Aluminum Association, Washington, DC)*

2. The difference between expansion rates of aluminum and a steel fixture amounts to about 0.005" per inch at 1000°F. When small parts are brazed, this difference in expansion is often ignored. When larger, nonflexing parts are brazed, springs, weights, and levers are used to hold parts in place. This aid also serves as an assembly fixture (Figure 24.3).

3. Small, thick, or vertical aluminum parts generally need no support during brazing. Long, thin, horizontal parts need support, which can be supplied in many ways. Interlocking tabs, resistance welding, or tack welding can serve the purpose.

4. Some heat-treatable alloys can be returned to their former temper without severe quenching. Also, some methods of quenching, such as proprietary solutions, reportedly preclude distortion. When distortion is expected and finish tolerances are closer than 0.005", make the parts slightly oversized. Then, the assembly can be finish-machined after brazing. By this means, a brazed structure can be produced with close tolerances and flat sides (Figure 24.4).

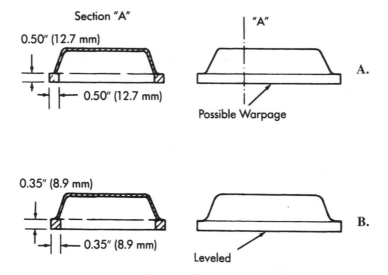

FIGURE 24.4 Machining after brazing. Highly accurate final dimensions are easily obtained by using slightly oversized stock and machining the assembly after the parts have been brazed: A. Assembly immediately after brazing. B. Assembly machined to a fine tolerence after the parts have been brazed. *(Aluminum Association, Washington, DC)*

JOINT PARAMETERS

The following conditions should be met if the highest-quality brazed joints are desired:

1. Faying surfaces and a small distance beyond should be chemically clean, and free of foreign adhesions, undesirable bumps, and dimples. In other words, for the best joints, faying surfaces should be clean and flat. Generally, brazing shops merely degrease, then braze.

2. The oxide, which is always present on aluminum, should be removed from the faying surfaces and beyond for ½", just prior to brazing.

3. The distance between faying surfaces should be correct for joint width, filler and parent metals, brazing method, and the time and temperature used.

4. The joint should be designed to allow entrance of flux and filler metal. Excess flux, oxide, and gases should be able to exit easily.

5. Faying surfaces and the adjacent fillet area should be fluxed prior to brazing, unless the joint is to be dipped or vacuum brazed.

6. The assembly and furnace, if one is used, should be vented to allow escape of trapped expanded air and gases generated during the process.

7. The joint should be brought to proper temperature and held there long enough for a good brazing job to be done.

8. A fixed relationship should be maintained between parts during the process.

9. If the brazing sheet is supplying filler metal, the clad surface of the brazing sheet should touch the part it is to join for the length of the joint.

Note: The brazing sheet is aluminum sheet of varying thicknesses clad on one or both sides with solder, also of varying thicknesses. Brazing sheet can be spun and formed as severely as the parent metal without danger of loosening the filler cladding. Although the use of a brazing sheet increases the cost of the filler metal in an assembly, it can eliminate the need to preplace filler metal, thus reducing labor costs (Figure 24.5).

10. If one or both parts are not made of brazing sheet, the correct quantity of filler metal is covered with flux and properly positioned prior to brazing, or is hand-fed during brazing.

JOINT CLEARANCE

Capillary force is determined by joint clearance. Hundreds of simultaneous brazed joints are made possible by the drawing of molten filler metal into each crack and crevice, around corners, up vertical joints, and overhead.

When capillary attraction acts on molten filler in a salt-dip brazing bath, it can draw liquid metal into and up a vertical joint 24" high. This force makes it unnecessary to place filler in its final position. In a dip brazing pot, the molten flux reduces the filler's surface tension, allowing liquid filler to follow a clean, fluxed path into a joint.

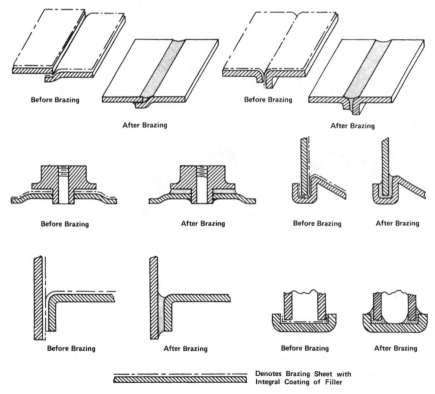

Before Brazing   After Brazing   Before Brazing   After Brazing

Before Brazing   After Brazing   Before Brazing   After Brazing

Before Brazing   After Brazing   Before Brazing   After Brazing

Denotes Brazing Sheet with Integral Coating of Filler

FIGURE 24.5 Typical joints made with aluminum brazing sheet, before and after brazing. *(Aluminum Association, Washington, DC)*

Capillary force is directly related to joint clearance. The smaller the clearance, the greater the force. Also, the longer it takes molten metal to traverse the joint, the greater the possibility that oxide, flux, gas, and foreign material will be trapped inside. It is also possible that the filler will stop flowing prematurely because alloying with the base metal it is less fluid at that temperature.

Overly large clearances pose problems of their own. Capillary action is reduced so that flux and filler might not follow the joint to its end. More stress is placed on the fillet and the joint will be weaker. Gaps might appear in the joint, a smooth fillet might not form, and filler metal will be wasted.

Fortunately, deciding on the correct clearance is easy. Satisfactory joint clearance can be determined by means of a few test joints made under actual production conditions. Once gap dimensions have been established, they will hold true as long as the other factors involved (brazing method, time, temperature, flux, and alloys) are not changed.

SUGGESTED JOINT CLEARANCE FOR VARIOUS BRAZING METHODS

Dip brazing and vacuum brazing	
Joint width	Suggested clearance
Less than 0.250"	0.002 to 0.004"
Over 0.250"	0.002 to 0.025"

Torch, furnace, and induction	
Less than 0.250	0.004 to 0.008"
Over 0.250"	0.004 to 0.025"

ESTABLISHING AND MAINTAINING JOINT-GAP CLEARANCE

In casual brazing, joint clearance can be established by a layer of flux between the mating surfaces. A strong, useful joint will result if the assembly remains fairly motionless, except for small thermal movement during the cycle.

A simple, but more positive, method of establishing gap clearance is to use correct thickness shims of brazing filler to separate the parts. The shims melt during brazing and they shrink about 5%. On cooling, they can't be relied upon for close-tolerance joint gaps. Additional support and fixturing are required (Figure 24.6).

IMPROVING BRAZING FILLER METAL FLOW

Molten metal flows best on clean, roughened, or etched surfaces. Each scratch and pore acts as a capillary to pull the metal along. Grooved and serrated surfaces help and a minimum of entrapped flux. Metal surfaces should not be smooth or mirror-like if they are to be brazed. There are proprietary chemicals for cleaning aluminum that also quickly etch aluminum surfaces that are too smooth.

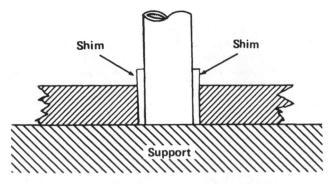

FIGURE 24.6 How brazing filler metal shims can be used to center parts, and provide escape routes for gas and flux. *(Aluminum Association, Washington, DC)*

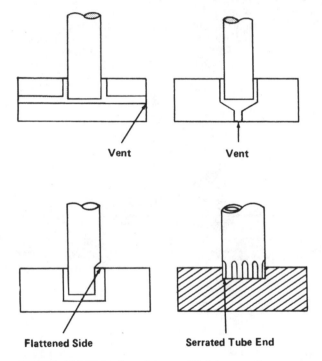

FIGURE 24.7 Examples of how solid joints can be vented. *(Aluminum Association, Washington, DC)*

AVOIDING FLUX ENTRAPMENT

Dry flux is chemically inert, but even in a dry joint there is no ensurance that moisture will not enter and contaminate. This would, of course, initiate corrosion and damage. Good design provides an escape route for molten flux, heat-generated gases, and expanding air. It also allows easy access to solidified flux, which must be removed after brazing (Figure 24.7).

Long joints should be vented along their lengths. Joints should not be filled with filler from both ends because that could lock flux and gas inside.

Earlier, it was mentioned that too small gaps could stop the flow of filler. They could also trap flux inside. Low brazing temperature or insufficient filler could also trap flux. At the brazing temperature, flux releases hydrogen and other gases. At the same time, air within the enclosed space expands. Together, they can produce pressures that would distort the assembly or even rupture it. Closed vessels must be vented. Afterward, the vent can be brazed shut. Blind holes also must be vented.

JOINT TYPES

Lapped joints are strong, easily brazed, and require no special care. These joints can easily have more strength than the base metals by making the lap length two to three times the thinner of the two base metals. See typical brazed joints, Figure 24.8.

Lap joints are best for pressure vessels (Figure 24.9). This is the strongest type of joint. It provides the longest braze path, thus reducing leakage possibilities. When an exact calculation of joint strength is needed, the formula given in Figure 24.10 can be used.

Another device is to bend one of the pieces into an offset that can determine gap clearance (Figure 24.11).

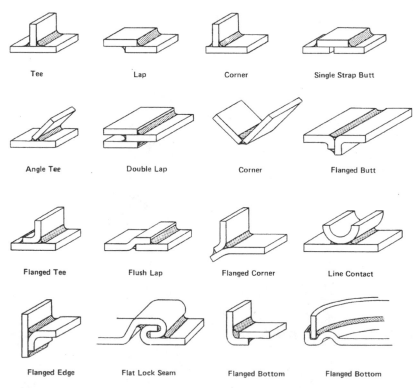

Tee   Lap   Corner   Single Strap Butt

Angle Tee   Double Lap   Corner   Flanged Butt

Flanged Tee   Flush Lap   Flanged Corner   Line Contact

Flanged Edge   Flat Lock Seam   Flanged Bottom   Flanged Bottom

FIGURE 24.8   Typical brazed joint designs. *(Aluminum Association, Washington, DC)*

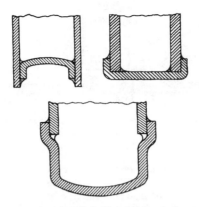

FIGURE 24.9 Typical brazed joints that have proven suitable for pressure-tight containers. *(Aluminum Association, Washington, DC)*

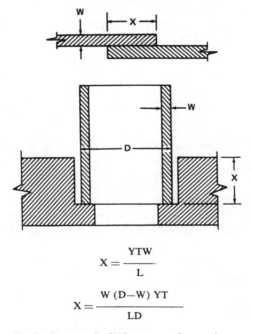

$$X = \frac{YTW}{L}$$

$$X = \frac{W\,(D-W)\,YT}{LD}$$

T   Tensile strength of thinner or weaker part in lbs/sq inch (kg/mm²)

X   Length of lap in inches (mm)

W   Thickness of the thinnest part in inches (mm)

D   Diameter of the area of shear in inches (mm)

Y   Safety factor—this is usually 4 or 5

L   Shear strength of the braze alloy in lbs/sq inch (kg/mm²)

FIGURE 24.10 Calculating needed overlap by formula. *(Aluminum Association, Washington, DC)*

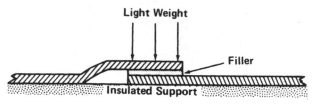

FIGURE 24.11   A simple offset, formed in one of the parts, can be used to establish lap-joint clearance. *(Aluminum Association, Washington, DC)*

Tee joints are easily made because there is no necessity for joint clearance if the vertical member is thin. The molten metal will flow under the edge of the vertical part, bonding to the butting edge, and forming fillets on both sides of the tee. Line contact joints, such as when a tube or similar shape is positioned on a flat surface, are easily made because they are like tee joints, too (Figure 24.12).

Strong butt joints can be made, but they lack ductility and require so much care that designers avoid them whenever possible. To produce a 100% efficient butt joint, the edges must be held in perfect alignment and mating surfaces must be parallel. When thick aluminum plate is to be butt joined, the faying edges can be machined to an angle producing a scarf joint, thus increasing the faying surfaces or the faying surfaces can be V'd or rounded (one edge concave, the other convex) so that one edge fits into the other (Figure 24.13).

Lap-joint clearance can be established and maintained in many ways. You could use a prick punch to dimple the thinner piece or you could form protrusions of the correct height on the thicker part. Filler metal of the correct thickness could be positioned between the parts and a weight paced on top to hold the assembly immobile (Figure 24.14).

If the vertical member of a tee joint is thicker than ⅛", the butting edge should be reduced somehow to improve filler flow. This can be done by rounding, serrating, roughing, or angling the butting edge. Any cut to this edge should not be larger than 0.010" and no where should a gap exist between the faying parts, larger than 0.010". If a thicker piece is used as the vertical member and angled, the edge should be kept within 5 to 8 degrees of the plane (see Figure 24.12 again).

Weep holes should be used to vent lap joints wider than ½" (Figure 24.15). When brazing aluminum, pressed joints should not be used because there is danger of flux and gas entrapment. In fact, it is conceivable that the flux will not enter the joint at all.

FILLER, PLACEMENT, SHAPE, AND QUANTITY

There are three ways to bring filler to the joint which is being filled. Before brazing, the filler can be placed in or near the joint. Filler metal can be hand-fed while brazing or a brazing sheet can furnish filler. If the first method is selected, there are hundreds of rods, sheets, washers, tubes, wires, and rings. For hand-feeding, filler is supplied in many wire sizes. For the third method, the sheet can be provided in many different thicknesses.

Generally, it isn't necessary to measure filler metal exactly. Nevertheless, it is better to provide a little more than not enough. Using an excessive quantity of filler should be avoided though because it is unsightly and could actually run where you don't want it, interfering with operations.

Maximum fillet height can be achieved when dip brazing; that is 0.5" high. With other brazing methods, the maximum fillet height is about 0.3".

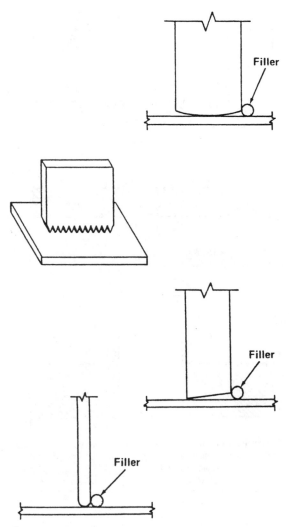

FIGURE 24.12   How the contacting edges of thick sections can be rounded, serrated, angled, or curved to promote the flow of filler metal into the joint. *(Aluminum Association, Washington, DC)*

If the assembly is removed too rapidly from a flux bath or if it is shaken accidentally, fillet height can very likely be reduced.

PROPER FILLER PLACEMENT

A good design will have the filler positioned as close to its ultimate resting place as possible. The shorter the travel distance for the filler, the less time there will be to alloy and the less oxide has to be removed.

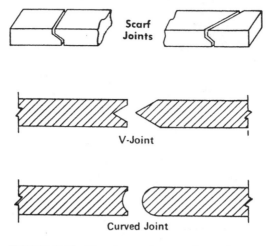

FIGURE 24.13 Alternatives to the square butt. *(Aluminum Association, Washington, DC)*

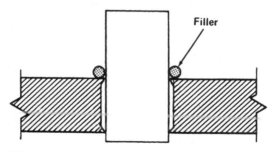

FIGURE 24.14 How prick punch indentations and protrusions can be used to center and hold the part in place. *(Aluminum Association, Washington, DC)*

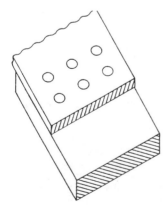

FIGURE 24.15 Weep holes cut into one joint member are used to vent flux, gas, and filler metal when making wide-lap brazed joints. *(Aluminum Association, Washington, DC)*

If the mass of filler metal is very small compared to the parts being brazed, the filler must be prevented from melting too soon. Otherwise, it might ball up and roll off. One preventative measure is to make filler grooves within a part.

Sometimes, when there is considerable mass difference between the parts being brazed, the thinner part will reach brazing temperature first. Therefore, it is best to place the filler in contact with the heavier part. It is imperative to protect filler metal from the source of direct heat when the heat is supplied from one direction. Otherwise, the filler might melt prematurely and roll away before the base metal melts.

DIMENSIONAL CHANGES DURING THE BRAZING CYCLE

When brazing an aluminum plug in a hole in a piece of aluminum, if the part masses are similar, the hole and plug should expand and contract at about the same rate as the temperature goes up and down. A correct joint clearance, in this case, with cold parts, will provide a good assembly. If the external part is so large that it remains cool through the brazing cycle, the inner piece will expand and reduce the joint's clearance as it heats up. One inch of aluminum will expand about 0.012" at brazing temperature (1000°F). Steel, normally used for fixturing, expands about half that amount. This difference must be considered by the designer.

FIXTURING

Simple weights and levers will frequently be sufficient to hold parts in alignment, except when dip brazing. Positive restraints are required when dip brazing. Although the media in a dip-braze tank has a specific gravity that is similar to that of aluminum and (for all practical purposes) supports the assembly, the parts are liable to separate as they are dipped into and removed from the bath. Self-fixturing assemblies are far less expensive to braze, although they might require more planning (Figure 24.16).

Many costs are caused by fixturing. First, is the initial cost of design and fabrication. Next, a quantity of fixtures are probably required for production. The fixtures pick up flux encrustations, as do the assemblies and they must be cleaned frequently. Fixtures add to both the heating and cooling load because each one adds to the assembly's mass. In dip brazing, fixtures contribute to contamination and add to the quantity of flux dragged out.

Parts can be held together by tabs, pins, crimps, staking, rivets, springs, clips, tie wires, and welding. Only the designer's imagination limits the self-fixturing methods.

If the brazed assembly sticks to the fixture, the parts could be coated with stop-off material, or the fixture could be constructed from oxidized stainless steel. Stainless steel fixtures can be oxidized simply by heating them, unloaded, in a furnace.

Whatever fixtures are used, or whatever arrangement of weights, springs, or levers is devised, the aluminum pieces must be held lightly, but firmly.

For short runs, low-carbon steel is generally satisfactory for fixture material. If the flux is removed after each cycle or at least often, the fixtures will serve you longer.

For production jobs, no bare or coated steel is recommended for dip brazing. Iron and steel will quickly contaminate the flux bath. The action of hot flux on nickel- or aluminum-coated steel will deteriorate these coatings in about 10 or 15 immersions.

For longer runs, it is advisable to use stainless steel or Inconel X-750 for such assignments. For very long runs, Inconel should be used because it will last much longer than stainless. With repeated exposure, the hot flux dissolves the nickel on the surface of stainless, and it is necessary to shot peen the dark color off the stainless. The dark color is ex-

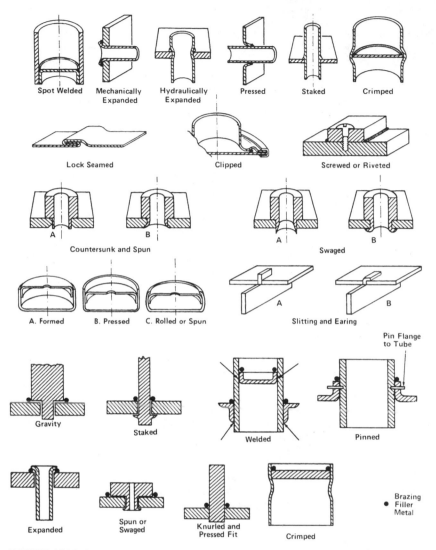

FIGURE 24.16 Twenty-one suggestions for making self-fixturing brazing assemblies. *(Aluminum Association, Washington, DC)*

posed iron, which would contaminate the bath. The fixture should be kept as small and light as possible to minimize heat loss.

COSTS

The cost factors are: precleaning, labor, energy (fuel), brazing alloy, flux (when used), and post cleaning. The design of the part, its tolerances, and fixture design must be considered to minimize scrap and reworking. The capital costs involved, such as dip pots, furnaces, torches, and automation devices (such as conveyors, robots, and computers) can be amortized properly. All of these costs can be closely determined.

APPLYING THE FLUX

All surfaces that are to be wetted by the filler and the filler itself must be fluxed, unless, of course, one of the fluxless brazing methods or dip brazing is to be used. Molten filler will ball up unless it is fluxed.

Because flux is very hygroscopic, it should not be exposed to air longer than necessary; certainly not more than 45 minutes. In other words, apply it as soon as possible before brazing. You want to ensure that flux is present within the joint. Do this by covering the filler metal and both facing sides of the joint with flux. If the filler cannot be positioned in the joint, the path it will take must be fluxed.

Because steel containers would contaminate flux, it should be kept in special vessels (such as glass, porcelain, or porcelain-lined vessels). It can be applied by brushing or spraying. Flux in powder form can be mixed with water to form a thick paste. Fresh flux should be mixed every 4 to 6 hours.

Excessive flux should not be used because hot flux will stain any metal that it touches. Yet, sufficient flux must be used to allow it to do its job. When using brazing sheet, the entire clad surface is normally fluxed to assist filler flow.

STOPPING THE FLOW OF FILLER

By eliminating flux from an area, the flow of filler can be stopped. However, preventing capillary attraction is a more positive method. This can be done by cutting short one of the brazing surfaces by 0.10", where you want to stop the flow. Many excellent commercial stop-off materials are on the market. These substances block the flow of filler. A few tests will indicate which stop-off is best for your job. The stop-off must not harm adjacent surfaces and it must wash off easily.

SEQUENTIAL JOINTS

If it is necessary to form a second joint adjacent to one already made, a higher melting point filler must be used for the first joint. Some products are made with a dozen adjacent joints. Each successive joint must be made with a melting point at least 50° lower than its predecessor.

HEAT-TREATABLE AND NONHEAT-TREATABLE ALLOYS

The physical and chemical properties of aluminum are altered by the addition of various elements. Certain elements produce alloys that can be hardened and strengthened only by cold work. These are known as *nonheat-treatable alloys*. The reminder of alloys, which can be strengthened by heat treatment, are called *heat-treatable*.

Both groups lose strength as they are heated. The nonheat-treatable alloys soften and return immediately to zero temper. The heat-treatable alloys must be maintained at annealing temperature for at least 20 minutes before they start to lose temper. These can be reheat treated. The nonheat-treatable alloys can only be tempered by more cold working.

Aluminum brazing temperature is between 1030°F and 1195°F. The annealing temperature is from 650 to 800°F. Brazing causes a slight annealing. The amount of annealing that occurs depends on the alloy, time/temperature, and mass. Recold working can't often be done. Therefore, brazing a nonheat-treatable aluminum generally results in an assembly that is close to zero temper.

In cases where high strength is not required, the selection of nonheat-treatable aluminum would reduce material cost. Most nonheat-treatable alloys (excepting those with high magnesium content) do not retain oxides as tenaciously as those that are heat treatable. When maximum quality is not required, nonheat-treatable alloys are often prepared for brazing by vapor degreasing alone. The chemical or mechanical cleaning is skipped.

When the aluminum leaves the quench tank, it is soft and workable. It can be kept in this workable condition for many hours by maintaining it at below freezing temperature. This can be helpful when there is much straightening to do and not enough time.

With the passage of time, these materials harden somewhat. If it occurs at room temperature, the alloy is said to have *aged naturally*. If hardening has been accelerated by heating, it is said to have been *artificially aged*. Time and temperature affect aging rate and temper, and vary from alloy to alloy. Some alloys require a month to fully self harden, others require many months. Much can be learned about the heat treatment of aluminum by reading brochures from the prime producers of aluminum.

An alloy that has already been hardened loses very little of its temper as a result of brazing. When some distortion is permissible, hot brazed parts can be dumped into boiling water to remove flux from the brazed areas. This is the best water temperature for flux removal.

Unless casting quality is very poor, castings can be brazed just as easily as wrought material. Brazeable aluminum casting alloys include 356, 357, 359, 443, A712, C712, and D712.

FURNACE BRAZING

Except for tip brazing, more assemblies are brazed in a furnace than by any other means. It is so popular because the equipment cost is low and many existing furnaces can be used for brazing aluminum then changed back for another use. The weight of many assemblies can make them self-jigging for furnace brazing.

Furnace brazing is excellent for assemblies that might trap flux if dipped or have pockets that could trap air. It is also good for highly polished parts that might etch if dipped in hot flux.

Before putting parts in the furnace, they should be cleaned and have excess oxide removed. Faying surfaces are fluxed, filler metal is positioned, and the pieces assembled and jigged. The assembly is then heated to about 300°F to drive moisture out of the flux. This is normally done in the preheat section of the furnace, perhaps on a conveyor. From there, the assembly travels through the main portion of the furnace for about 3 to 5 minutes. The assembly is finally removed, cooled or quenched, then cleaned. The total furnace time seldom exceeds 15 minutes.

TWO FURNACE SYSTEMS

Two types of furnaces are commonly used for brazing today: the batch furnace and the continuous furnace. The cost of the batch furnace is low, as is its maintenance. It can be run intermittently. Brazed assemblies can be made economically in batch furnaces. Cycling can be accomplished in 10 to 20 minutes, depending on the type of furnace and the mass of the parts. Batch furnaces are ideal for a large variety of parts per day and any production quantity whatsoever. Most any part that can fit into such a furnace can be brazed.

The continuous furnace, on the other hand, could be one long unit or several joined units, which normally have different temperatures in each section. The parts or assemblies move through the separate units on a conveyor. The parts are gradually brought to brazing temperature and finally moved through a cooling zone. This type of furnace can handle the highest hourly production rates and the highest hourly weights of aluminum of all methods of brazing.

When furnace time approaches 30 minutes, the quality of the brazed joints fall off. This is because of three basic reasons and in their order of significance: flux change, liquation, and diffusion. When flux is exposed to air, it absorbs water and reacts with adjacent metal. This hygroscopic characteristic of flux is something to beware.

If the temperature is raised very slowly, it could melt some of the lower melting-point constituents of the filler material. That could leave the chemistry of the filler changed and inadequate to do its job. This change is called *liquation*.

If assembly parts are very unequal in mass, the filler might turn liquid in one portion of the assembly before the rest of the assembly is ready for brazing. This is called *diffusion*. Nevertheless, any new job that, for whatever reason, requires more than 30 minutes should be tested because some assemblies that require much longer than 30 minutes can be brazed successfully.

SELECTING THE PROPER FURNACE

If your furnace can be brought to brazing temperature and maintained there within ±5°F, it can be used for aluminum brazing. The heating source can be oil, gas, electrical resistance, or infrared heating lamps. Although any of these energy sources can be used for brazing, it is best to avoid the oil-fired furnace. Oil combustion is never complete. Soot could deposit on the aluminum parts. As a matter of fact, direct combustion furnaces, which are the least expensive, should be avoided. Moisture, always a by-product of these furnaces, is also likely to degrade brazed joints.

Hot-wall furnaces, in which hot combustion gases pass through tubes or behind furnace walls, provide a cleaner, dryer, atmosphere for brazing. Electrically heated hot-wall furnaces with unexposed heating elements are also recommended. Alloys 6061, 6063, and 6951 do not braze well in direct-combustion furnaces.

TORCH BRAZING

The method generally used for one-of-a-kind brazing jobs, short production runs, and repairs, is hand-held torch brazing. If the occasion requires it, the torch can be used for any brazing situation (Figure 24.17 and 24.18).

There are many sources of heat for torch brazing. However, most people use the same type of equipment, torch, controls, and gases that are used for fusion welding. For aluminum brazing, the operator merely changes goggle lenses and torch nozzles.

The art of torch brazing can be learned quickly and easily by most mechanically able people. Of all the gases used to fuel the torches, oxyacetylene produces the highest temperature. Other gases are cooler and, thus, easier to use with light-gauge aluminum. By maintaining a low gas pressure (under 4 psi), the flame can be easily controlled. Visual inspection can be used to adjust oxyacetylene.

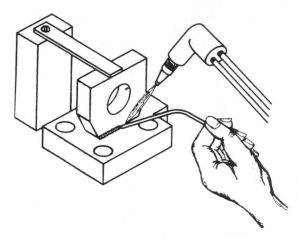

FIGURE 24.17 Typical torch brazing set up. A simple fixture holds the parts in place. The torch is moved around the work, as necessary. *(Aluminum Association, Washington, DC)*

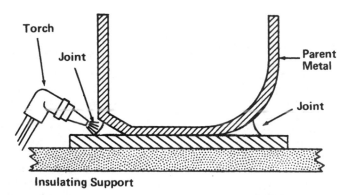

FIGURE 24.18 Faying surfaces that meet at a small angle or a gentle curve can be brazed without joint clearance for short distances. *(Aluminum Association, Washington, DC)*

The white cone produced by the mixture, should be extended from the torch tip for 1 or 2". The inner cone should be half this length. Oxygen pressure should always be about half of the accompanying gas. You should try to produce a flame that is slightly reducing. This type of flame tends to protect the aluminum parts from oxygen because this flame consumes oxygen faster than the oxygen is delivered. A reducing flame is also less intense.

PROCEDURE

After the parts are cleaned, joint clearances are established. For torch brazing, joint gaps are normally set between 0.004 and 0.025", when possible, parts should be designed for self-jigging. Aluminum at brazing temperature is soft. The gas pressure exiting the torch could move it. Long sections need support. After flux is applied, care must be taken with application of the torch to prevent premature melting of the filler.

MEASURING JOINT TEMPERATURE

Three common methods are used to determine approximate joint temperature. Flux indicates temperature by its appearance. As the temperature increases, it dries out and turns white. With a continuing rising temperature, flux melts and turns gray. At brazing temperature, the flux becomes transparent and the aluminum subsurface glows with a silvery sheen. Some proprietary fluxes indicate temperature by showing color.

Special crayons are used to make a mark near the joint. When the desired temperature is reached, the mark changes color or melts. The brazing filler itself is the third method. The tip of the filler wire is tentatively applied to the work being brazed from time to time. When the filler's tip softens and starts to melt, the brazing temperature has been reached. Temperature overshoot can be avoided at this time by withdrawing the flame for a moment.

HANDLING THE TORCH

The torch should be held a few inches from the work at any convenient angle from 5 to 45 degrees to the part. Never permit the flame to remain on any one portion of the part for more than a moment. Move it constantly either in circular motion or from side to side. The outer edge of the cone provides maximum heat.

When parts of varying mass are being joined, the torch flame is moved rapidly over thin sections and more slowly over thick. Aluminum is such a good conductor of heat that heating unequal pieces is not as difficult as it might appear. If the parts are very thin, it might be necessary to play most of the heat over the filler, rather than the parts.

Supplying filler from one point is a good habit. This way, the filler will flow its own way down the joint. If the joint is too long, the filler might have to be fed from more than one position. Being sure that the bright metal flows the full length of the joint is a good way for the operator to ensure complete joints. If the joint is clean and fluxed, capillary action will draw molten filler into all cracks and form the fillet by itself. It is best to heat a joint once and not do any reheating. Reheating a fillet sometimes causes collapse of the matrix metal from overheating.

If the heat of a single torch is insufficient to braze a large part, a second torch wielded by a second operator might be required. Multiple-head torches are available to heat large assemblies. Assembly heat losses through conduction can be reduced by reducing contact between parts and fixture. Losses from convection or radiation can be reduced by reflective walls of some sort. OSHA frowns on the use of asbestos for that purpose.

If a quantity of assemblies must be made, they can be brought to the operator on a conveyor. Fans are commonly used to cool brazed assemblies so that they can be quickly dropped into a wash tub for defluxing (Figure 24.19).

Because hot flux creates toxic fumes, brazing and cleaning should be done in a well-ventilated area. The operator should always wear protection; clothing, gloves, and goggles. Wear the class and type of goggles recommended by the American Welding Society.

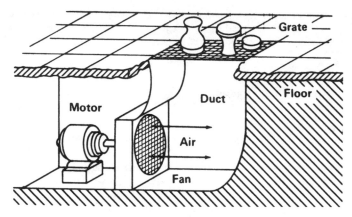

FIGURE 24.19 Parts are frequently quenched with a blast of ambient-temperature air. *(Aluminum Association, Washington, DC)*

AUTOMATIC TORCH BRAZING

Automatic torch brazing is quite similar to manual torch brazing. The only difference is that the assembly or the torch is automatically moved toward the other. The torch might be programmed to move or the assembly might be moved past or around the torch.

First, the operator changes adjustments on the controls until the correct parameters for that specific braze is determined. When the torch-to-work distance, gas mixture, and heating time are correct, there is no need for further experiments. The flux and filler must also be positioned. Additional torches can be used to handle large masses.

In automatic brazing, care must be taken not to shake the assembly or rotate it too fast because of the danger of losing some of the fillets to gravity. Assemblies that cannot be dip brazed or furnace brazed often can be economically brazed by automatic torch.

DIP BRAZING

As in all other types of brazing, assemblies to be brazed by this method should be cleaned, freed of oxide, assembled, and jigged. The assembly is next preheated to almost 1000°F in a preheat furnace. Finally, the assembly is submerged in a dip-braze tub for one or two minutes. During this time period, one or many joints can be made simultaneously.

The hot molten flux is maintained at a temperature slightly above the liquidus point of the filler metal. It fluxes the joints, serves as a heat-transfer agent, and partially supports the submerged assemblies. Hot flux has the consistency of thickened water. It readily contacts every inch of surface and enters each joint and cranny unless prevented by dirt or air. The fluxing action is similar to the action of flux placed on a faying surface.

The temperature of dip-brazing furnaces can be maintained very close so that small, large, or heavy pieces can be consistently dipped without pause. Blind or inaccessible joints can be dip-brazed easily and simultaneously.

Dip furnaces are heated by ac current flowing between electrodes immersed in flux. Electrodes wear and have to be replaced occasionally. Proper clearances must be maintained between electrodes and parts, and between parts and the dip-pot walls. Self-jigging parts are ideal for dip brazing. Brazing sheets work very well in this method.

One problem with dip brazing is economics; the pot should be kept fluid at all times. That means you must have enough work to keep the pot busy; otherwise, maintenance would be too expensive. You would then have to use the services of a vendor with dip-brazing facilities.

Salt pots (dip brazing pots) are steel-reinforced vessels, lined with acid-proof fire brick (about one foot thick). The large pot covers can be power operated on rollers and they are also well insulated. These units generally last 10 to 20 years. To start the pot, cold flux is placed in the pot and heated by a torch or portable electric heater. Once molten, the flux becomes electrically conductive and is kept at brazing temperature by the passage of electricity between electrodes immersed in the flux.

CHAPTER 25
WELDING

BACKGROUND

Welding contributes to the production processes of most manufactured products, and to the maintenance of many other businesses. This chapter contains references to a variety of welding methods, but it is not much more than that—a reference. Entire volumes have been written about each method, so if you need more information, it can be located easily at most libraries. The only intention of this chapter is to display the menu of welding processes available, thus helping you to select the best one for your purpose.

The Gas Metal Arc Welding (GMAW) process was developed and made commercially available in 1948, although the basic concept was actually introduced in the 1920s. In its early commercial applications, the process was used to weld aluminum with an inert shielding gas, giving rise to the term *"MIG" (metal inert gas)*, which is still commonly used when referring to the process. And in its turn, Gas Tungsten Arc Welding (GTAW) is the process very often called by its old name, *tungsten inert gas (TIG) welding*.

The GMAW process uses either semiautomatic or automatic equipment and is principally applied in high-production welding. Most metals can be welded with this process and can be welded in all positions with the lower energy variations of the process. GMAW is an economical process that requires only a moderate amount of cleaning of the weld deposit. Warpage is reduced and metal finishing is minimal compared to stick welding.

THE GMAW PROCESS

GMAW is an arc-welding process that incorporates automatic feeding of a continuous, consumable electrode that is shielded by an externally supplied gas. Because the equipment provides for automatic self-regulation of the electrical characteristics of the arc and deposition rate, the only manual controls required by the welder for semiautomatic operation are gun positioning, guidance, and travel speed. The arc length and current level are automatically maintained. See Figure 25.1 for a GMAW system installation.

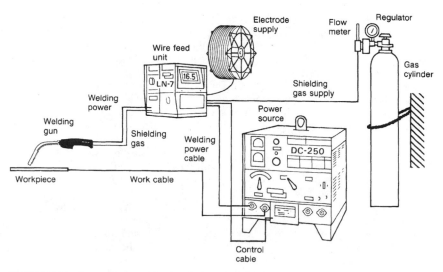

FIGURE 25.1 Gas metal arc-welding installation. *(Lincoln Electric, Cleveland, OH)*

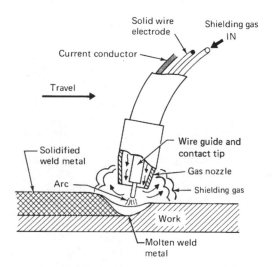

FIGURE 25.2 A gas metal arc-welding (GMAW) gun.
(Lincoln Electric, Cleveland, OH)

The cable assembly performs three functions. It delivers shielding gas to the arc region, guides the consumable electrode to the contact tip, and conducts electrical power to the contact tip. When the gun switch is depressed, gas, power, and electrode are simultaneously delivered to the work and an arc is created. See the GMAW gun in Figure 25.2. The wire-feed unit and the power source are normally coupled to provide automatic self-regulation of the arc length. The basic combination used to produce this regulation consists of a constant-voltage power source in conjunction with a constant-speed, wire-feed unit.

The characteristics of GMAW are best described by four basic modes of metal transfer, which can occur with the process: axial spray transfer, globular transfer, short-circuiting transfer, and pulsed spray. Axial spray and globular transfer are basically associated with relatively high arc energy. Both of these methods are normally limited to the flat and horizontal welding positions, with material thicknesses greater than ⅛". Short-circuiting transfer is a relatively low-energy process that is generally limited to thicknesses under ⅛", but is used in all weld positions. Pulsed spray transfer, in which the average energy level is reduced, is an exception to the GMAW process. In fact, it is called *GMAW-P*.

In axial spray transfer, a gas shield with a minimum of 80% argon is used. In this mode, metal transfer across the arc is in the form of droplets of a size equal to or less than the diameter of the electrode. The droplets are directed in a straight line from the electrode to the weld puddle. The arc is smooth and stabile, with little spatter and a fairly smooth weld bead. Penetration is less than can be obtained with the high-energy globular-transfer mode of GMAW, but more than that of shielded metal arc welding (stick welding).

The globular-transfer mode uses carbon dioxide or helium as a gas shield. In this mode, metal transfer is in the form of irregular globules randomly directed across the arc, resulting in a considerable amount of spatter. This spatter could be minimized when using a carbon-dioxide shield by positioning the electrode's tip below the surface of the molten metal.

In short-circuiting transfer, all metal transfer occurs when the electrode is in contact with the molten puddle on the workpiece. In this mode, the power-source characteristics control the relationship between the intermittent establishment of an arc and in the short circuiting of the electrode to the work. Because the heat input is low, weld penetration is shallow and care must be exercised to ensure good fusion in heavy sections. However, these characteristics permit welding in all positions and the mode is adaptable to welding thin-gauge sections.

PULSED ARC WELDING

In addition to these three basic modes of metal transfer in the GMAW process, there are several uniquely significant variations, of which the pulsed current transfer has become most important. This mode is capable of all position welding at a higher energy level than the short-circuiting transfer. In this variation, the power source provides two current levels: a steady background level, too low in magnitude to produce any transfer, and a pulsed peak current, superimposed upon the background current at a regular interval. The combination of the two currents produces a steady arc with a controlled transfer of weld metal in the pulsed peak current.

The proper terminology for the process has been established by the American Welding Society as GMAW-P. This power pulse produces a spray arc at procedures that would normally be glob transfer or short-arc transfer with conventional power sources. The advantages of pulsed arc are:

1. Better arc characteristics at deposition rates below full spray.
2. Lower spatter levels because of the improved arc characteristics.
3. Reduced cold casting over short-arc welding.
4. Lower heat input at a given wire-feed speed than spray arc or glob transfer.
5. Reduced smoke levels.
6. A wide range of materials can be welded.
7. Good operator appeal.

	Argon (Ar)	98Ar/2Oxygen	95Ar/5CO$_2$	90Ar/10CO$_2$	80Ar/2OCO$_2$
Mild Steel				X	X
Low Alloy Steel		X	X	X	
Stainless Steel		X			
Aluminum	X				
Silicon Bronze	X				
Inconel	X				

FIGURE 25.3 Pulsed arc gas blends. *(Lincoln Electric, Cleveland, OH)*

Pulsed arc welding requires a shielding gas of either pure argon or pure helium or a mixed gas with high percentages of argon or helium. The parts that are welded must be clean, which means a clean wire and an oxide-free plate. Dirt affects the pulsed arc more than conventional arcs both in arc character and weld puddle wetting. Pulsed arc welding requires more equipment maintenance (liners and contact tips) than conventional welding. The high-current pulse needs good contact and smooth wire feeding to provide uniform arc characteristics and low spatter.

The advantages of controlled or reduced heat input with pulsed arc can also be a disadvantage in some applications. For many applications, a minimum heat input is required to obtain the proper bead shape, edge wetting, and follow. If the heat necessary to make the weld is more than the pulsed arc setting can deliver, the weld cannot be made. The arc characteristics might be good with no spatter, but the weld will be humped, poorly wet, over welded, or slow. The most useful range for pulsed arc welding is the upper half of the glob-transfer range and the lower part of the spray-transfer range.

Pulsed arc can often be used where other processes are marginal. If another process is doing well, pulsed arc might not have any advantage at all. Use pulsed arc when spatter must be minimized. It is a good mode for welding out of position, without cold casting. This mode has been used with great success in butt-weld applications with low-alloy steels. The controlled heat input will permit welding in the vertical up position and lower the cost of welding in areas that have been dominated by stick welding.

The controlled heat input of pulsed spray is excellent for aluminum welding and welding silicon bronze. Higher-alloy (monel and inconel) electrodes require minimum heat input for good bead wetting. See Figure 25.3 for suggested gas blends for pulsed arc welding. Be careful in the following areas when pulse arc welding:

1. Good ground connections are required.
2. Good, tight tips are preferred.
3. The process can be "finicky" and equipment can be more complex than conventional, which is why experience is helpful.

GMAW EQUIPMENT

The GMAW process can be used either semi-automatically or automatically (see Figure 25.1 again). The basic equipment for any GMAW installation consists of the following:

1. A welding gun.
2. A wire feed motor and associated gears or drive rolls.
3. A welding-control unit.
4. A power source.

5. A regulated supply of shielding gas.

6. An electrode supply.

7. Interconnecting cables and hoses.

The welding gun, which is used to introduce the electrode and shielding gas into the weld zone and to transmit electrical power to the electrode, also has a number of accessories (see Figure 25.2 again):

1. Contact tip.

2. Gas nozzle.

3. Electrode conduit and/or liner.

4. Gas hose.

5. Water hose for water-cooled guns.

6. Power cable.

7. Control switch.

The contact tip, usually copper or a copper alloy, is used to transmit welding power to the electrode and to direct the electrode to the work. The contact tip is connected electrically to the welding power source by the power cable. The inner surface of the contact tip is very important because the electrode must feed easily through this tip and also make good electrical contact. The literature typically supplied with every gun will list the correct contact tip for each electrode size and material. The contact tip must be held firmly by its collet and it must be centered in the shielding gas nozzle.

Different types and sizes of welding guns have been designed to provide maximum efficiency, regardless of the application, ranging from heavy-duty guns for high-current, high-production work, to lightweight guns for low-current or out-of-position welding. Water- or air-cooled guns and curved or straight nozzles are available for both heavyweight and lightweight guns.

SUBMERGED ARC WELDING

In this process, the arc melts the electrode, base metal, and flux into a common pool. The molten flux acts as a cleansing agent and floats to the top of the weld to form a protective slag while the weld solidifies. Deposit rates of more than 100 pounds per hour are possible using up to 600 amps on a single $\frac{3}{32}$" wire. Submerged arc welding features outstanding fast-fill characteristics. This mode of welding can approach a 100% operating factor, sometimes for an hour or more—especially on long, large welds, multipasses and overlays.

The ripple-free, no-spatter, submerged arc welds seldom require machining or grinding. Slag peels off with little or no chipping. Easy cleaning minimizes finishing costs. The ability to penetrate deep into the joint minimizes the amount of weld metal required for full strength. Using standard procedures, submerged arc welds easily meet code requirements, including stringent radiographic, ultrasonic, and other testing procedures. The ability to consistently produce sound welds of code quality has been a major reason for the wide acceptance of the process.

Submerged arc deposits are low in hydrogen, feature excellent crack resistance, and have generally superior weld quality. Machine control of welding parameters and arc placement eliminates numerous opportunities for operator error. Once the weld procedure is established, submerged arc welds of the same size and quality will be consistently reproduced.

This process provides more operator comfort than any other welding process. The granular flux almost completely covers the arc, which minimizes flash, glare, heat, smoke, and fumes. Operators can weld without cumbersome helmets and heavy gloves. They are able to work longer with better accuracy and less fatigue. Other employees and sensitive equipment are automatically shielded from welding glare and flash. Even the sound is muffled under the flux bed.

WELDING TECHNIQUES

Although there is a variety of welding methods, as this chapter points out, one common requirement does exist; the need for positioning the pieces and holding them accurately and securely. Usually, welding fixtures are designed for specific jobs. It could be special lathe chuck jaws to hold parts that require a circular weld. Or a "Lazy Susan" might hold parts for circular motion. Normally, special fixtures are justified only when production quantities warrant the expense. The question is how to handle the small quantity jobs.

These jobs are usually done on a steel table using C clamps, De-sta-co clamps, and large metal blocks. Several companies are marketing modular workholding systems. These are tables mounting a variety of blocks, angles, and other accessories. The tables are flat within a fairly close tolerance and have perforations to accept screws, and slots for T bolts that secure standard straps. In many cases, such a system would not be a necessity, but more of a convenience.

Some welds on both high-strength and low-alloy steels should be made with only one or two types of stick electrodes. It is important to select the best of the alternatives.

1. Classify the joint as fast-freeze, fast-fill, fill-freeze, or a combination of these.
2. Choose the electrode group as fast-freeze, fast-fill, fill-freeze, or low hydrogen from the information listed in the literature from each supplier of welding consumables.
3. Review the electrodes in the appropriate group to select the best for the specific job.

The following general information is for the fast-freeze group, whose characteristics are:

1. Truly all purpose. Particularly good for vertical and overhead.
2. Light slag with little slag interference for easy arc control.
3. Deep penetration with maximum admixture.
4. In appearance, the beads are flat with distinct ripples.
5. Made especially for welding carbon steel.
6. For general purpose and maintenance welding.
7. Capable of x-ray quality welds out of position.
8. Pipe welding cross country, in plant, and noncritical small-diameter piping.
9. Best choice on galvanized, plated, painted, or dirty, greasy steel that cannot be completely cleaned. The weld quality might be lower than on clean steel.

The following general information is for the "Fast-Fill Group:"

1. For multiple pass welds, including fillets and deep-groove butts.
2. Single-pass welds—especially production fillets and laps.

3. Medium-carbon, crack-sensitive steel when low-hydrogen electrodes are not available. Preheat might be required.

4. Highest deposition rates of all electrodes.

5. Good bead appearance: smooth, ripple-free beads with minimum spatter.

6. Easy slag removal. Heavy slag tends to peel off.

7. High iron powder content (50%) by weight in the coating gives fast-fill electrodes their major advantage of high deposit rate.

8. Weld heat is more efficiently used to melt coating, rather than excess quantities of base metal.

9. The iron powder is added to the deposited weld metal.

10. Optimum currents are higher than with conventional electrodes.

The charts that can be located in the stick-electrode handbooks will tell whether to use ac or dc polarity with each type of stick. When welding on the flat, hold a ⅛" (or shorter) arc or touch the work lightly with the electrode tip. Move fast enough to stay ahead of the molten pool. When making vertical welds, use ³⁄₁₆" (or smaller) electrodes. Vertical-up is used for most plate-welding applications. Vertical-down drag techniques are used by pipeliners and for single pass welds on thin steel. For overhead and horizontal butt welds, use a ³⁄₁₆" (or smaller) electrode. For sheetmetal edge and butt welds, use dc electrode negative.

The following general information is for the "Fill-Freeze-Group," whose characteristics are:

1. Good for welding carbon steel.

2. Good for irregular or short welds.

3. Good for sheetmetal lap and fillet welds.

4. Good for joints with poor fit.

5. All position welding, but mostly for downhill or level positions.

6. General-purpose welding in all positions.

Use dc polarity for best performance, except when arc blow is a problem. Touch the tip of the electrode to the work or hold a ⅛" (or shorter) arc. Use a ³⁄₁₆" (or smaller) electrode for vertical and overhead welding. When welding vertical-down, point the electrode up so that arc force pushes molten metal back up the joint. Move fast enough to stay ahead of the molten pool. When welding vertical-up, point the electrode slightly upward so that arc force assists in controlling the puddle and travel slowly enough to maintain the shelf without spilling. When welding overhead, travel fast enough to avoid spilling.

The following are general applications for the "Low-Hydrogen Group:"

1. They resist cracking in medium- to high-carbon steels, hot short cracking in phosphorus-bearing steels, and porosity in sulfur-bearing steels. However, x-ray quality and mechanical properties might be lower.

2. Thick sections and restrained joints on mild and alloy steel plate, when shrinkage stresses tend to cause weld cracking.

3. Alloy steels that require 70,000-psi tensile-strength deposits.

4. Low-hydrogen electrodes can produce dense, x-ray quality welds with excellent notch toughness and ductility. The low hydrogen content in the deposit reduces the danger of underbead and microcracking in low-alloy and high-carbon steels and on thick weldments.

5. Be sure that all low-hydrogen electrodes are shipped in hermetically sealed boxes, which can be stored indefinitely in normal storage conditions without danger of moisture pickup. Moisture pickup is significant because it could cause hydrogen-induced cracks, particularly in steels with 80,000 psi and higher strength. Moisture-resistant electrodes can be left exposed for up to 10 hours.

6. Use dc polarity with ⁵⁄₃₂" (and smaller) electrodes and ac with larger sizes.

7. When x-ray quality is required, drag the electrode lightly or hold as short an arc as possible: ¹⁄₁₆" maximum. Do not use a long arc at any time because this type of electrode relies principally on molten slag for shielding.

8. Weld vertical-up with ⁵⁄₃₂" (or smaller) electrodes. Do not use a whip technique or withdraw the electrode from the molten pool.

9. For overhead welding use ⁵⁄₃₂" (or smaller) electrodes. Move fast enough to avoid spilling weld metal, but don't be alarmed if some slag spills.

There are specific electrode applications listed in most handbooks for all jobs. It pays to read all about your choice before using it. Check out the typical applications, the chemistry, and the mechanical properties as deposited.

Any company that does much welding should consider robotic welding if only to know when it becomes economical to purchase a robot. You should study the entire spectrum of welding; the quality issues, welding consumables used, maintenance schedules, equipment depreciation, cost of external (vendor) work, annual scrap rates, and the intangibles involved in robot-versus-manual welding. Flexibility in switching jobs is also a factor. Then, you should consider where the company is heading. What are the projections for next year and for five years?

Compared to manual welding, robots could improve quality, decrease spatter, use less consumables, and reduce scrap, with an estimated pay-back period of one year. Twenty years ago, this would have been a hopeful projection. Now, experience has demonstrated the accuracy of these projections.

To avoid the expense of electron beam or laser welding, especially if the company does not possess that type of equipment, many manufacturers will attempt to do the job by means of GMAW. Much experimentation has been done with what is classified as micro-arc welding. Currents as low as 0.1 amps are being used and the pulsed arc method is also being used.

STAINLESS STEELS

Stainless steels are defined as iron-based alloys, which contain at least 11.5% chromium. A thin, but dense, chromium-oxide film, which forms on the surface of a stainless steel, provides corrosion resistance, and prevents further oxidation. When welding stainless steel, it is important to make the right selection of electrode. The base metal could be austenitic, ferritic, martensitic, precipitation hardening, or a duplex stainless steel. The alloying elements besides chromium could be nickel, carbon, molybdenum, columbium, titanium, phosphorus, sulfur, selenium, manganese, and silicon. Hardening these materials involves different procedures. Naturally, the electrodes used to weld these metals should be selected with an eye to the final condition and use of the parts.

Several problems could be caused by the wrong selection of electrodes. Sensitization, which is caused by chromium carbide precipitation to grain boundaries in a heat-affected zone, reduces corrosion resistance at these local areas. A filler material should be chosen that would reduce the amount of carbon available to combine with chromium. Hot cracking is caused by low-melting compounds of sulfur and phosphorus, which tend to penetrate grain boundaries. When these compounds are present in the heat-affected zone, they will penetrate grain boundaries and cause cracks as the weld cools and shrinkage stresses develop. The problem could be eliminated by reducing the amount of sulfur and phosphorus, but this would increase the cost of the steel. A more realistic approach would be to use an electrode that would provide a microstructure that contains a small amount of ferrite in the austenite matrix.

The selection of electrodes to weld stainless steel depends on what is required by the application. In many cases, the primary consideration is corrosion resistance at an elevated temperature. Then, some minimum mechanical properties will be desired. Other factors to consider, which are mentioned previously, are: resistance to pitting, crevice corrosion, and intergranular attack. If you follow the charts that can be found in many handbooks, matching filler materials can be found for any situation.

Most stainless steels are considered to possess good weldability. They can be joined by several welding processes, including: the arc-welding processes, resistance welding, electron-beam and laser welding, friction welding, and brazing. For any of these processes, joint surfaces and electrodes must be clean. And once again, the selection of filler material is a significant step—especially for permanence of integrity.

HARDFACING

There are stick electrodes for welding most materials. For welding stainless steels, there is a complete line. Cast iron has two classifications of electrodes. Nickel electrodes are used for cast-iron repair work and have either limited or good machinability. A steel electrode, which is nonmachinable, is also available for cast iron. An extruded electrode is used to weld aluminum. But when it comes to hardfacing, Lincoln Electric, one of several welding equipment suppliers, features at least nine different electrodes under the name of "Wearshield," which are used for that purpose.

One is used as an underlay prior to hardfacing with a second "Wearshield" electrode. Some are martensitic deposits and others are austenitic. Some are machinable and others nonmachinable. Some are for use on alloy steels, some on plain-carbon steels, and some for manganese steels. Some can be used to resist abrasion or high impact, and some are similar to type M-1 air-hardening tool steel. The point is that stick electrodes are available for all kinds of hardfacing.

As an alternative to stick welding hard coatings to steels of various types, there is the process of plasma-welding hard metals in powder form to a substrate. Plasma welding of metal powders is accomplished at melting temperatures between 5000 and 50,000°C. Using this method, practically all high melting-point materials could be overlaid. The Tucson Transatlantic Trading Company (among others) markets a series of powders, some self-fluxing, for use on iron base. They produce copper, bronze, brass, aluminum-bronze (with anti-friction properties), and stainless-steel powders. They also make a spectrum of nickel-based powders for overlays to provide corrosion or abrasive-wearing qualities. These welding powders give new life to worn parts.

PLASMA WELDING

All arc-welding processes use an ionized gas or plasma, consisting of free electrons, positive ions, and electrically neutral atoms. However, plasma arc is associated with a process that uses a constricted arc. The main differences are the special features of the plasma torch and how it is used.

A pilot arc is initiated between the electrode and the inside of the plasma torch orifice. This arc is generally less than 10 amps, and it bridges the orifice and the part surface with a narrow plasma column. This column conducts current as it is switched on and off during the weld process. Inside the torch is a low flow of inert gas to conduct the arc current and to shield the weld pool. The power source is strictly dc.

The plasma arc orifice is small, forming a constriction that results in a higher voltage across the arc, which concentrates its power. This provides a narrower, deeper weld than can be obtained by the GMAW method. The plasma arc weld is fast, cool, and has a low heat input, compared to the GMAW method. In the same comparison, the plasma arc is less sensitive to variations in orifice to work distance.

The depth of the welds can be varied by adjusting the orifice gas flow. Narrow, single-pass welds can be made in metals up to ⅜" thick by what is called *keyhole welding*. This process uses a higher gas flow across the orifice, which melts a hole completely through (like in laser welding). The molten metal flowing behind the hole fills it, and although filler material might be used, it isn't required. There is another small advantage to this process. The electrode inside the plasma torch is protected from the environment of weld-surface contamination, which extends its useful life.

There is an area where this process reigns supreme. Some companies that have not invested in an intense energy type of welding system (EB or laser), use the plasma to weld assemblies that probably would be better welded by EB or laser. Knife handles, automatic gun magazines, medical needles, and sundry small components are a few of the parts welded by plasma arc. The ideal type of components for this process are thin sections.

Several companies market plasma cutting systems for cutting steel plate used in ship building, bridges, and other large projects. This plasma cutting equipment moves on rails in a semi-automatic mode.

ARC SPOT WELDING

Spot welding capability can easily be provided by adding an arc timer to standard GMAW equipment and special nozzles to the gun. The major functional difference between arc spot welding and resistance spot welding is that in arc spot welding, the nugget begins to form from the outside of one of the members being joined, rather than at the interface between the two members. This has both advantages and disadvantages when comparisons are made. If arc spot welding is desired for any of a number of reasons, the voltage, electrode diameter, type of gas shielding, electrode extension, and part fit-up are significant factors that must be considered.

ELECTRON-BEAM WELDING

Electron-beam welding (EB) is a fusion process that focuses a narrow, high-velocity beam of electrons onto the welding surface. The resulting molten area is narrow and, depending on the current used, the weld can be exceptionally deep, as much as 6" in a single pass (Figure 25.4). With a reduced current, the weld would, of course, be shallower. The electron beam cannot exist outside of a high vacuum. All EB welding must occur within this vacuum because of the functional requirements of the electrode and the focusing system, and the high temperature of the electron source.

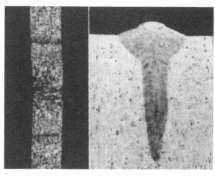

A. **B.**

Electron beam welding as performed by this machine is probably the most versatile metal joining system available today. Unique capabilities include:

−Infinitely variable power changes.

−Ability to produce deep narrow welds and join thin foils with thick sections (up to ¾″ with 6kW) on the same machine.

−High vacuum environment allows clean, high integrity welds on reactive metals.

−Welding a wider range of metals than possible with other fusion welding techniques.

−Continuous beam operation at welding speeds up to 360 inches per minute.

FIGURE 25.4 Electron beam welding. *(Ebtec East, Agawam, MA)*

Small components are often held in compound fixtures to weld many parts per evacuation. The maximum use of the chamber volume should be a goal. It would be expensive to have to pump a vacuum for several minutes for each weld. These fixtures generally require tight tolerances for such details as concentricity and locations. The manipulation of the fixture is done either remotely or by computer program.

The electron-beam (EB) system (see Figure 25.5) is composed of an electron-beam gun, a power supply, control system, motion equipment, and a vacuum chamber. Fusion of base metals eliminates the need for filler metals and the vacuum chamber eliminates the need for shielding gases.

The electron-beam gun has a tungsten filament that is heated, freeing electrons that are attracted to a positive pole or anode. Passing through a hole in the anode, the beam is directed by magnetic forces of focusing and deflecting coils. The beam is directed out of the gun column and it strikes the workpiece.

The potential energy of the electrons is transferred to heat upon impact with the workpiece and it cuts a perfect hole at the weld joint. Molten metal fills in behind the beam, creating a deep, finished weld. The gun and workpiece are manipulated by means of precise, computer-driven controls within a vacuum chamber, thereby eliminating oxidation, contamination, and the need for shielding gas.

The advantages of the electron-beam system are:

1. Maximum amount of penetration with the least amount of heat input, which reduces distortion.

2. Reduces the need for cleaning and finishing (Figure 25.6).

3. Repeatability is achieved through an electrical control system (Figure 25.7).

4. A cleaner, stronger weld is produced in a vacuum, such as that of an EB system. This environment precludes contamination.

5. Exotic metals and dissimilar materials can be welded.

6. Extreme precision because of the programming and the magnification of operator viewing.

7. Approved for salvage and repair by the Federal Aviation Agency.

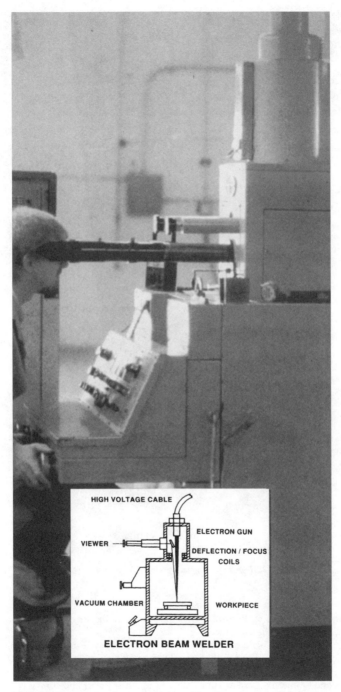

FIGURE 25.5 An electron-beam welder. *(Ebtec East, Agawam, MA)*

FIGURE 25.6 Welding comparison: Conventional welding (left) and electron-beam welding (right). *(Ebtec East, Agawam, MA)*

FIGURE 25.7 A high-production electron-beam welded product. *(Ebtec East, Agawam, MA)*

The limitations of this system are high capital cost and the size of the vacuum chamber, which limits the size of the workpieces.

Most materials available today are weldable by the EB system, including most ferrous metals, such as steels up to and including 40 points of carbon, most light metals (such as aluminum and magnesium), and most alloys (such as titanium, tantalum, molybdenum, zirconium, and hafnium). Even dissimilar metals (such as kovar and copper) can be joined. The following joint designs can be accommodated: all kinds of can welds, bellow welds, feed throughs, lap welds, butt welds, plug welds, tube welds, tube-to-flange and external fillet welds. Because EB is a fusion process, joints can be designed specifically for each application.

To get an idea of the cost of EB systems, Ebtec Corp. in Agawam, MA, lists their electron-beam systems for $50,000 and up. Their laser-welding systems sell for about $75,000. So, price alone should not determine which system to purchase.

LASER WELDING

LASER is an acronym for "Light Amplification by Stimulated Emission of Radiation." This is a device that produces a concentrated, coherent beam of light by stimulating molecular or electronic transitions to lower energy levels, causing the emission of photons. The solid-state laser utilizes a single crystal rod with parallel, flat ends. Both ends have reflective surfaces. A high-intensity light source or flash tube surrounds the crystal. When power is supplied by the pulse-forming network (PFN), an intense pulse of light (photons) will be released through one end of the crystal rod. The light released is of a single wavelength, thus permitting only a minimum divergence (Figure 25.8).

One hundred percent of the light will be reflected off the rear mirror and 30 to 50% will pass through the front mirror, continuing on through the shutter assembly to the angled mirror and down through the focusing lens to the workpiece. The coherent light beam furnishes energy to make narrow welds. Many substances are used in making lasers, including crystals of yttrium, aluminum, and garnet (known as *YAG*). Gases like carbon dioxide can also be used to make laser energy (Figure 25.9).

There are two significant characteristics of this laser light. It is coherent, meaning that all the lightwaves are in step, and it consists of only one wavelength. These factors permit it to be focused on a narrow spot. The edges of the parts to be joined must contact along the entire length in order for the weld to have integrity. For instance, when parts are to be butt welded, it is common to hold them together against a grinding wheel to ensure carefully mating edges. A laser weld is appropriate when a low heat input is necessary.

Recently, the automotive industry has been using what they call "laser blank welding" to butt weld flat sheetmetal pieces of various thicknesses to create tailor-made steel blanks, which are then drawn or stamped. In this type of welding, a coherent beam is focused on a small spot (generally less than ½₂" in diameter). No filler metal is used and mechanical properties are excellent. This is virtually a spot weld, and currently is producing about 12 different parts for automobiles.

Restrained circular joints might present a problem to laser welding. Circular joints might crack if the weld is shallow. If the weld is shallow, the temperature at the top of the weld is much higher than at the bottom immediately after solidification. As the weld metal cools, it begins to shrink, but is restrained by the cooler metal at the bottom of the joint and severe stresses build up. To minimize the stress, it is recommended that the weld penetrate to the bottom of the joint.

There are situations that are tailor-made for laser welding. Parts with complex shapes or exotic materials (such as hasteloy, titanium, or kovar) can be handled nicely by laser welding. Several laser models are not much more difficult to operate than a common drill press. It is certainly not more complicated than conventional welding.

Several large international companies laying pipelines are now using laser systems for the welding and find that they have several worthwhile advantages. Laser Machining, Inc. of Somerset, WI, has specialized in laser welding for a number of years. Now that they have so much laser equipment on hand, they use it for other jobs, too, such as cutting,

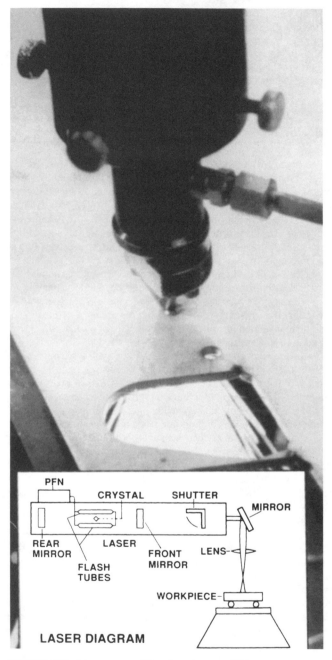

FIGURE 25.8   A laser system. *(Ebtec East, Agawam, MA)*

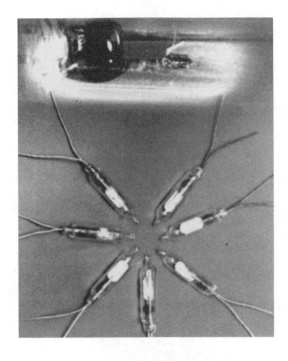

A.

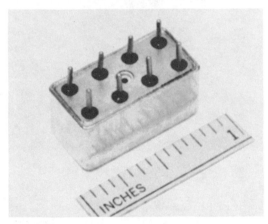

B.

FIGURE 25.9 Two problems solved by laser welding. *(Ebtec East, Agawam, MA)*

scribing, heat treatment, and drilling (holes at an angle). There are quite a few companies in the same position and their names can be found in the Thomas Register.

Laser welding is widely used for welding medical and dental components, where both cosmetic appearance and superior joint integrity are essential. Lasers can easily weld entire joints, sealing off any open gaps where contamination might enter. The control of weld

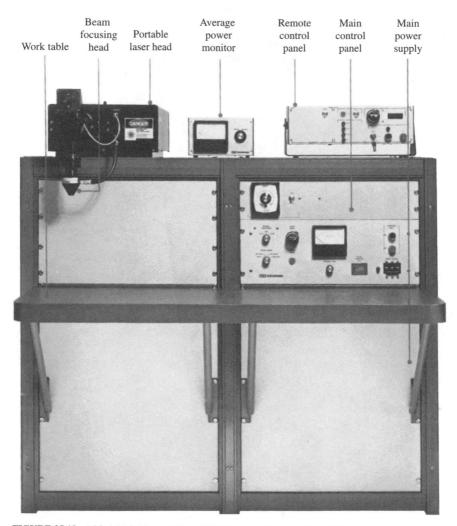

Work table | Beam focusing head | Portable laser head | Average power monitor | Remote control panel | Main control panel | Main power supply

FIGURE 25.10 A Model 1610 laser system *(GTE Sylvania, Mountain View, CA)*

placement makes the laser ideal for precision welds, such as thin metal strips, small hypodermic tubes, and endoscopic components. Little, if any, post-weld finishing is necessary. This paragraph could also have been written about electron-beam welding because the beams they both project are so similar in operation.

Cardiac pacemakers are now being made by both EB and lasers. This product demands a high-integrity weld and no third method has been used. The similarity of the power densities and the reliabilities of the processes make both EB and lasers ideal choices. Other products that are being welded by both processes are small, hollow pistons of 52100 steel, miniature electromechanical assemblies, positive sealing of microelectronics, pressure-sensing devices, welded bellows, and diaphragms of stainless steel and/or beryllium copper.

Two interesting cases of laser welding are shown in Figure 25.9. Figure 25.9A was a production problem in how to trim a miniature thermostat. Conventional welding altered the delicate thermostatic balance of the bi-metallic plate in this thermal switch, causing a high scrap rate. The solution was to fine-tune the temperature range, then spot weld the tungsten wire to the bi-metallic plate after final assembly—by laser welding through the glass enclosure.

Figure 25.9B shows a miniature relay, which (in its original procedure) had been super-heated, damaging the delicate circuitry, creating a severe low-yield problem. Hermetically joining the kovar header on the Cu-Ni body further complicated the production problem. Laser welding solved the problems by producing nearly 400 units per hour. Both of these problems were eradicated on a GTE Sylvania model 1610 Precision Laser (Figure 25.10).

A few years ago, a lot of investigation went into the welding of print bands for computer printers, and both EB and lasers were selected. EB advantages were greater beam power per investment dollar and the ability to weld a wider variety of metals, which are processed in a totally inert environment. The laser advantages were a greater choice of tooling designs, the potential for greater productivity, and the nonvacuum conditions.

Some companies had to be dragged "screaming" to laser welding. They resisted the change until the volume of unusable production increased precariously. When tolerances, materials, or undesirable conditions cause management woes, it pays to consider laser or EB welding as a substitute for a more conventional form.

CHAPTER 26
INDUCTION HEATING

BACKGROUND

For about 70 years, American induction heating companies have been supplying advanced solutions to complex heating requirements. These noncontact heating techniques have yielded superior results in production efficiency, energy savings, and final product quality, when compared to other industry methods. One induction heating specialist, Lepel, in New York City, maintains a 24-hour service hotline, a state-of-the-art laboratory equipped with advanced measuring instruments, metallurgical analytical equipment, and power supplies of varying powers and frequencies (Figure 26.1).

Here, you can try the industry's widest selection of 100% solid-state oscillator-tube medium- and high-frequency induction-heating power supplies. Generating power from 0.5 to 2000 kW with a frequency range of 3 kHz to 30 MHz, this broad range of equipment can be matched to virtually any customer requirement.

Induction heating can be applied to a variety of applications, from heat treating and joining to crystal growing and fiberoptics. The correct power supply will be recommended and you will also get the correct coil design for your application. Induction heating is a noncontact method, delivering localized heat to a variety of products through custom-designed coils. In a typical application, the coil surrounds, but does not touch the customer's product. As the coil is energized by a power supply, the heat is induced in the product via an electromagnetic field with no contamination to the product (Figure 26.2).

Typical applications include: brazing, soldering, tempering, levitation melting, optical-fiber production, deep hardening, surface hardening, annealing, stress relieving, epitaxial growth, crystal growing, zone processing, sputtering, plasma processes, vapor deposition, glass-to-metal seals, plastic coating, plastic joining, curing of coatings, adhesive bonding, shrink fitting, fatigue testing, and melting.

SOLDERING

A primary advantage of brazing and soldering by induction heating is its ready adaptation to production-line methods, permitting strategic arrangements of equipment, and, if necessary, remote control.

FIGURE 26.1 Standard bench-model induction heaters. *(Lepel, Inc., Edgewood, NY)*

Basically, brazing and soldering involve fusion of a joining alloy between the surfaces of metal parts to be joined. If the metal parts are clean, intimate contact is established and the joining materials alloy with each surface. The two methods of joining differ only in the type and melting temperature of the alloy used to make the joint. If the alloy's melting point is more than 800°F, the joint is brazed; if it is less than 800 degrees, the joint is soldered. In soldering, the alloy materials are tin and lead (Figure 26.3)

Soldering alloys for soldering aluminum must be selected very carefully. Because all solders have a potential that is significantly different from aluminum alloys, soldered joints might disintegrate rapidly as a result of galvanic action when exposed to moisture. Most solder alloys are available in wire, strip, and powder. Different sizes are available, which conserves material. In addition, preforms are available in the shape of rings, washers, and any shape desired, for preplacement where they are to be used. Powders are also available, but they are not as popular because of the difficulty of controlling the amount of solder and the displacement of the alloy as the flux spatters during heating.

Thorough cleaning prior to and during heating is basic for successful soldering. Many joint failures can be traced to poor cleaning and inadequate fluxing. Surfaces to be soldered should be chemically cleaned prior to heating and the joint areas fluxed as soon as possible to avoid contamination from handling or exposure. Suitable fluxes prevent oxidation and also help dissolve residual oxides. They improve wetting characteristics, thus promoting free flow upon melting.

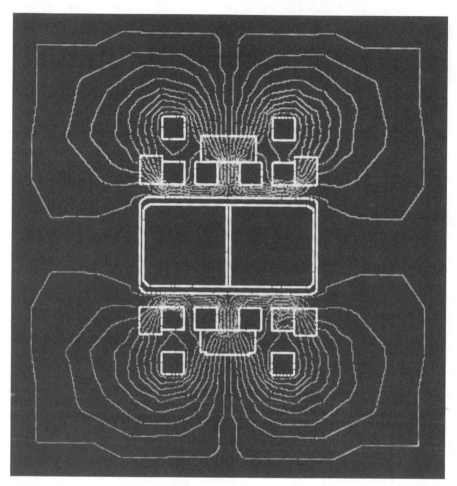

FIGURE 26.2 Advanced computer modeling is used to design custom coil applications. *(Lepel, Inc., Edgewood, NY)*

Alloy	Tin	Lead	Silver	MELTING RANGE °F. Liquidus	Solidus
1	99.8	—	—	450	450
2	62	38	—	361	361
3	60	40	—	370	361
4	50	50	—	420	361
5	40	60	—	460	361
6	30	70	—	500	361
7	—	97.5	2.5	590	590
8	—	96.5	3.5	603	590

* Additional alloys are available. For special problems relating to alloy selection, it is recommended that prospective users consult their suppliers of alloys.

FIGURE 26.3 Chemical composition and melting range of alloys commonly used in induction soldering. *(Lepel, Inc., Edgewood, NY)*

Zinc chloride, aluminum chloride, and ammonium chloride fluxes in paste form are most commonly used in induction soldering because they are the most active. Unfortunately, they also leave residues that are corrosive, electrically conductive, and hygroscopic. These residues must be removed. When this is impossible, use rosin-core solder.

Hydrazine derivative fluxes offer a good solution for induction soldering copper, brass, and other copper alloys. During soldering, while the temperature is in the range of 375 to 600°F, the hydrazine rapidly decomposes and vaporizes. Special proprietary fluxes are available to facilitate soldering of stainless steels, printed circuits, and brasses subject to season cracking. These fluxes are distributed by manufacturers of soldering alloys and fluxes. To avoid any flux spattering, the flux should not be highly diluted.

Experimental work indicates that ultrasonics is an aid to fluxless soldering. Ultrasonic vibrations remove oxide films and clean aluminum alloy parts quite well without flux.

BRAZING

Precleaning is no less important in brazing than in soldering. Surfaces to be brazed should be chemically cleaned prior to heating and the surfaces should be fluxed as soon as possible to avoid contamination. It is interesting that brazed joints involving cast-iron parts are more dependably gas-tight, liquid-tight, and possess higher strength if the parts are first electrolytically treated to remove graphite particles from the joining surfaces before fluxing and heating.

Fluxes containing fluorides and alkali salts, preferably potassium, are generally used for induction brazing. These fluxes, normally used in paste form, become fluid and active below 1100°F, protecting the metal surfaces to be joined. Because most of the fluxes commercially available are of a proprietary nature, it is most satisfactory to use the flux recommended by the manufacturer for their specific alloy.

Techniques have been developed for induction brazing in a reducing atmosphere to avoid the problems of flux and thorough flux removal. In these cases, a purified, dry, reducing gas (such as hydrogen) enters the bell jar at the top, displaces air and surrounds the parts to be heated.

Soldered or brazed joints made by induction heating require special consideration of the heating pattern, the preplacing of the joining alloy, the tolerances between mating parts, the thermal conductivity, and the expansion characteristics of the materials to be joined. Several principles serve as an aid in joint design. First, it is important that all areas adjacent to the joint are above the melting point of the joining alloy. Furthermore, it is desirable to confine the heat to the joint areas in such a way that the area arrives at the joining temperature first. Coil design is thus of paramount importance. See the various coils in Figure 26.4. In situations where the component parts are stronger than the joining alloy, a sandwich braze, utilizing a filler of copper clad on both sides with silver brazing alloy, should be used.

The generators required for induction brazing and soldering are generally small. That is because the joining temperatures are low, the volume of material heated is generally small, and the rate of heating is low to permit the diffusion of heat throughout the joint area. The minimum size of generator for a particular job, can best be determined by trial. However, an estimate of the size can be made by considering the power absorbed by the work. Actually, improper selection of power level is a frequent source of trouble in induction brazing and soldering. The difficulties encountered in the field are traced to too rapid heating of the parts. If one of the parts has poor thermal conductivity, such as stainless steel, you can anticipate a problem.

Induction heating coils are generally coated with Glyptal or a similar substance to protect them against spattered flux. Coils must be cleared periodically in water to remove the flux and prevent electrical shorting between turns. Hand-operated induction soldering and

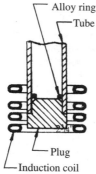

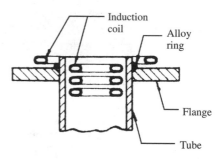

Plug-to-tube joint. Turns of solenoid coil may be spaced for suitable heat pattern depending upon the materials being joined. Tight fit desirable for maximum heat transfer.

Flange-to-tube joint. Internal-external coil. Flange chamfered to hold preplaced alloy.

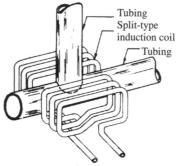

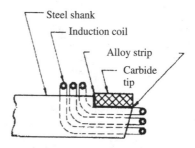

Joining carbide tips to shanks. Open end coil. Alloy strip preplaced as shown.

Tube-to-tube T joint. Split solenoid coil. Alloy ring preplaced internally to provide uniform fillet without excess joining alloy along the sides of the tube.

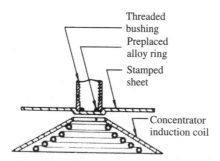

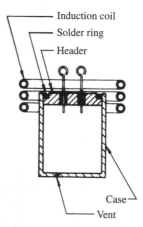

Threaded bushing joined to metal stamping. Concentrator coil localizes heat. Alloy preplaced at base of bushing.

Header-to-case seal. Simple solenoid coil for uniform heating. Preplaced alloy recessed. Vent permits escape of expanded air to avoid leaky joint.

FIGURE 26.4 Typical joint and coil designs for induction soldering and brazing. *(Lepel, Inc., Edgewood, NY)*

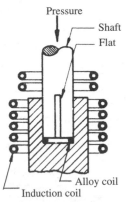

Shaft to fitting joint. Solenoid coil with varying diameter. Alloy preplaced internally at base of recess. Pressure used to seat shaft. Small flat permits escaped of gases.

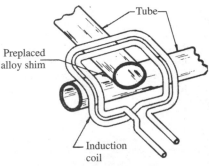

Tube-to-tube joint. L-shaped pancake coil.

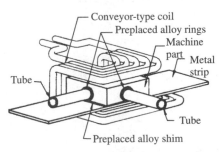

Assembly of components. Conveyor type coil. Alloy shim preplaced between machine part and metal strip. Alloy rings preplaced on tubes.

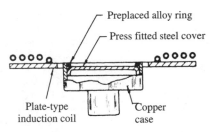

Assembly of formed metal shapes. Plate coil for localized heating. Alloy preform partially shielded from electromagnetic field by case.

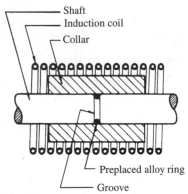

Shaft-to-collar joint. Solenoid coil. Alloy ring preplaced in machined groove.

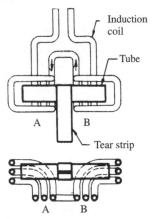

Tear strip to tube assembly. Bucking type coil i.e. reversal of coil current between sections A and B prevents overheating tear strip (.010"). Alloy preplaced beneath strip.

FIGURE 26.4 Typical joint and coil designs for induction soldering and brazing. *(Lepel, Inc., Edgewood, NY) Continued*

brazing fixtures can be extremely simple, although fixtures for mass production might become complex. Because metal fixture parts can heat inductively if placed too close to the coil, nonmetallic materials are used extensively in solder fixture design. These materials include transite, Mycalex, glass-filled epoxy, Diamonite, lava, and ceramics. All are good insulators and have been used successfully, but they vary extensively in their abilities to be formed or machined, and in their costs.

Brazing alloys melt at much higher temperatures than solder alloys and provide high-strength joints that can resist reasonably elevated temperatures without failure. Cutting torches, carbide-tipped tools, ice skates, jewelry, musical instruments, and structural frames are frequently fabricated by brazing. The metals to be joined include carbon and alloy steels, stainless steel, cast iron, copper and copper alloys, nickel and nickel alloys, and aluminum alloys.

Figure 26.5 lists the common alloys used for induction brazing. The selection depends on the melting temperature range, ability to wet the metallic surfaces, the tendency of the material to oxidize or volatilize, and, in some cases, unfavorable metallurgical reactions. With the exception of controlled-atmosphere brazing, copper and brass are not preferred as joining alloys in induction brazing because the high temperature required for making the joint slows production and causes a breakdown of the common fluxes.

Because soldering is a lower temperature process, several tools can be used that wouldn't stand up in brazing. Silicone rubber bands are used to hold parts together and melamine (G-5) or silicone fiberglass laminate (G-7) are used as nests to hold parts during assembly. Nests are used to hold components in alignment. In high-production fixtures, turntables are used in combination with conveyor-type coils.

Most induction heating generators use water to provide cooling for the oscillator tube, as well as for the working coil. Medium-sized generators require 4 to 8 gallons of water per minute for satisfactory cooling. To avoid difficulty, this water should be soft (less than 10 grains per gallon) and have high electrical resistivity (more than 5000 ohm-cms). For situations where high-resistivity soft water is not available, a closed-system distilled-water recirculator has been designed for use with Lepel generators.

Alloy No.	Silver %	Copper %	Zinc %	Cadmium %	Other %	Solidus %	Liquidus %	Color
1	72	28	—	—	—	1435	1435	White
2	15	80	—	—	5P	1185	1445	Gray
3	50	15.5	16.5	18	—	1160	1175	Yellow
4	45	15	16	24	—	1125	1145	Yellow
5	35	26	21	18	—	1125	1295	Yellow
6	50	15.5	15.5	16	3 Ni	1170	1270	Yellow
7	54	40	5	—	1 Ni	1340	1575	Off Wh.
8	65	20	15	—	—	1235	1325	White
9	10	52	38	—	—	1410	1565	Yellow
10	—	99.9 min.	—	—	—	1981	1981	Copper
11	71.8	28	—	—	0.2 Li	—	—	White
12	85	—	—	—	15 Mn	1760	1780	—
13	—	Rem.	—	—	Au 37.5	1770	1825	—
14	(Ni 65/75, Cr 13/20, B 3/5, Fe+Si+C. 10 max.)					1850	1950	—
15	(11/13 Si, remainder Al)					1070	1080	—

* Additional alloys are available. For special problems relating to alloy selection, it is recommended that prospective users consult their suppliers of alloys.

FIGURE 26.5 Chemical composition, melting range, and color of alloys commonly used in induction brazing.* *(Lepel, Inc., Edgewood, NY)*

In the generator, the water is distributed into three channels. One to provide cooling to the oscillator tube, another to provide cooling for the coil, and the third to permit cooling for the circuit components. It is important to remember that induction heating is not just for brazing or soldering materials together. Some things having to do with metallurgy are processed by induction heat also.

Induction heating is used in zone refining, zone leveling, and crystal growing. The machine used by the RCA Semiconductor Plant in Somerville, NJ, has carbon boats containing germanium bars, pulled through successive heat zones to achieve a high degree of purity. Basically the zone processes are related to a fact that has been well-known to metallurgists for years (i.e., the segregation found in alloy castings). Generally, this segregation is a troublesome problem that foundrymen have to live with, rather than being a useful tool. Pfann visualized a method of using segregation to move alloying elements about within an ingot. Such movement is brought about and controlled by a small portion of the ingot, called a *zone*, which is heated to a molten state.

This liquid zone is made to traverse the ingot and has a profound influence on the distribution of the alloying elements. To better understand how alloying elements can be redistributed and controlled by moving liquid zones, examine segregation more closely. As an alloy is slowly cooled from the liquid state, it reaches a point at which a solid begins to form. The solid that forms has a different alloy concentration than the original liquid.

Frequently, the amount of alloying element in the solid is less than that in the original liquid. As a result of this segregation during freezing, it is possible to view the interface at which freezing occurs as a net that has one of two properties. If the alloying element lowers the freezing point, the net acts to pass only a fraction of the alloying element into the solid, causing the solid to have a lower alloy concentration. If, on the other hand, the alloying element raises the freezing point, the net acts to pass the alloying element more easily than the solvent metal, causing the solid to have a higher alloy content than the liquid. This one fact, that a liquid and solid phase in equilibrium with each other, have different alloy concentrations, is the basis of all zone processes.

It is common for the solidification interface, the net, to decrease the concentraton of the impurity by a factor of 100, and, in some cases, the factor is as high as 10,000. It is easy to see why repeated zone passes have resulted in such phenomenal elimination of impurities.

Zone refining is carried out in the same manner if the impurity is of the type that raises the melting point and tends to concentrate in the solid, rather than the liquid. Now, the impurity is swept to the opposite end, and the purist part of the ingot is in the tail, rather than in the head end.

For studies of dislocations and the bulk properties of metals without the complications of grain boundaries and the zone processes of grain boundaries, the zone properties of preparing specimens are especially attractive because they readily lend themselves to the growth of single crystals. The only modification of the zone-leveling process needed to obtain alloy single crystals of uniform composition is to place a single crystal of the same material at the head end of the polycrystalline ingot. The zone is formed by melting some of the single-crystal material, called a *seed crystal*, and a portion of the ingot including the alloy addition. As the molten zone moves, the solid that forms solidifies on the single crystal and has the same crystallographic orientation as the seed crystal, thus extending the seed crystal to the other end of the ingot.

As the melting points of the materials increase, it becomes more difficult to find a suitable boat to contain the liquid without contaminating it. For these materials, the high melting point, reactive metals, a crucible-less method was developed—the floating zone refining. This technique eliminates the boat in the normal horizontal zone processing by floating the zone between the solid portions of a vertically held ingot. This technique greatly extends the usefulness of the zone processes, permitting a wide range of materials to be subjected to the zone-refining and zone-leveling methods.

INDUCTION HEATING AND PLASTICS

Only materials that are electrically conductive can be heated by induction. Because plastics are insulators, it is pertinent to ask what part induction heating plays in the plastics industry. In operation, metals are frequently associated with plastics in the manufacture of end products. For example, many plastic components have metal inserts and plastic coatings are applied to metal wires. Also, plastic adhesives are used to bond metals together.

Plastic applications using induction heating fall into two categories. Heating to develop sufficient plasticity for flow upon application of pressure, and heating of plastic adhesives for bonding by suitable curing. In each instance, time and temperature are important. Because induction-heating cycles are measured in seconds, it is apparent that special consideration must be necessary in the selection of the thermoplastic for optimum plastic flow.

Thermosetting plastics have no place in this arena. All we can work with are thermoplastics. The plastic-to-metal bond here is primarily mechanical, involving the flow of plastic around metal protrusions, and into holes and indentations. It is of benefit to use interference fits between metal and plastics. The plastic adhesives themselves are generally of the thermosetting variety and include accelerators to control the rate of curing. Bonding is primarily by chemical action, but it depends somewhat on secondary chemical forces, such as the van de Waals type. The epoxy resins seem admirably suited for this job, but urea and phenol resins have also been used successfully. Metal inserts are often used in plastic components, and sometimes the bond is plastic to plastic.

INDUCTION SURFACE HARDENING

Lead screws and racks, used extensively in the machine tool industry, require hardening of the threads to minimize wear and maintain accuracy. Unfortunately, the long length and relatively small cross-section subject these parts to distortion during heat treatment. Progressive surface hardening by induction permits rapid production of these parts with a high degree of accuracy. The former method of hardening lead screws was much too long and caused too much distortion. To avoid these problems, the hardening was changed to progressive surface hardening by induction and oil quenching. By this method, each screw was heated in the threads only and immediately quenched. This maintained a relatively cool core in strong condition. See Figure 26.6 for an induction-hardening system, a vertical scanner with dual spindles.

Hedervan, a new water-hardening type of tool steel, developed by Latrobe, is heated to higher temperatures to compensate for the short heating times. The new temperature, anything above 1500°F, did a fine job. It provided case depths of more than 0.072" at Rc 68, while the core remained at Rc 21.

Selective heat treatment of parts is shown in Figure 26.7. The design of the heating coils is an assignment for experts. Anybody can bend a copper coil to fit around a special part, but it requires an expert to foresee what will happen and prepare for it.

Induction heating has many advantages over competitive techniques, such as radiant or convection heat, laser technologies, electron beam, and iron-nitriding methods. Figure 26.8 shows a vertical scanner with dual spindles. Following are some of the many benefits of induction heating:

- *Optimized consistency* Accurate temperature control provides uniform results consistently and with less scrap.

- *Minimized distortion* Site-specific process delivers heat exactly where it is needed, and as rapidly as needed, so that other parts are not exposed to distortion.

FIGURE 26.6 VSS-20 vertical scanner with dual spindles. *(Lepel, Inc., Edgewood, NY)*

- *Improved quality* The contact-free process induces heat in the product without touch-ing it, reducing reject rates.
- *Increased profits* The energy-efficient process converts up to 80% of the expended en-ergy into useful heat to save costs.
- *Maximized productivity* Instantaneous heat permits increased production and reduced distortion.

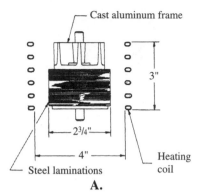

Cast aluminum frame

3"

2³/₄"

4"

Steel laminations

Heating coil

A.

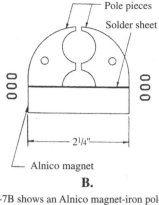

Pole pieces

Solder sheet

2¹/₄"

Alnico magnet

B.

In Fig. 26-7A, the cast aluminum rotor was heated to approximately 350° F while the steel laminations reached 500° F. The coil was loosely coupled in an effort to provide more uniform heating of the rotor.

This method of assembly is rapid and clean, and is ideally suited to an automatic indexing set-up for high production. Assembly by shrink-fitting has a wide application in industry.

Figure 26-7B shows an Alnico magnet-iron pole piece assembly for a dc milliammeter joined by soft soldering. Sheets of soft solder are placed between the pole pieces and the Alnico magnet, fluxed, and then heated locally by induction as shown. Rapid heating, confined to the vicinity of the joint avoids overheating of the permanent magnet.

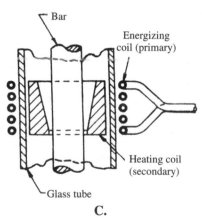

Bar

Energizing coil (primary)

Heating coil (secondary)

Glass tube

C.

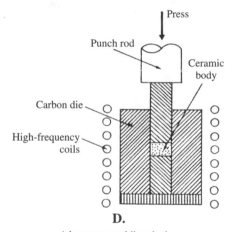

Press

Punch rod

Ceramic body

Carbon die

High-frequency coils

D.

Figure 26-7C shows an arrangement designed to permit heating by a physically isolated but electromagneticaly coupled heating coil.

The heating coil (secondary) is maintained within a ceramic tube which may be provided with a special atmosphere or evacuated during heating of the bar. The heating coil is energized by the primary winding placed outside the glass tube and connected to the generator.

A hot-press molding device.

FIGURE 26.7 Coils used to achieve specific actions. *(Lepel, Inc., Edgewood, NY)*

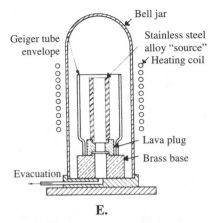

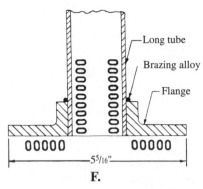

The relative positions of the stainless steel source, glass envelope to be coated, and induction heating coil.

E.

An interesting aspect of this operation is the ease of inspection. Molten alloy drawn into the tube-flange joint will appear at the bottom of the flange signifying adequate flow and a sound joint. A uniform small fillet remains at the top of the flange.

F.

FIGURE 26.7 Coils used to achieve specific actions. *(Lepel, Inc., Edgewood, NY) Continued*

Features/Benefits:

- Medium frequency (10 kHz) output power for small and medium-size parts requiring case depth of 0.060″ to 0.150″ (1.52-3.81 mm).
- Provides flexibility for general purpose heat treating tasks, as well as brazing, shrink fit and other applications.
- Compact, stand-alone system for small shops and workcells.
- Self-contained solid state power supply, cooling system, tuning capacitors and transformer.
- Four tooling options for maximum production flexibility.
- User-friendly control with off-line programming capability.
- Digital timers for heating, quench cycle and quench delay.
- Quality and performance equal to custom-designed induction systems.
- Backed by the world's leading manufacturer of induction systems.

FIGURE 26.8 An Unimat general-purpose induction heater. *(Lepel, Inc., Edgewood, NY)*

- *Elimination of contamination with atmospheric control*   The workpiece can be isolated in an enclosed chamber with a vacuum, inert gas, or reducing atmosphere.
- *Environmentally sound*   Clean, nonpolluting process produces no smoke waste heat, noxious emissions, or noise.

The induction-heating process produces clean, reliable joints in just seconds. It can be integrated into your line with continuous automatic feeds and designed to provide heat at multiple locations simultaneously.

The localized induction-heating process pinpoints the area to be brazed while protecting other areas. Because there is no product contact, there is little opportunity for breakage. In addition, oxidation and discoloration are minimized, often eliminating the need for a controlled atmosphere. Special circuitry can be added: one for rapid heating of the filler to the melting point and the other to allow material flow without overheating. Temperature monitoring and control is available for better process accuracy and repeatability.

- *Soldering*   The accurate temperature control provided in induction heating is a key benefit in soldering applications, which often involve softer materials (if overexposed to heat, these materials can disintegrate).
- *Shrink fitting*   Induction accomplishes shrink fitting much faster than traditional machining or brazing methods. Distortion is also minimized because of site-specific heat.
- *Bonding*   Another application well-served by induction heating, bonding, lends itself to a high degree of automation. A wide range of power and frequency levels can accommodate a variety of bonding applications, including glass-to-metal, metal-to-metal (using epoxy), plastic-to-metal (with or without epoxy), and plastic-to-plastic (using epoxy with magnetic filler).

Induction heating is also ideal for controlled-atmosphere bonding. Metal assemblies can be joined without flux, and the resulting joints are strong and corrosion-free. The process leaves smooth fillets and eliminates fluxing and cleaning costs.

The ability of the induction-heating process to provide heat exactly where it is needed, for the exact amount of time necessary for optimum quality, makes it ideal for a variety of heat-treating applications. Instances of distortion, warpage, and decarburization that can occur with other heating techniques are rare.

- *Annealing*   Providing exceptional accuracy and control, induction heating is perfect for annealing cold work materials. Parts can be annealed with or without a controlled atmosphere, and can even be annealed right on the production line. Localized annealing can be accomplished with special coils.

  Wire and tube annealing can be easily done in an inert (argon, nitrogen, or reducing atmosphere, such as nitrogen-hydrogen or pure hydrogen) by using specially designed chambers. Cooling can be done in the same type of atmosphere.
- *Tempering*   The rapid heating and controlled temperatures of induction heating produce the same hardness as traditional tempering, but in much less time.
- *Other heat-treating applications*   Induction heating allows ease of normalizing, stress relieving, and hardening. Localized heat relieves stress caused by welding, hardening, or brazing to restore ductility and prevent cracking. Use of induction for preheating before welding can minimize such stresses.
- *Hardening*   In hardening applications, induction heating effectively hardens the outer layer to a specified depth while retaining the rest of the part's originally ductility. The depth of hardness can be precisely controlled with various methods (such as changing

power density, frequency, and part-feeding rate). Shafts, tubes, gears, etc., can be hardened without the need for finishing after quenching. A wide variety of power supplies enables the optimum concept for through hardening, case hardening, contour hardening, single-shot hardening, and scan hardening.

CRYSTAL GROWING

Induction heating is particularly suitable for the production of ultra-pure materials, such as the thin films and crystals used in solid-state and laser technology. The ability to provide high temperatures, localize heat, and operate in a controlled atmosphere are crucial for crystal growing, semiconductor applications, and other high-technology requirements.

One line of power supplies is specially designed and adapted for crystal-growing applications. Highly accurate feedback restart and override circuitry, coupled with high reliability, allow uninterrupted system operation, even through short, periodic, power-line interruptions. The frequency can be selected to contain the magnetic field inside of the crucible wall, thus minimizing or eliminating stirring.

- *Epitaxial deposition* A line of power supplies to be used in the epitaxial deposition field for any substrate size from less than 25-mm to more than 760-mm diameter susceptors. Coils are modeled by a computer program for optimum temperature distribution across the susceptor (see Figure 26.2 again).

- *Zone refining* Induction heating is ideal for horizontal zone processing, in which multiple-coil arrangements can produce several molten zones simultaneously in one ingot. Horizontal zone processing is widely used for relatively unreactive materials, which can be melted without contamination in graphite or ceramic boats. More reactive materials are processed with the floating-zone method.

The advantages of induction in continuous heating are many. They include rapid heating, highly accurate temperature control, and the ability to process moving parts as they pass through a coil. The coil design can also be much more compact than competitive furnaces or infrared tunnels.

- *Wire and cable continuous* Normalizing, stress relieving, annealing, hardening, curing, and coating can all be accomplished both quickly and economically on such materials as carbon and stainless steel, copper, aluminum, brass, titanium, superalloys, tungsten, molybdenum, etc. These processes can be performed in conditions from room temperature up to 3000°C, as well as in air or controlled atmospheres.

- *Fiberoptics* In the heating of optical fibers, induction heating excels over other methods, including conventional radiant furnaces. It provides the high temperatures necessary to melt silica glass, 2000°C, using zirconia susceptors, which eliminates the need for protective furnace atmospheres that can damage the fibers. It is also ideal for the furnaces using graphite susceptors inside of protective atmospheres. Resulting fibers have greater strength and better optical properties.

High-frequency induction heating is also ideal for the processing of tubes, rods, strips, blades, and pipes. It is exceptionally effective for experimental work and metallurgical research. For small-scale melting, it can quickly melt ferrous and nonferrous metals, common or precious metals, or alloys with high or low melting points. Melting can be done in any chosen atmosphere—even in a vacuum. Induction heating is well suited to centrifugal, precision casting, vapor coating thin films, and levitation melting. In addition, induction heating's instantaneous heat, localized applications, and cleanliness make it ideal for many specialized applications. In fact, this equipment provides solutions to a variety of unique requirements.

- *Plasma torch* Induction-coupled plasma melting provides the temperatures necessary for scientific research for high temperature, highly refractory metallic materials in contained environments. Induction-coupled plasma produces low gas velocities, minimizing turbulence during crystal growth and providing favorable conditions for chemical reaction. Plasma can also provide light for spectrographic analysis without contamination.

  Lepel's product line covers the entire range of equipment necessary for plasma torch applications. Typical applications are powder metal manufacturing, spray coating, waste disposal, and high-temperature melting of tungsten, tantalum, niobium, iridium, molybdenum, titanium, and a variety of ceramics.

  Induction-heating technology easily adapts to standard mechanical equipment, providing uniform temperatures over the test section for high-temperature fatigue tests.

- *Levitation melting* A specially shaped levitation induction coil melts small quantities of conducting materials while suspending them electromagnetically. Induction levitation offers a quick method of preparing melts of reactive or high-purity metals without the danger of crucible contamination. Melting can proceed in either a vacuum or an inert atmosphere.

- *Catheter tipping* Induction heating allows the edges of plastic catheters to be rounded, or tipped, within seconds. Tips can be formed from 26 gauge (0.018") to 15 mm (0.590") and are used to prevent patient discomfort during insertion.

- *Getter flashing* Getter flashing used in the manufacture of all kinds of vacuum tubes, including CRT and X-ray tubes, is quickly accomplished through induction heating. When getter material is heated to a high temperature, its coating is vaporized. This process effectively scavenges any residual gas in the tube.

- *Thermoshocking* This process is used in the manufacture of "squirrel cage" rotors for induction motors. The rotor, consisting of aluminum bars and magnetic steel laminations is heated to several hundred degrees F, then quenched in cold water, thermally shocking the dissimilar metals. Because of the different thermal coefficients of expansion, the metals break away from each other, thus reducing or eliminating shorts.

- *Power output controls* The power control is of the thyristor type, which is capable of varying the ac main line input voltage to the plate transformer automatically or manually from zero to maximum. Devices are incorporated to prevent damage to the system because of voltage transients. The impedance of the system also limits surge currents during a fault condition to well below the ratings of all solid-state devices. This system contains solid-state rectifiers, which eliminates the need for fuses. Response time is less than 50 milliseconds to prevent saturating the transformer core.

The primary saturable core reactor permits a range of control of the RF power from approximately 0.5% to 95% of generator capacity at full load. This reactor system is connected to the primary side of the plate transformer and provides a smooth wave shape of the RF power output for applications where peak voltages would be undesirable. This power control incorporates solid-state rectifiers. The reactor can be operated either manually or with the addition of a small solid-state driver (SCR), automatically, through an external signal, such as a temperature controller. The water-cooled tank capacitors, tank coil, and tank circuit components are all selected to provide high KVAR service. Numerous devices have been incorporated for protecting the equipment from damage caused by overloading, main-line phase failure and inadequate air or water cooling. Heavy-duty transformers are equipped with primary taps to accommodate a wide range of line voltages. This feature permits proper adjustment of filament and plate voltages, which produce uniform generator performance and optimum tube life.

These units are furnished with a readily accessible tank coil with multiple taps that permit rapid and convenient adjustment to match a wide range of load coils. A wide-range externally variable grid control is furnished, which permits external regulation of the grid current under various load conditions while RF power is on. To activate power, a footswitch and a single-circuit pushbutton is available for automatically timing the predetermined heating cycle. Remote power-control stations are available. A 115-V power supply circuit is provided for control and timing.

All units are furnished with grid and plate current meters, filament voltage, kilovolt, and RF output current meters, which enable the operator to adjust and check the performance of the generator at all times under all load conditions. All access panels are equipped with safety interlock switches that automatically de-energize the power when any panel is removed. For additional safety, one side of the load coil is permanently grounded. All water cooling for the oscillator tube, tank circuit, and load coil is provided through one supply line and one drain line. In areas where the water supply is unsuitable for cooling purposes because of low electrical resistivity, high mineral content, or insufficient pressure, it is highly recommended that a closed-system water-to-water recirculator with heat exchanger be installed within the cabinet.

VSS-20 (Figure 26.6) is a unitized vertical scanning system with dual spindles. It features maximum scan length of 30". It is a self-contained system easily integrated into a workcell. It has power ratings and frequency ranges available for application-matched precision hardening. It requires minimum floor space.

The Unimat system, Figure 26.8, was designed for general-purpose induction heating for heat-treating and quenching applications. Unimat systems can be used for selective hardening to a case depth of 0.060 to 0.150" for processing parts (such as gears, shafts, and cams), as well as for brazing, shrink fit, and similar applications. This user-friendly programmable system is easy to set up and operate.

The model LTF is an induction heating system for catheter tip forming, welding plastic to plastic, or plastic to metal. The system can be equipped with quick-change tooling for multiple-sized catheters. It will radius both tip ID and OD. It will close or taper tubing ends. It can be used to punch holes in tubes (one side or both). It can punch round or irregular shapes, and single or multiple holes at one time.

Lepel water-cooling recirculators are designed for a variety of conditions that are so vital in the operation of induction-heating generators. These closed systems are completely self-contained mobile units using distilled water through a radiator-type heat exchanger. They can be operated anyplace in the plant without the need for external plumbing connections.

The Uniscan IV (Figure 26.9) is a high-production vertical-scan hardening system. It is comprised of two dual-spindle machines, each with independent control integrated through a solid-state transfer switch to a common power supply. Four sizes are offered. Each dual-spindle unit contains its own heat station and NC scanner. It can be programmed to process one or two parts at a time on each tower. Process parameters are entered via keyboard.

The design of the coil in your induction-heating process is a significant factor in achieving maximum product quality. Once a coil is designed, you have a wide choice of power supplies to fit your application. At Lepel, your coil will be designed and built using sophisticated computer modeling. The state-of-the-art software for this design includes finite

Features/Benefits

• Twin dual-spindles scanning units integrated for high productivity.
• Common power supply with solid state SCR transfer switch.
• Four power ratings available with dual frequency capability.
• User-friendly controls simplify setup, changeover and diagnostics.
• 42 in. long x 50 lb./part/spindle flexible workpiece capacity.
• Built to machine tool standards for reliable performance.
• Operate as stand-alone workcell or integrated in FMS system.
• Backed by the world's leading manufacturer of induction systems.
• Worldwide parts and service ensure maximum system uptime.
• Compact. Requires 20-50% less floor space.

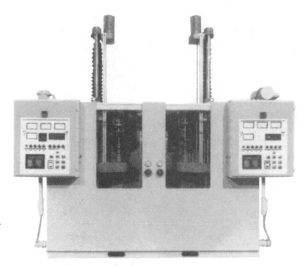

FIGURE 26.9 A four-spindle, vertical-scanning precision induction-hardening system. *(Lepel, Inc., Edgewood, NY)*

element analysis to produce 2-D and 3-D models. Lepel will take your drawings and enter specific process parameters, such as current and frequency, generate numerical and graphic plots predicting electromagnetic fields, and design a coil of optimum size and shape for your needs (see Figure 26.2 again).

Solid-state equipment using MOSFET technology for high frequency ranges are available in several configurations:

• *LSS line* The line (with a frequency range of 50 to 200 kHz) features six tap selector switches for easy tuning and ample power ratings.

• *MR line* Self-oscillating power supplies that operate at 100 to 450 kHz with output power ratings of 500 W to 100 kW. The principle of operation is very much like a vacuum-tube generator, except that MOSFETs are used for power conversion.

• *LST line* This line uses a constant-current circuit. Output powers are available from 35 to 2500 kW at frequencies from 50 to 800 kHz.

• *LCG line* This line is composed of power supplies designed for crystal growing and fiberoptic wire draw applications, with emphasis on precise output control and restart circuits, in case of a power loss.

• *Vacuum-tube floor-model lines* There are two types of floor model lines. One is called *low power*, and the other, although quite similar, has a different power input and RF power output.

There is a vacuum-tube bench model and four types of RWWEX series. There are six types of LSP 12 models and nine types of LUP models. There is, indeed, a model to suit every requirement.

SPECIFIC USES

When Sperry Gyroscope brazes one of the many joints in their Klystron tubes, the joints must meet rigid tests. Most of the joints are furnace brazed, using temperature-graduated assembly methods. But, in the latter stages, numerous vacuum seals and other joints must be made with an eye to localizing heating to the greatest possible degree. For these critical joints, preforms are used in bell-jar brazing. The bell jar eliminates flux, which is a contaminant.

High-temperature fatigue testing using either an R.R. Moore Rotating Beam Machine or a Baldwin Fatigue Tester is done with induction heaters adapted for uniform temperature over the test section.

Shafts and tubes are selectively surface hardened by induction heating. Spindles for Cone automatic lathes are being done this way on a fixture, built by Lepel, which resembles a lathe standing on its head-stock end.

Wave guides are readily assembled using induction heating. Alnico magnet-iron pole pieces are assembled by induction heating. Another common use of shrink fits occurs when motor rotors are fitted to their shafts. In this application, steel-laminated rotors are shrink fitted onto rotor shafts by induction heating. The bore of the rotor is expanded by heating, the cold shaft is pressed into the hot rotor, and the assembly is allowed to cool. Contraction of the rotor upon the shaft provides a shrink fit with ample strength.

Millions of tons of iron castings are produced each year by American foundries and most of that tonnage requires no hardening. Many of these castings would perform better if one or more surfaces could be hardened. Among these machine parts are cams, gears, shafts, and lathe beds. Several types of iron castings are improved by induction heating. The principal difference in these castings is the form in which the carbon is present in the microstructure, whether in the combined state or as graphite. This, in turn, depends on the composition of the casting, particularly with respect to carbon and silicon content, the cooling rate, and prior addition agents to the liquid metal before casting. Perhaps the greatest difficulty in induction hardening is the variation in combined carbon content, which affects the pearlitic matrix of the microstructure.

Specific methods are used when induction-hardening spring steel strip, fine wire, and shafts. Also specific steps are taken in almost any induction-heating process. For instance, bearings can be heated by two coils, one around the outer race and one just inside the inner race. Then, the bearing would be lowered onto the shaft and pressed. Many induction-heating problems disappear when a suitable coil is used.

Metal vapor coating of nonmetallic parts has been on the increase. Very thin, uniform coatings of gold, silver, stainless steel, aluminum, and other metals onto metals, plastics, or glass parts to increase corrosion and wear resistance, or to provide conductive surfaces.

During construction, glass Geiger tubes are coated with a thin film of stainless steel, which has two functions. It serves as a cathode and also acts as a shield to keep out light. Light entering the tube causes ionization and registers in the same manner. The general arrangement for vacuum coating Geiger tubes is shown in Figure 26.10. The bell jar is evacuated to about 0.1 micron to facilitate the transfer of metallic vapor by heating the source.

Some peculiar-shaped special coils are used to heat equally peculiar-shaped products. The coil arrangement used to selectively harden the edges of cutter blades used in agricultural equipment is odd in that the coil assumes the shape of a triangle. The triangle is the blade whose Rc 60-62 edges fall away to a 20-30 interior. A chain manufacturer in

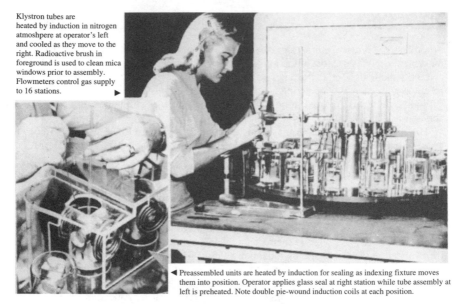

Klystron tubes are heated by induction in nitrogen atmoshpere at operator's left and cooled as they move to the right. Radioactive brush in foreground is used to clean mica windows prior to assembly. Flowmeters control gas supply to 16 stations. ▶

◀ Preassembled units are heated by induction for sealing as indexing fixture moves them into position. Operator applies glass seal at right station while tube assembly at left is preheated. Note double pie-wound induction coils at each position.

FIGURE 26.10   An indexing feature simplifies the sealing operation. *(Lepel, Inc., Edgewood, NY)*

western Massachusetts was having a problem with failures in hardened links. Finally, he read the chapter on selective hardening and tried that approach. He surface hardened each link, leaving the center soft (around Rc 30-25). That was the end of his troubles. The metallurgical heating and cooling should be thoroughly understood.

CHAPTER 27
LASERS

HISTORICAL BACKGROUND

LASER is an acronym for *Light Amplification using Stimulated Emission of Radiation.* In the 35 years since the first semiconductor laser, the industrial ramifications of the tool has grown exponentially. Soon after the invention of the laser in 1960, researchers all over the world began investigating the use of this laser radiation to weld, drill, and cut materials. It took about 30 years to develop the vigorous market share it now holds within the field of metalworking. It has also invaded the field of medicine; ophthalmology has been using it very successfully, and dermatologists and surgeons are now trying it out.

Around this same time, 1960, there was an awakening demand for electron-beam (EB) welding. This exotic process was required for special welds that were deep and narrow, stronger, and more reliable than any other method. EB welding produced a joint with minimal contamination. Its heat source resulted from the collision of a tightly focused stream of electrons, traveling about ⅔ the speed of light, with the workpiece. The depth-to-width ratio is very high. It can weld up to a 1-foot thick material down to a thin sheet or even foil.

But the disadvantages are also significant. The weld must be done in a vacuum of about 0.0001 torr, meaning, of course, that a vacuum chamber is required. A small chamber about 4" × 4" × 8", large enough for one small assembly, would take about 10 minutes to evacuate. A larger chamber with fixtures to hold six or eight small assemblies to be welded at the same time would probably require an hour or so to evacuate. And, of course, that chamber must be well-lined with lead to protect nearby personnel.

Yet, there was a constant demand for EB welding by both the military and commercial industry. This was the niche in the manufacturing arena that fueled the experimentation with lasers as a substitute for EB welding. As we approached the 1970s, many small businesses that performed EB welding as contracted vendors, were beginning to weld, drill, and cut with lasers. Through experience, they soon became as proficient with lasers as they had been with EB welding.

Experts in manufacturing believe that not even half the potential applications for laser technology have been exploited. They feel that the noncontact aspect of this tool is of tremendous significance. There is no wear on the tool, which is the laser beam itself. The workpiece is not stressed by any mechanical force, so dimensions cannot change. The versatility and reliability of the laser are now well-known. And, of course, because they are computer controlled, they are flexible and efficient.

One company that supplies several different types of lasers is A-B Lasers of Acton, MA. The many uses include:

1. Marking and engraving
2. Drilling and perforating
3. Solder reflow
4. Metallurgy surface hardening
5. Cutting
6. Welding
7. Trimming
8. Laser chemical vapor deposition

I remember the first time I tried to laser weld two pieces of $\frac{1}{16}$" thick × 1" wide stainless steel strip. After burning holes along the 1" width, I was disgusted and carried the two pieces to a grinder. Holding them together against the flat side of the grinding wheel, I removed the damages and tried to weld again. This time the weld was excellent, which taught me that the edges had to be flat, parallel, and touching. We progressed by experience; nobody told us how to do the job. We muddled through and acquired the processing knowledge gradually.

Both EB and laser welding are high-energy density-welding processes. Some purists feel that because EB is completed in a vacuum, it must be at least slightly higher in quality. But laser welding cannot be far behind; besides, laser welding, not requiring a vacuum, is much quicker (on the order of 3 to 10 times as fast).

The laser is focused on the work surface to obtain the unique, deep weld, which is typical of laser welding. This is quite similar to the EB type of weld. The laser welds are extremely narrow and, because of the small heat-affected zone, they have excellent mechanical characteristics. Generally, there is no need for filler material.

Lasers have been one of the most misunderstood and underutilized automation tools at our disposal. Engineers, designers, and industrial leaders should know that the potential from 20 years ago is now here. We are currently using the laser for many tasks, and the purpose of this chapter is to relate some of the more common uses. Hopefully, this will instigate some original thinking and solve some industrial problems.

WHAT IS A LASER

A laser device generates an intense, directed, and coherent beam of light. Normal daylight, by contrast, is an incoherent mixture of wavelengths at different frequencies. Today's laser substances are solid-state crystals, gases, and liquids. All lasers work on the same basic principles. A laser consists of the laser substance, an energy source (pump), and a laser resonator. In its simplest form, it is an optical path defined by two mirrors.

An energy source pumps electrons in the laser medium to a certain energy level so that when they return to a lower level of excitation, photons are emitted. Different lasing media produce different wavelengths and all produce monochromatic light. The light waves are then reflected back and forth between the two mirrors on the resonator. This amplifies the light energy and lines it up in parallel and in phase to produce the intense, coherent light.

Of the two resonator mirrors, the rear mirror reflects almost all the laser light, while the output mirror reflects part of the light and outputs the rest. This output is the laser beam, which can be focused, directed, and controlled with great precision.

SAFETY STANDARDS FOR LASER-BEAM PRODUCTS

Because of its high energy density, laser beams can affect eyes, skin, and other parts of the human body. For safety handling, lasers are classified according to the degree of laser intensity. Lasers are perfectly safe when under control and there are standards that must be maintained to ensure this control. European standards are often used in the literature of foreign companies. In the USA, Standard 21 CFR of the National Center for Devices and Radiological Health, is the main requirement that should be read and followed.

21 CFR is the bible for laser products, and it covers any and all applications. When you use a laser, the power required drops it into a classification, and each class has a power limitation. Some classes require the operator to wear goggles and some even have a precaution for protecting the skin. But many applications don't require goggles. It is imperative, however, to read and follow the precautions that every manufacturer sticks on their products.

LASERS FOR WELDING

Since 1990, the quantity and diversity of CO laser applications between 10 and 250 W has grown tremendously. They have gas lifetimes approaching 30,000 hours and cost about $150 per watt. Their small size is also a plus. Consequently, the sealed CO laser has become one of the fastest expanding technologies to hit the marketplace recently. Japan is currently using about 2000 of them. The Ford Motor Company's automatic transmission plant in Livonia, MI, has been using welding lasers for 10 years. They do a lot of round part welding. High-powered CO lasers are welding automobile frames and welding roofs to bodies and are coupled with robots for many operations.

The Department of Defense's (DOD) Advanced Research Projects Agency is using diode-pumped, solid-state laser technology to create high-speed laser machine tools. These tools should raise the quality and lower the cost of manufacturing military products, such as planes, ships, and all sorts of heavy equipment. The Agency is particularly interested in using a diode-pump YAG laser that will deliver its power through fiberoptics. Naturally, all industries will benefit from these trials as soon as they are successfully completed.

Marrying fiberoptics to laser technology will provide new opportunities because the fiberoptics will transmit the beam to the work. The DOD's Agency has introduced new manufacturing means to American industry for more than a decade, and now it is hoped that the investigation of the fiberoptic delivery system will lead to such a demand that the price of lasers will decrease significantly.

Nd:YAG lasers are compatible with a wide range of fiberoptic beam-delivery systems. On the current market are rugged, compact, and sealed laser heads, incorporating state-of-the-art components. Some permit the selection of fibers with smaller diameters than conventional. They have control consoles that contain the microprocessor, laser status indicators, and CRT monitor.

CUTTING MATERIAL BY LASER

Today's industrial production is characterized by ever shorter development cycles and increasing product diversity. Flat-formed and welded sheetmetal is being used more frequently because of its inherent freedom in design. Sheetmetal processing, which meets high technical standards, is being used more often to make inexpensive, lightweight parts. There is also a trend toward smaller lot sizes, which creates an ideal situation for computer-controlled, laser sheetmetal processing.

The Trumpf Laserpress was probably the first combined punching and laser machine when it was produced in 1979. This company continued to combine quality machinery with state-of-the-art laser technology to make flexible, productive, CNC laser-cutting centers. Important features of these centers are:

1. A large enclosed work area.
2. Wide sheetmetal thickness range.
3. Short setup times.
4. Nonwear tooling.
5. Efficient chamber exhaust system.
6. The cutting head moves over the surface of the workpiece.
7. Fast acceleration and speed, regardless of sheetmetal weight.
8. Efficient material utilization.
9. Soft and pliable workpieces can be cut, which is not possible on moving-workpiece systems.
10. The need for clamping is eliminated.
11. Stationary workpieces eliminates most scratches on material surfaces.
12. Pallet exchange system increases machine utilization.

In addition, a few other features are available. When high-pressure cutting is used, it produces non-oxidized burr-free cut edges. A multichamber suction removal system guarantees effective suction removal of cutting residues directly at the cutting point. A capacitance-type height-regulation system maintains a constant distance between the cutting nozzle and the metallic workpieces.

Lasers are now commonly used instead of punches to cut metals. They produce a high-quality finish, and, when driven by a computer program, they preclude the changing of tools. With the ordinary punch press, a dozen or more individual punches might be necessary to complete a sheetmetal chassis. Using the laser, those punches are not required. The computer-controlled beam can create any complicated contour, round hole, or elongated slot or any shape desired, and with high accuracy.

Cutting with a laser causes only localized heating. The part is not stressed and it remains flat, which is important for sheetmetal work. You do not have to purchase punches and maintain them. The laser used for cutting is economical for both small- and medium-production lots. The cutting edge produced by the laser is perpendicular to the surface and the surface is left smooth.

Recently, I visited a machine shop that did contract work for many manufacturers. There were several 3-axis laser cutters with cutting tables up to 48" × 96", which could cut stainless steel up to 0.250" thick and mild steel up to 0.500" thick. There was a 5-axis machine with 72" X-axis, 120" Y-axis, and 32" Z-axis travel. This was used principally to run automobile body panels, and to trim prototypes. It was CAD/CAM compatible.

A lot of this kind of prototype work is being done by vendors for manufacturers because these independents have had thousands of hours of laser experience. Many machine shops have purchased general-purpose lasers and have developed the capability to do laser welding, cutting, and drilling. The Casey Tool & Machine Company of Casey, IL, recently cut a ¼"-thick helical gear in 8 minutes on their 5-axis Laserdyne 890 BeamDirector. This job had previously taken several hours of work on a 4-axis machining center. With the moving-beam system, the parts didn't need much clamping during cutting. A large sheet of 1045 steel was laid on the table and multiple parts were cut from the sheet.

Quasar Industries of Rochester Hills, MI, is a complete machine shop that has 30 lasers. Their first prototype order was from the Ford Motor Company for a plate with eight holes in it. It took the shop three hours to manually program a machine to cut 20 pieces. Later, when 400 additional parts were ordered, the original programming time was recovered. The shop saved 30 to 40 minutes of processing time per part, compared to conventional machining. A few years later, this same shop used a robot with mirrors on an arm to cut three-dimensional parts with a 5-axis laser system. Lasers have evolved into flexible process centers.

MARKING WITH LASERS

Laser marking is noncontact, flexible, permanent, and automatic. The process is computer-controlled and can provide any fonts, logos, or figures desired. CAD system data can also be used. Numerous facilities are available to do contract laser marking if your company doesn't possess the equipment. They will do graphics of all kinds, logo designs, bar codes, date codes, tool specifications, alphanumerics, and serialization. They can mark most materials, such as metals, ceramics, carbides, glass, wafers, IC's, packages, leather, wood, fabric, plastics, and even stone and diamonds.

A variety of parts and products have been marked. Electrical contactor housings have been marked with a wiring diagram and rating information. Reusable containers and barrels (no longer with labels and glues), key caps for mathematical calculators and computers, along with countless plastic products, have been marked. When lack of permanence poses a danger, laser marking is called for. Micrometers and calipers must be marked with absolute accuracy in order to be able to measure in tenths of thousands of an inch. Injection nozzles and saws, chassis numbers on cars and hardened gears, twist drills and ceramic parts are all identified by laser markings.

Each laser-marking workstation, if fully integrated, is capable of performing several functions, such as running rotary work tables, manual or motorized Z-axis tables, and X and Y movements on a standard table. It will have software to program any type of marking required, on-line or at a separate workstation. It will be able to mark around the circumference of a part just as easily as it can mark on a flat surface.

When production quantities require a high-speed marking process, a beam-steered laser is just the thing. The computer's ability to quickly change serial identities works very well married to a laser system. Read/write heads for computer hard-disk drives are about $0.04" \times 0.1"$ in size and they must be serialized with numbers that are $0.007" \times 0.004"$. The noncumulative tolerances of the digits are less than $0.001"$. For the laser, this is a simple and reliable marking job that is being done consistently.

Just-In-Time (JIT) manufacturing has resulted in smaller job lots for many companies. Improved marking capabilities have become important to overall production improvements necessitated by JIT. Utilizing either an Nd:YAG or a CO laser-beam steered system, a wide range of metallic and nonmetallic materials can be marked. The mirrors mounted on high-speed, high-accuracy galvanometers direct the focused laser beam in response to computer commands, which create any text or graphic desired. The process is like writing on paper, but uses a laser beam instead of a pencil.

The heat generated by the focused laser permanently alters the material it hits, often changing the color of the engraving. It is possible to permanently mark products by older, more conventional means than the laser. Some of these traditional systems can also be directed by computer. Some can operate as fast as the laser, but none can provide all three characteristics, except the computer. Also, the requirement to serialize products could jeopardize production schedules if a marking system other than the laser is contemplated. An important consideration is the timing of marking. The laser is used after the product/part is completed, whereas some of the other marking methods are done earlier and this might cause a problem in salvage attempts of parts that don't pass inspection.

The depth of most laser beam marking is only 0.001 to 0.002". But deep engraving of metals sometimes leaves a slight residue of vaporized material along the marks, which can be easily cleaned.

LASER SCANNERS

Around 1965, the railroads were having a difficult time locating railroad cars that were missing, some for a year or more. Generally, they had been parked on sidings near plants and forgotten. This wasted a lot of money for the railroads, building extra cars to make up for those missing. A GT&E scientist, working on a military program, invented the concept of bar code identification while investigating another problem and it was immediately tried by the railroads. Identity stripes were painted on railroad cars and what we now know as *bar-code readers* were established at selected geographical locations. With this method, the position of all cars that passed that point were reported to computers. The computer printouts listed the whereabouts of all cars daily.

In a short time, the concept was tried by supermarkets and other retail establishments. It is now a well-known and accepted identification and is also heavily used in warehouses. Formerly, warehouses were sectioned off and material was kept in definite places. Of course, this wasted a lot of space because most sections were seldom full, although space had to be set aside for possible shipments.

Now, as the materials are received, they are marked with a bar code and simply placed at any convenient spot, without regard for the old sections. When the parts are desired, a computer printout provides the part's bar code and its location in the warehouse. Hand-held laser scanners have been designed to withstand the wear and tear of industrial scanning applications found in warehouses, distribution centers, automotive plants, utilities, and other demanding environments. This type of heavy-duty scanner is durable enough to withstand multiple 6-foot drops. It has features, such as a shock-absorbing rubber boot, spring return trigger, and high-impact optics.

Other desirable features to look for are:

1. Operating temperatures from −22 to 122°F.
2. Memory scanning option for walk-around scanning and downloading to a computer.
3. Hands-free scanning option.
4. Marker beam option for aiming at long-distant labels or scanning in bright ambient light.

5. Models are available for scanning up to 5 feet.

6. An RS-232 computer interface option is available.

7. Direct connection to several different computers or to cash registers.

8. Stands that provide hands-free automatically triggered scanning. These laser scanners are being used in commercial, medical, transportation, education, and other fields of commerce besides retail and warehousing.

LASER DAMAGE-THRESHOLD TESTING

Laser damage-threshold testing is a destructive test that provides the level at which a particular coating or substrate begins to exhibit laser-induced damage. The procedure involves measuring the percentage of sites that exhibit optical damage at various fluence levels. Below the damage threshold, 0% of the sites will fail. At a location well above the damage threshold, 100% of the sites will fail. Within a finite range, however, the percentage of failed sites is a linear function of the incident fluence. By measuring the percentage of failures at various fluences, a linear regression fit can be provided to determine where there is a zero possibility of failure, referred to as the *damage threshold*.

This test is useful in coating research-and-development applications, including evaluations of new coating designs and materials. Laser Damage Certification Measurement is a test to a component, where it is exposed to a predetermined laser fluence, as dictated by the eventual use of the part. These tests are best performed by a full-service independent laboratory.

UNUSUAL LASER APPLICATIONS

Unusual applications are being found for lasers. People are doing surface machining with lasers. The high power density permits fast heating and cooling that can alter a small area metallurgically, producing new types of surfaces with special desired qualities.

THERMOMETER WITH SIGHTING BY LASER

The Raytek Company of Santa Cruz, CA, markets a large range of compact, infrared, non-contact, thermometers with laser sighting. The various models are used to check hot spots on all kinds of engines and perform automobile diagnostics. They are used with HVAC installations, electrical and plant maintenance; asphalt, concrete, and roofing applications; and food processing, storage, and plastic-forming work.

All models feature backlit digital displays for easy viewing in dim conditions. They are useful from −4 to 932°F. Fire departments find these units helpful in determining when a fire is truly out. By locating hot spots, they can avoid having a fire rekindle itself six hours after they think it is out (quite common, both in house fires and forest fires). Locating faulty ballasts in high, overhead lights or smoldering insulation in walls can be expedited by using these thermalert products (Figures 27.1 and 27.2).

FIGURE 27.1 Measuring temperature by laser sighting. *(Raytek, Inc., Santa Cruz, CA)*

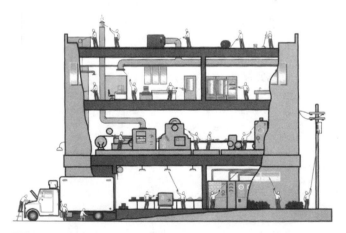

FIGURE 27.2 The uses of a hand-held laser thermometer. *(Raytek, Inc., Santa Cruz, CA)*

COMMON APPLICATIONS

Some of the common laser applications are:

1. Drying, curing, production, and converting of plastics.

2. Glass processing, including tempering, annealing, forming, sealing, and bending.

3. Ferrous and nonferrous metal processing, heat treating, and foundry work.

4. The processing of asphalt, ceramics, pharmaceuticals, textiles, foods, chemicals, and petrochemicals.

5. Research and development.

Figures 27.3 through 27.12 show a variety of uses for the laser. Hopefully, they will trigger original thinking on the part of designers and engineers. Many personnel are reluctant to consider the use of lasers because they lack familiarity with the concept. Remember, you don't have to be an automotive engineer to drive a car. That little old lady from Pasadena makes good use of her car, although she probably doesn't understand how it works.

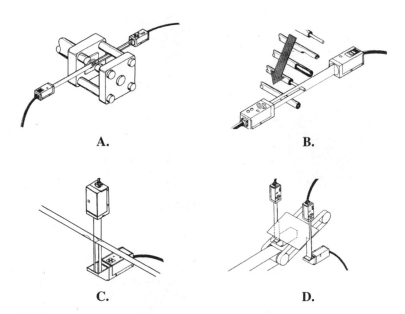

FIGURE 27.3 What can laser beam sensors do? A. Detect remaining plastic at injection machine. B. Detect if wire harness is still on the wire. C. Detect uneven coating on video tapes. D. Detect misplacement of workpieces. *(Ramco Electric Co., West Des Moines, IA)*

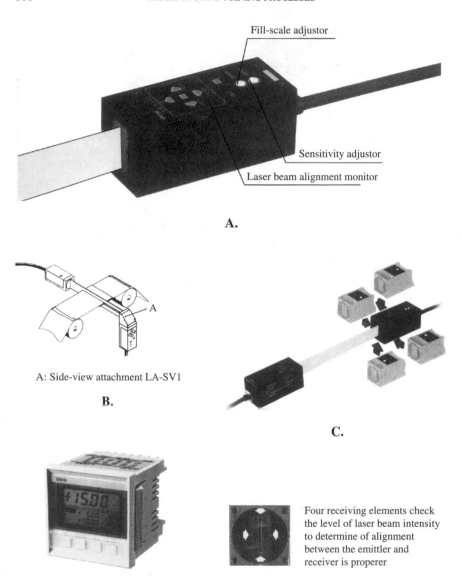

A.

A: Side-view attachment LA-SV1

B.

C.

D.

Four receiving elements check
the level of laser beam intensity
to determine of alignment
between the emittler and
receiver is properer

E.

FIGURE 27.4 Laser beam alignment: A. Laser beam alignment monitor. B. Various mountings are possible with the use of the side-view attachment. C. The laser beam alignment monitor makes alignment easier. D. An optional component. E. Four receiver elements check the level of laser beam intensity to determine if alignment between the emitter and receiver is proper. *(Ramco Electric Co., West Des Moines, IA)*

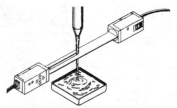

Suitable amount of paint is detected by measuring the
time duration while the sensor is ON.

A.

Object with many holes can not normally be detected by the
through-beam type of sensors.
Since laser beam of **LA511** has 15mm of sensing width, these
objects can be detected reliably.

E.

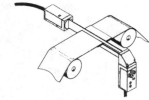

Detects overlapping filter of cigarettes

B.

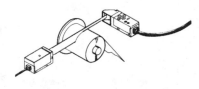

Remaining quantity of sheet material is detected.

F.

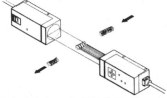

Overlapping spring can be detected by the difference in
laser beam interruption.

C.

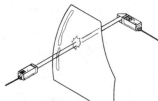

Colors can be differentiated by the difference of laser
beam interruption.

G.

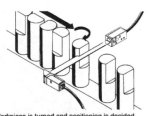

Workpiece is turned and positioning is decided
according to the designated direction.

D.

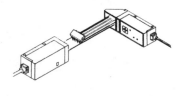

Two types of IC, with 2.54mm pitch and half pitch, can
be differentiated.

H.

FIGURE 27.5 Detection by laser beam: A. Painting detection of dispenser. B. Detection of folding filter of cigarettes. C. Detection of overlapping spring. D. Detection of workpiece positioning. E. Detection of lead frame. F. Detection of remaining quantity at drum. G. Detection of glass colors used in automobiles. H. Detection of ICs. *(Ramco Electric Co., West Des Moines, IA)*

FIGURE 27.6   A CCD laser micrometer. *(By permission Keyence Corp. of America, Woodcliff Lake, NJ)*

Parallel laser beam detects minute targets at long distances.
■ FEATURES
• Parallel laser beam able to detect minute bumps or breaks at 300mm.
• 5 sensor head types available according to application.
• Compact design.
• Analog output voltage.
• High and low output tolerance settings.

■ APPLICATIONS
• Detection of double-fed or mispositioned resistors (A).
• Detection of incorrectly positioned tablets (B).
• Detection of loose bottle caps.
• Detection of differences in screw sizes.

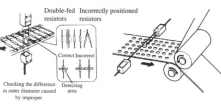

A.   **B.**

FIGURE 27.7   Laser thrubeam photoelectric sensors. *(By permission Keyence Corp. of America, Woodcliff Lake, NJ)*

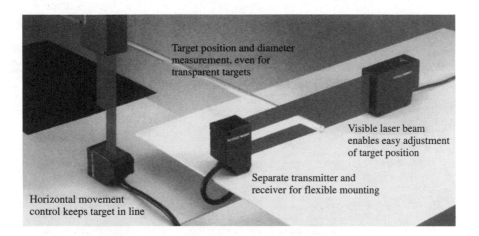

Target position and diameter measurement, even for transparent targets

Visible laser beam enables easy adjustment of target position

Separate transmitter and receiver for flexible mounting

Horizontal movement control keeps target in line

35 x 300 mm measuring area

The visible parallel beam has a wide measuring area enabling edge detection even if target is vibrating or sliding. Targets of various sizes can be measured.

5 µm resolution and 780 c/s sampling rate

Using a CCD image sensor as the receiver gives the VG series superb resolution. The high sampling rate enables high-speed measurement of targets travelling at high speeds.

No external synchronous input is required.

Measurement begins when a target interrupts the beam or at a preset time using the SELF-TIMING function. This useful feature saves time and money.

Easy-to-see display with two-color LEDs

Two-color LEDs (red and green) indicate if the product is acceptable or not. The measurement position display function enables the currently measured part to be confirmed at a glance.

Simplifies product changeover.

TEACH | **Teaching function simplifies tolerance setting.**
–Teaching function
Set a target at the upper or lower limit within the measuring range and press the TEACH button. It's that simple. That limit is now set.

SEL | **8 sets of upper/lower limits can be stored.**
–Tolerance memory function
Upper/lower limits can be easily set or changed using external equipment.

ZERO | **Quick zero setting**
– AUTO-ZERO function
Press the ZERO key and the current displayed value is set to "0". ZERO setting is simple using a reference target.

GAIN | **Corrects fluctuations in light distribution.**
– AUTO-GAIN function
The AUTO-GAIN function quickly corrects fluctuations in light distribution on the CCD image sensor caused by dust on the lens surface or a misaligned optical axis. AUTO-GAIN can also be activated externally. This function maintains stable measurement accuracy.

External I/O Flexibility

Supports intelligent system configuration.
The VG series has RS-232C and analog voltage outputs for communicating with external devices such as personal computers and PLCs. Measurement modes and tolerance settings can be changed externally. This keeps the number of steps to a minimum.

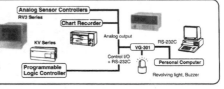

FIGURE 27.8 The VG-301 controller. *(By permission Keyence Corp. of America, Woodcliff Lake, NJ)*

Regular-reflective: LC-2420 & 2430

Ultra-high accuracy, super-small beam spot

Diffuse-reflective: LC-2440 & 2450

High accuracy, long detecting distance, wide range

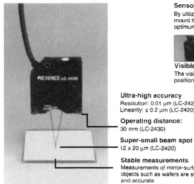

Sensor head position LED
By utilizing this LED, you can easily mount the sensor head at the optimum position.

Visible beam spot
The visible beam spot makes positioning of minute targets easy.

Ultra-high accuracy
Resolution: 0.01 µm (LC-2420)
Linearity: ± 0.2 µm (LC-2420)

Operating distance:
30 mm (LC-2430)

Super-small beam spot
12 x 20 µm (LC-2420)

Stable measurements
Measurements of mirror-surfaced objects such as wafers are stable and accurate

Wide range, long distance detection:
50 ± 8 mm (LC-2450)

Resolution: 0.2 µm
Linearity: ± 3 µm (LC-2440)

Sensor Head

Type	Regular-reflective		Diffuse-reflective	
Model	LC-2420	LC-2430	LC-2440	LC-2450
Measuring range	±200 µm	±500 µm	±3 mm	±8 mm
Operating distance	10 mm	30 mm	30 mm	50 mm
Light source	Visible laser, class 2			
Minimum spot diameter	20 x 12 µm	30 x 20 µm	35 x 20 µm	45 x 20 µm
Resolution (Averaging)	0.01 µm (@512)	0.02 µm (@512)	0.2 µm (@512)	0.5 µm (@512)
Linearity	±0.2 µm	±0.5 µm	±3 µm	±8 µm

New Applications

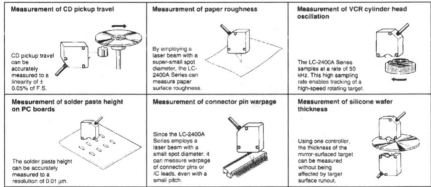

Measurement of CD pickup travel	Measurement of paper roughness	Measurement of VCR cylinder head oscillation
CD pickup travel can be accurately measured to a linearity of ± 0.05% of F.S.	By employing a laser beam with a super-small spot diameter, the LC-2400A Series can measure paper surface roughness.	The LC-2400A Series samples at a rate of 50 kHz. This high sampling rate enables tracking of a high-speed rotating target.
Measurement of solder paste height on PC boards	**Measurement of connector pin warpage**	**Measurement of silicone wafer thickness**
The solder paste height can be accurately measured to a resolution of 0.01 µm.	Since the LC-2400A Series employs a laser beam with a small spot diameter, it can measure warpage of connector pins or IC leads, even with a small pitch.	Using one controller, the thickness of the mirror-surfaced target can be measured without being affected by target surface runout.

FIGURE 27.9 Several different sensor heads. *(By permission Keyence Corp. of America, Woodcliff Lake, NJ)*

Measurement of target diameter or width
NORMAL

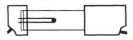

Measures the width (segment) of the shadow projected by the target.

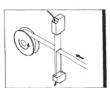

Measurement of resin tube outer diameter

Measurement of video tape width

Measurement of clearance
EDGE #1 LIGHT

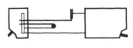

Measures the width of the bright area between the first top edge to the next edge.

For EDGE #2 LIGHT: Measures the bright area between the first bottom edge and the second bottom edge.

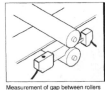

Measurement of gap between rollers

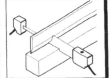

Measurement of cutter blade height

Measurement of amplitude of edge control or shaft eccentricity
EDGE #1 DARK

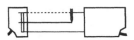

Measures the width of the shadow between the top end edge and the second edge.

For EDGE #2 DARK: Measures the dark area between the bottom and the second edge.

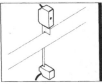

Film edge control

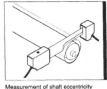

Measurement of shaft eccentricity

Measurement of position or dimensions of complex shapes
SEGMENT

Measures the width between specified edges. This mode is used for measuring complicated shapes or unstable bright/dark edges.

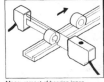

Measurement of bearing inner diameter

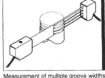

Measurement of multiple groove widths

Operating principle

The red laser beam emitted from the visible laser diode in the transmitter is converged into a parallel beam by the transmitter lens. When the laser beam is interrupted by a target, a shadow with an area proportional to the target size is projected on the receiver. The one-dimensional CCD image sensor scans and calculates the shadow area and position at a 780 c/s sampling rate to measure target dimension or position.

Each CCD bit stores an electric charge proportional to the received light quantity. By scanning the CCD, the sensor detects the shadow area on the CCD and determines the target edge position. The controller calculates the target dimension based on the number of bits between the specified edges. The VG series employs a 5000-bit CCD image sensor (1 bit: 7 μm). Using an original segment calculation method, the VG series offers a 1 μm resolution internally, enabling a 5 μm stable display resolution through averaging.

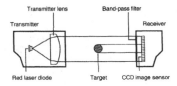

FIGURE 27.10 Four different modes of use for the Keyence VG series. *(By permission Keyence Corp. of America, Woodcliff Lake, NJ)*

340° rotation of marking head
The head can be rotated through 340°. This enables the top, bottom, front and sides of the target to be marked.

Built-in visible red pilot laser for positioning

Up to 100 memory functions can be stored in the main unit

Minimum distance from target for marking: 130 mm

150 cps marking speed
On stationary targets, a 50 x 50 mm area can be marked at up to 150 characters per second.

Sharp letters on moving targets
KEYENCE's unique Galvanometer Scanning System produces the clearest characters ever on moving targets.

Fine character/line adjustment
Character height and line spacing can be set in 0.1 mm increments. Up to nine lines can be marked simultaneously.

Marking direction adjustable
KEYENCE's unique Automatic Scanning Control System enables the marker to print perpendicular to the direction of target travel, a function impossible with conventional markers.

Automatic numbering function
The laser marker is equipped with four independent counters which automatically update the individual numbers.

Automatic date/time functions
The current date and time is updated automatically. And the standard model has a SHIFT CODE function for marking numerals and alphabetical characters which correspond to the desired preset time period.

Change marking data externally through control terminals

External control through RS-232C port

Store marking data externally using an optional memory card

Complete set of safety features
The standard model is equipped with a laser beam shield, laser remote input function, and self-diagnostics.

FIGURE 27.11 A compact laser marker. *(By permission Keyence Corp. of America, Woodcliff Lake, NJ)*

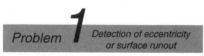

Problem 1 *Detection of eccentricity or surface runout*

"How can we detect eccentricity and surface runout? We've had trouble detecting them with our conventional photoelectric sensors. Do you have an answer?"

Yes. The LX detects eccentricity and surface runout accurately. It detects minute changes in the laser's interrupt area, and its wide beam makes alignment easy.

■ **Major applications**
• Detection of eccentricity in gears
• Detection of surface runout in CDs

Problem 2 *Edge control of sheet*

"To provide uniform edge control of our sheet material, we have to use multiple sensors. This adds greatly to our costs. Do you have a solution?"

Yes. Because the LX's analog voltage output changes linearly with the interrupted area of the beam, one LX sensor can do the work of multiple conventional sensors to ensure uniform edge control.

■ **Major applications**
• Accurate measurement of sheet material width
• Detection of edge parallelism of boards being cut

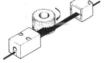

Problem 3 *Multilevel workpieces*

"We have to use several photoelectric sensors to differentiate close tolerance levels of workpieces. Each sensor must be delicately adjusted, otherwise even the slightest error in positioning can cause interference. Do you have a solution?"

Yes. The LX has an internal comparator which allows accurate 3-level differentiation, according to the difference in workpiece size. And inputting the LX's analog signal to an RV3 controller lets you differentiate 4 or more levels.

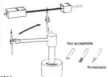

■ **Major applications**
• Differentiating screw diameter
• Differentiating lead length

Problem 4 *Transparent workpieces*

"Our present photoelectric sensors cannot differentiate between varying levels of target transparency. Do you have a sensor that can solve this problem?"

Yes. The LX can accurately differentiate target transparency. It does it by precisely measuring the intensity of the laser beam passing through the transparent target.

■ **Major applications**
• Differentiation of variations in glass color
• Detection of density variations in plastic bottles

FIGURE 27.12 Problem solving with photoelectric sensors. *(By permission Keyence Corp. of America, Woodcliff Lake, NJ)*

P · A · R · T · 5

INJECTION-MOLDING PROCESSES

CHAPTER 28
POWDER-INJECTION MOLDING

Often, a young engineer, when witnessing the plastic-injection molding operation for the first time, would have the thought, "Wouldn't it be great if we could do that with steel?" Well, now we can!

Powder-Injection Molding (PIM), formerly called *Metal-Injection Molding (MIM)*, offers designers a new process for producing steel and ceramic products. Currently, the combination of plastic-injection molding equipment and the powder metallurgy process can produce complex and irregular small shapes in steel or ceramics at lower cost than is possible by any other method. PIM is especially effective for parts smaller than 2" × 1" × 1".

PROCESS OVERVIEW

PIM is the process of mixing elemental or alloyed powders with thermoplastic binders. The binders are selected with two thoughts in mind. First, the binder should deliver the optimum flow characteristics to facilitate uniform distribution of the material in the mold. Second, it should evaporate completely and smoothly during the debinderizing step.

Under moderate pressures and temperatures, the mixture of metal powder and plastic binder is extruded into a mold. The material hardens as it cools and is removed from the mold in what is called the *green condition.*

At this point, the part is larger than the desired shape by the amount of binders used in the mixture. These binders are preferentially and selectively extracted from the "green" part. Generally, a low-temperature furnace treatment slowly evaporates the binder over a 1- to 3-day period.

The parts are then sintered in a high-temperature furnace under controlled atmospheric conditions and a temperature profile selected to remove the remaining binders and present a finished part with the desired physical properties. Obviously, the entire process can be well adapted to microprocessor control (Figure 28.1).

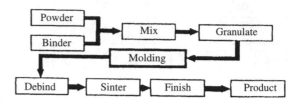

What are the key steps?
What are some of the variants?
Why are there variants?

FOCUS
Materials
Powders
Binders
Mixing
Molding
Debinding
Sintering
Properties

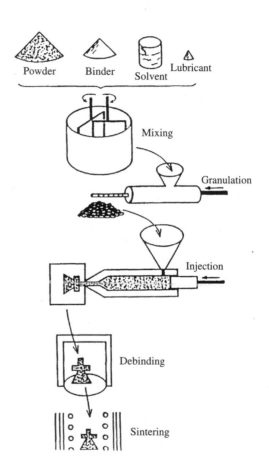

FIGURE 28.1   The PIM process. *(By permission of Prof. R. M. German, Rensselaer Polytech Institute*

PROCESS DETAILS

Manufacturing is a dynamic institution. New means of producing parts, such as PIM, are continually surfacing. This process provides a new freedom in part geometry, which eliminates the restrictions of the more conventional metal-working processes. Sometimes two or more parts can be molded as one.

Although PIM had progressed for about 15 years in the laboratory, it finally entered the arena of competitive production during the 1980s. The point is that, although this is a newly emerging metal-working technique, it has endured a long, arduous birthing procedure.

This process makes intricate parts from a mixture of finely divided metal powders (5 to 10 microns in size) and a binding agent that could be one of several thermoplastics. The mixture ordinarily contains 10 to 35% binders. Because the binders are later removed by heat, solvents, or both, the molds should be precisely oversized to accommodate the 10 to 35% binders of the molded part.

As could be expected, portions of this process are proprietary and are closely guarded secrets for each vendor. At least four different companies license their processes.

Dr. Ray Weich, of Multi-Material Molding in San Diego, CA, holds three patents on the use of thermoplastic binders. One patent uses a solvent-extraction process to remove binders prior to sintering. Several manufacturers have been licensed to use the Weich patents.

New Industrial Techniques Inc., Coral Springs, FL, is using the Rivers Process, licensed by Hayes International, Kokomo, IN. Some people identify this as the *Cabot Process*. This process uses water-soluble methylcellulose as the binder. A 0.25" thick part could be debinderized in less than four hours and possibly enable a full cycle (molding, debinding, and sintering) to occur in one day (one shift).

Two of the PIM process features that create expense are the length of time to debinderize and to sinter. If this 4- to 5-day cycle could be cut to one day, much expense would be avoided. That is why so much experimentation is going on in this area. There have been problems in attempts to accelerate debinderization.

Amax Specialty, Ann Arbor, MI, owns a process that removes the binder with a solvent. This permits debinderization for a ½" section to be completed in four hours. Most binderization processes in current use are based on these processes customized for the individual vendor.

Multi-Material Molding of San Diego, CA, has introduced a continuous binder-removal process.

Recently, a list of 14 different binder formulations was published (Figure 28.2). Each binder has advantages and disadvantages. Most vendors experiment to enhance the former and eliminate the latter.

Several problems have been solved, but some still remain. Because of this, there seems to be specific areas where each type of binder excels. Several PIM vendors use more than one basic type of binder for this reason. But experimentation continues, just as it does in most of the conventional manufacturing methods. The latest work in binder systems raises the debinderization temperature as high as 1250°F.

PROCESS DESCRIPTION

PIM is a recently developed technique that is applicable to small, complex, steel, or ceramic parts. Special machines are used to accelerate and efficiently complete each step. Each part processed must undergo four steps: mixing, molding, debinderizing, and sintering.

Binder attributes	**Example PIM binder formulations**	

Flow characteristics
  Viscosity below 10 Pa s at the
    molding temperature
Low viscosity change with
  temperature during molding
Strong and rigid after cooling
Small molecule to fit between
  particles

Powder interaction
Low contact angle and good
  adhesion with powder
Capillary attraction of particles
Chemically passive with respect to
  powder

Debinding
Multiple components with differing
  characteristics
Noncorrosive, nontoxic
  decomposition product
Low ash content, low metallic
  content
Decomposition temperature above
  molding and mixing temperatures

Manufacturing
Inexpensive and available
Safe and environmentally
  acceptable
Long shelf life, nonhydroscopic
  without volatile components
Not degraded by cyclic heating
  (reuseable)
Lubricity
High strength and stiffness
High thermal conductivity
Low thermal expansion coefficient
Soluble in common solvents
Short chain length, no orientation

Binder 1
  70% paraffin wax
  20% microcrystalline
    wax
  10% methyl ethyl ketone
Binder 2
  67% polypropylene
  22% microcrystalline wax
  11% stearic acid
Binder 3
  33% paraffin wax
  33% polyethylene (wax)
  33% beeswax
  1% stearic acid
Binder 4
  69% paraffin wax
  20% polypropylene
  10% carnauba wax
  1% stearic acid
Binder 5
  45% polystyrene
  45% vegetable oil
  5% polyethylene
  5% stearic acid
Binder 6
  65% epoxy resin
  25% wax
  10% butyl stearate
Binder 7
  25% polypropylene
  75% peanut oil

Binder 8
  50% carnauba wax
  50% polyethylene
Binder 9
  35% polyethylene
  55% paraffin wax
  10% stearic acid
Binder 10
  58% polystyrene
  30% mineral oil
  12% vegetable oil
Binder 11
  96% polyethylene
    (wax)
  4% palm oil
Binder 12
  56% water
  25% methyl
    cellulose
  13% glycerine
  6% boric acid
Binder 13
  72% polystyrene
  15% polypropylene
  10% polyethylene
  3% stearic acid
Binder 14
  4% agar
  3% glycerine
  93% water

FIGURE 28.2 Binder attributes and examples of PIM binder formulations. *(By permission of Prof. R. M. German, Rensselaer Polytech Institute*

Mixing

Metal powders are mixed with binders in a small mechanized bucket, like a cement mixer. The material is pelletized and poured into the injection-molding machine hopper.

Molding

This is done in a standard plastic injection-molding machine, just as though it was a glass-filled nylon part being made. The feedstock is heated, then injected into a mold. Complex geometries are handled routinely because the feedstock fills the volume of any design, just as though it was a plastic part. The same type of multiple slide molds and adaptors, used in plastic-injection molding, extend the capabilities of the process. The part cools and solidifies the same way the nylon part would. Then, it is ejected as a "green" part.

A simple single-cavity mold might cost $4000. But, if production quantities warrant a higher mold cost, a multiple-cavity mold with 2 to 12 cavities (or more) could be utilized. $20,000 spent for a multiple-cavity mold could lower a piece price from $3.00 to $1.50. If volume was 100,000 per year, it would not take long to amortize the additional tooling cost.

The parts are loaded onto large ceramic shelves, each shelf holding hundreds of parts. The shelves, loaded with "green" parts are placed in the debinderizer.

DEBINDERIZING

Debinderizing is done in a low-temperature furnace, which seldom goes higher than 400°F to do its job. If more than one plastic makes up the binder, the temperature will sequentially evaporate the various constituents. The parts (now called *brown*) should by now have shrunk down to nearly the final size. The list of temperatures used and the soak times required make this work ideal for computer control. The ceramic shelves, loaded with parts, are now placed in the sintering furnace.

SINTERING

Here, the parts attain condition *white*. They assume their final critical shape and properties. This is another area that should be computer controlled to maintain proper atmospheric conditions with a temperature profile designed to extract any remaining binders and produce the required density, mechanical properties, and surface finish.

In this step of the process, the final temperature is close to the melting point of the steel material. This diffuses the powder and increases density to about 94 to 97% of wrought material. The shrinkage is uniform along X, Y, and Z axes and, as a result, critical dimensions and complex shapes can be attained.

SECONDARY OPERATIONS

If the parts could be produced without secondary operations, it would be most economical. A good MIM mold designer knows what to expect when he views the part drawing. It is always a good idea to discuss the part configuration with him before the drawing is finalized. This step is worthwhile, whether the part is PIM, sheetmetal, a casting, or even a forging.

But, if some undesirable design characteristic cannot be changed, secondary operations should be considered. Internal threading can be facilitated by positioning the hole and countersinking it in the PIM process. This leaves only a tapping operation, which would not require deburring. In some cases, both internal and external threads can be economically molded.

Normally, a ±0.002" tolerance can be held in a 1" length. But if a tighter tolerance is specified, make the part slightly oversized, then machine that surface. Some parts will require coining or sizing to maintain a consistent quality. During the sintering step, cavities might be found slightly egg-shaped or tapered, instead of round. This situation can be corrected by opening the hole to size with a punch while supporting the part in a cradle (Figure 28.3).

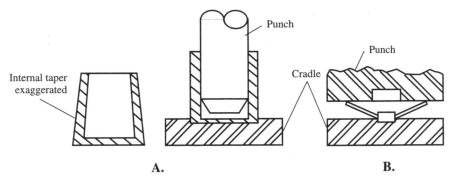

FIGURE 28.3   Coining examples.

Uneven cross-sections will cause thin areas to deform. Once again, a punch/cradle combination will help the situation (Figure 28.3). These two jobs might require some heat before straightening.

Most PIM parts require some type of finishing operation. Simple deburring or breaking sharp edges can usually be accomplished by tumbling in rotating tubs. Occasionally, the material is so soft in its annealed condition that the burrs simply flatten, instead of being removed. In these cases, it might be prudent to quench the parts first and give them enough hardness to preclude flattened burrs. The final use and the material of the part dictate what else will be done.

Sometimes heat treatment can be done immediately after parts are removed from the sintering furnace, while they are still in the high-temperature condition. Some can be hardened simply by quenching. Other materials can be nitrided or carburized. Welding or brazing can be performed.

Usually, there is no need to passivate the stainless steels. Some ferrous parts can be treated with a corrosion-preventative oil. Most conventional surface finishes can be used, such as nickel or chrome plating.

DESIGN CRITERIA

As mentioned, it makes sense to confer with the PIM mold designer before going into production. But often modifications to part design cannot be accommodated. Let the PIM vendor quote your job anyway. Many steps can be taken to meet your specifications—even the difficult ones.

He understands the process so well that he will be able to hold a few tight tolerances one way or another. When his first samples deviate from design dimensions, he will modify the tooling.

It has been said that any part smaller than a golf ball might be a candidate for PIM. Yet, larger pieces have been processed. One manufacturer showed parts ¾" square × 2" long. Theoretically, there is no limit to size. You are limited only by your imagination.

Any geometry that can be achieved by plastic-injection molding can be produced by PIM in steel or ceramic. Visualize a glass-filled nylon part. If the shape can be made in nylon, it can also be done by PIM.

If the design requires a minimum radius, it could actually be made sharp. Holes and undercuts can also be handled the same way you would do it in nylon.

When parts machined from bar stock or forgings are too expensive and where investment castings are too weak, PIM can often get the job done for you.

In a molding process, parting lines cannot be avoided. Wherever the mold halves come together, there will be some extrusion of material. By designing a small flat at this parting line, the flash will not affect the part function (Figure 28.4). Ejector pin marks can be hidden the same way, by using an undercut (Figure 28.5).

Before threading a hole, chamfers should be made at both ends of the hole to eliminate the burrs that occur after tapping (Figure 28.6).

In all types of molding or casting, good designs avoid large differences in material cross-section and long, thin areas. General tolerances in PIM are the same as for plastic-injection molding:

Angular: 0.5 degrees

Linear: 0.003" per inch of length

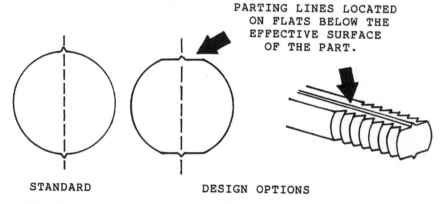

PARTING LINES LOCATED
ON FLATS BELOW THE
EFFECTIVE SURFACE
OF THE PART.

STANDARD   DESIGN OPTIONS

FIGURE 28.4 Parting lines: In many cases, the utilization of flats on contours and external threads can eliminate the possible adverse effects of parting lines without resorting to costly secondary operations.

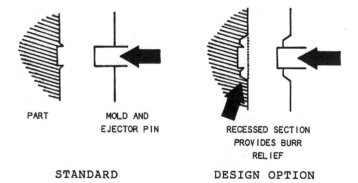

PART   MOLD AND
              EJECTOR PIN

RECESSED SECTION
PROVIDES BURR
RELIEF

STANDARD   DESIGN OPTION

FIGURE 28.5 Ejector pin marks: In a similar fashion, the area around an ejector pin can be recessed or flattened to eliminate any irregularities caused by flash.

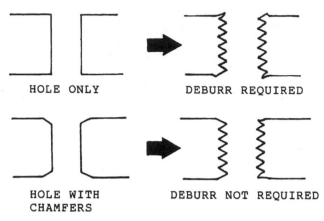

HOLE ONLY   DEBURR REQUIRED

HOLE WITH   DEBURR NOT REQUIRED
CHAMFERS

BURR RELIEF OF TAPPED HOLES

FIGURE 28.6   Internal threads generally must be tapped as a secondary operation. The starting hole, however, is molded to the proper size. Incorporating molded burr relief chamfers or slots along with the molded hole can eliminate the need to deburr for flush fits.

It might be possible to hold one dimension closer, but for consistency, a machine finish might be required.

The following parts have been made by MIM:

- Stainless-steel orthodontic braces
- Tungsten-carbide grinding burrs
- Kovar electronic packages
- Inconel and aluminum-oxide gears and spray nozzles
- Nickel-steel firearm parts
- High-speed printer hammers
- Automobile parts and consumer products
- Niobium rocket thrust chambers. This niobium part is 6" long and works in a 2500-degree F environment.

Production equipment is sometimes designed or modified for specific alloys. The 300 and 400 series stainless steels are being successfully produced by PIM. Nickel iron seems to be the most popular (easiest to handle) alloy in current use. AISI 4340 and tool steels are now being processed where high-strength and hardness requirements predominate. A 3% silicon steel is available for magnetic applications.

The alloy capabilities of PIM seem endless. Apparently, vendors are willing to try any alloy you might need, provided that the metal powder is available. Not all metals are available in fine powder form. However, any powders manufactured can be mixed in any desired combinations.

Most PIM vendors will show tables and charts that compare physical data, such as densities, chemical composition, tensile and yield strengths, elongation, hardness, and magnetic qualities of materials used in PIM production. These will be shown to guide your material selection.

Reading these is a sensible first step. However, for the sake of quality control, you must subject the first PIM samples that you receive, to the laboratory tests that concern you. As in any production, especially those not dealing with wrought material, it is possible to obtain an inferior product. In PIM work, that simply means that some step in the process needs adjustment; this is the time and place to learn that.

In the early days of PIM open competition, one company had some samples made. It would mean a change in manufacturing methods, if successful. It would also reduce part cost from $32 to $2 each. The samples looked beautiful and all dimensions were satisfactory. Nevertheless, some parts were sent to the laboratory for routine physical tests.

The project engineer was surprised to learn that there was a large scatter in the value of elongation, and ultimate and yield strengths. Apparently the 5% porosity of the part was segregated in one area that required great strength. The specimen's break surfaces indicated the existence of a slight sintering problem. In this case, a modification of the sintering step eliminated the problem. But without the tests, this correction wouldn't have been made.

ECONOMICS

This technology favors small, intricate sections. In fact, when small steel or ceramic parts have to be made, PIM has a cost advantage over some conventional processes (such as powder metallurgy, investment casting, and plain machining). It conserves time and material—even the mold material trimmings can be ground and recycled.

PIM material powders cost much more than powder metallurgy powders (PM), because they are finer in size and the quantity made and sold is so small, in comparison to PM. The small size of these powders promotes rapid diffusion during sintering, which permits superior homogenization, thinner walls, sharper edges, and better surface quality (always 63 microinches or less). Ceramic powders are available that are much finer than the 5-microinch steel powders. Because of these fine powders (1 microinch), mirror-like surfaces are now being produced at much lower prices.

Ferrous parts with parallel faces can be made by punch press or conventional PM cheaper than by PIM. Complex steel shapes, however, are natural candidates for PIM.

A local defense contractor had 24 parts cost-estimated by an PIM vendor. None of the tooling exceeded $20,000 and most of those were for multiple-cavity molds. The piece part prices averaged 100 to 600% less by PIM than by any other method. Naturally, the tool cost rises when you go for multiple-cavity molds. But, the piece part cost falls. This is where the economies of mass production must be compared with the cost of tooling, so an intelligent decision can be made.

One of those 24 parts required a quantity of only 1000 pieces. But the tooling cost was easily amortized and money was saved by eliminating difficult machining. Tooling costs and their design are comparable to those of plastic-injection molding. PIM sales in the 1990s are expected to exceed $100,000,000.

HEAT TREATMENT

Ferrous MIM parts of medium- or high-carbon content (0.3% or more) can be quenched directly from the sintering furnace for improved wear resistance and strength. The various other alloys present determined hardenability and, perhaps, a preferred method of heat treatment. An oil quench is often recommended because oil is not corrosive, if absorbed. If the steel parts have a low carbon content, most could easily be carburized.

PIM MATERIALS

Most metals or ceramics available in powder form, smaller than 10 microns, can be used in this process. Many PIM vendors have experience with the following materials:

1. Iron nickel alloys, 2 to 50% nickel
2. Stainless steels, 300 to 400 series and others
3. Pure metals: iron, nickel, molybdenum, cobalt, tungsten, columbium, and niobium
4. High-strength steel alloys, 4130 to 4340
5. A variety of tool steels
6. Kovar, Invar, and Inconel
7. Tungsten-carbide alloys
8. Aluminum oxide (99.9%)
9. Zirconia oxide
10. Silicon carbide
11. Silicon nitride
12. Ceramic and ceramic-metal composites

EQUIPMENT FOR PIM OPERATIONS

In a production facility, the following equipment is generally found:

Raw-Material Treatment:

- Ball mills
- Mixers
- Material granulizers
- Material dryers
- Precision scales

Molding Department:

- Injection-molding machines
- Material granulizers
- Trimming and handling equipment
- Precision scales
- Microscopes

Processing and Sintering Department:

- Batch processor
- Debinderizing furnace
- Sintering furnace
- Ceramic kilns

- Microscopes
- Precision scales
- Handling equipment

Quality-control Department:

- Comparators
- Microscopes
- Precision scales
- Density tester
- Carbon analyzer
- Tensile tester
- Computer
- Measurement equipment

Development Laboratory and Material Testing:

- Thermographic analysis
- Material mixer
- Centrifuge
- Ball mill
- High-temperature vacuum atmospheric sintering furnace
- Sedimentation-measuring device
- Debinderizer
- Metallurgical sample-preparation equipment
- Bench metallograph
- Carbon analyzer
- Scanning electron microscope
- Precision scales
- Viscometer
- Dewpointer
- Gas analyzer

GENERAL OBSERVATIONS

Injectalloy is the Remington Arms Company name for their PIM process. Similarly, *Injectomet* is the Engineered Sinterings and Plastics Company name for their process.

It is a good idea to work with more than one PIM vendor. Most have fairly small operations and their facilities would be swamped by a few large orders. Consequently, to safeguard your schedule, it might be necessary to investigate several vendors.

To avoid disagreements over the delivery of quality parts, it is also a good idea to make the PIM vendor responsible for the completed parts, including all secondary machining, heat treatment, and surface finishing.

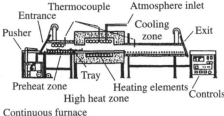

Thermocouple   Atmosphere inlet
Entrance
Pusher   Cooling
                          zone   Exit

                    Tray
Preheat zone   Heating elements
    High heat zone   Controls
Continuous furnace

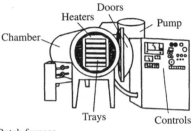

            Doors
    Heaters   Pump
Chamber

        Trays   Controls
Batch furnace

• There are several variables and several approaches
• Interaction between variables powder, component, binder, debinding, chemistry, sintering
• For a given material and component there are many viable pathways, multiple combinations
• No clear demonstration of optimal pathways in terms of properties, quality, or economics
• Need for objective, critical evaluation of processes
• Similar to factorization
  $12 = 12 \times 1 = 6 \times 2 = 4 \times 3$

FIGURE 28.7   Summary on PIM processes. *(By permission of Prof. R. M. German, Rensselaer Polytech Institute*

Each step of the PIM process is constantly being evaluated for improvement possibilities. Originally, the debinderizing and sintering steps caused about 75% of the PIM process expense. Now, this percentage has dropped dramatically and it approaches 30% in some cases. Of course, this depends on the powders used; some are very expensive. Another area of investigation is size. Although small parts are a natural for this process, some companies are trying to expand their size limitations. See Figure 28.7 for a summary of the PIM process.

Currently, about 30 vendors in the USA are available to do PIM work for you. A current list can be obtained from the Metal Powder Industries Federation in Princeton, NJ. The interest in PIM is worldwide. The meetings of the PIM Federation are attended by people from the following countries: England, France, Germany, Italy, Holland, Japan, Canada, India, Taiwan, Poland, Denmark, and Finland.

CHAPTER 29
INJECTED-METAL ASSEMBLY

WHAT IS IT?

Injected-metal assembly (IMA) is a widely accepted process that uses proven technology to assemble small parts using a specialized die-casting technique. Components to be joined are held in position within a tool while a small volume of molten metal, usually zinc alloy, is injected to form a hub at the intersection of the parts. The alloy quickly solidifies, forming a strong permanent lock. With very few limitations, the parts can be of most any structural material and shape. As soon as the assembly is completed, it requires no finishing steps and can be used immediately.

This is a successful, cost-effective replacement for many more-conventional processes, such as crimping, adhesive bonding, press and/or shrink fits, staking, swaging, riveting, soldering, brazing, and welding. All of these processes are excellent in their areas of expertise and need. But, any place where a simple assembly of two or more parts can be replaced by the quick-and-easy, almost-carefree, IMA step, it should be considered.

Injected-metal assembly overcomes many of the nuisances associated with the other conventional processes while providing several benefits. This is a 5- to 10-second process that can produce hundreds, perhaps even a thousand assemblies an hour with no secondary operations required. Machining, tumbling, filing, scraping, cleaning, are no longer done to the assemblies. Some tolerances can be relaxed and component material can often be saved. Additional shapes can be formed simultaneously with the assemblies, and parts can also be combined. Inspections can be accelerated while stocking and handling would almost certainly be reduced. Figure 29.1 shows typical applications. All photographs in this chapter have been supplied by the manufacturer of IMA equipment, the Fishertech Division of Fisher Gauge Ltd., Peterborough, Ontario, Canada.

The IMA process can be performed manually in a semi-automatic fashion or it can be automated with a programmable controller and feed mechanisms (Figure 29.2).

- disc and shaft
- gear and shaft
- electric motor rotors
- pins and plates
- laminated stampings
- metal hub in elastomer or ceramic disk
- abrasive points
- cable terminations
- interval positioning of parts along wire or shaft
- bridged mountings

FIGURE 29.1 Typical applications of injected metal assembly. *(By permission of Fishertech Div., Fisher Gauge Ltd., Peterborough, ON*

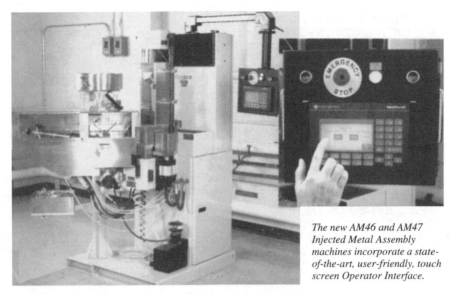

The new AM46 and AM47 Injected Metal Assembly machines incorporate a state-of-the-art, user-friendly, touch screen Operator Interface.

FIGURE 29.2   The injected metal assembly process. *(By permission of Fishertech Div., Fisher Gauge Ltd., Peterborough, ON*

THE SEMI-AUTOMATIC INJECTED-METAL ASSEMBLY SYSTEM

In this mode, the IMA system is comprised of a tool, an operating head, and a machine. The tool holds and positions the components during assembly. The components are loaded manually and the assemblies are removed the same way. The operating head holds and operates the tool, which is made in two halves, one stationary and the other movable. Component size and configuration determine the type of operating head to be used. The machine holds and controls the motion of the operating head and tool. It also holds, heats, and provides a means for injecting the molten metal.

The system cycle is managed by a programmable logic controller (PLC). Pneumatic circuits power the injection unit and the operating head. Each circuit has an independent pressure adjustment, lubricator, flow control valve, and a directional control valve. All power functions, except injection, can be initiated manually for set up and test.

This semi-automatic system is well suited for the assembly of complex or fragile components. These systems are versatile and adaptable to a wide range of assembly tasks and production volumes. By changing inserts, a single tool is able to assemble groups of similar components. The operating heads and tools can be easily and quickly changed to accommodate totally different assemblies.

THE FULLY AUTOMATIC
INJECTED-METAL SYSTEM

In this system, the machine has an automatic control unit with devices to control and monitor the stem functions. Components are automatically fed from vibratory feeders, hoppers, magazines, reels of wire, and pick-and-place units. The automatic cycle of the system is controlled by a PLC, which is contained in a dust-tight cabinet. An integral variable time control allows timer adjustment without interrupting the system. All in-cycle electronic control is done using solid-state devices. Each power function except injection can be manually initiated during set up. An assembly sensor ensures that a system cycle cannot be initiated until the previous assembly has been ejected from the tool.

POSSIBILITIES OF THE PROCESS

Assemblies can be made of both similar and different materials: most metals, glass, ceramics, plastics, and fibers. IMA machines are easy to operate, they are clean and relatively quiet, and the process is nontoxic. But size could be a limiting factor; 6" in the largest dimension is typical. Yet, there are cases where larger assemblies have been accommodated.

Although parts can be assembled to shafts and cables of virtually unlimited length, there is a limit to the size of the holding fixtures for part assembly. Shapes and sizes of an almost-infinite variety are possible, but a basic consideration must be the amount of molten metal the IMA machine can inject. Current models have capacities of one, two, and four ounces.

THE PROCESS

1. The parts to be assembled are accurately positioned in their correct relationship and held firmly by a tool mounted on the operating head of the IMA machine.
2. The machine injects molten metal into a cavity where the parts meet.
3. The molten metal flows in and around physical features (such as grooves, knurls, undercuts, splines, keys, ridges, and other shapes).
4. The metal quickly solidifies in a few milliseconds and shrinks to create a strong and permanent lock.
5. The machine ejects a ready-to-use assembly that can usually be handled safely without gloves. It is suggested that when contemplating this process, visualize the steps of the operation and consider opportunities to improve the design, cut costs, and solve problems.

SPRUE HOLE AND SHEAR

During the operation of an IMA system, the molten metal flows into the cavity through the sprue hole, a tapered passage in the tool. The injected metal in this hole solidifies at the

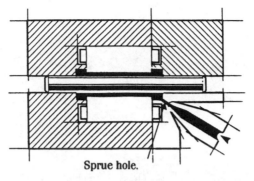

Sprue hole.

FIGURE 29.3 Sprue hole and shear. *(By permission of Fishertech Div., Fisher Gauge Ltd., Peterborough, ON*

same time as the injected metal in the cavity. Therefore, the sprue and injected metal form one piece of hardened metal. Fishertech equipment shears the sprue from the assembly during the operation, resulting in a ready-to-use assembly (Figure 29.3).

SHRINKAGE

As the injected metal cools and hardens, it shrinks slightly, creating a strong lock around the components being assembled. This shrinkage is always toward the center of the injected material. Shrinkage for zinc alloys used in the IMA process is 0.7%. As this shrinkage occurs, the injected metal pulls away slightly from the walls of the cavity, facilitating release. When a disk or gear is being assembled to a shaft, we take advantage of this shrinkage to gain rigidity and resistance to torque forces. Small grooves, knurls, and other physical features are incorporated into the components to be joined. The injected metal shrinks around these features to create a strong, permanent assembly.

FEASIBILITY

This process is almost certain to be the best choice in most of the following assembly operations:

1. If a component can be replaced by a shape that is formed as part of the joint.
2. If the inside diameter of a stamping or the outside diameter of a shaft must be held to close tolerances to achieve concentricity.
3. If several parts must be positioned in close relationship to each other—particularly if they are currently joined in separate operations.

These situations "cry out" for IMA (Figure 29.4).

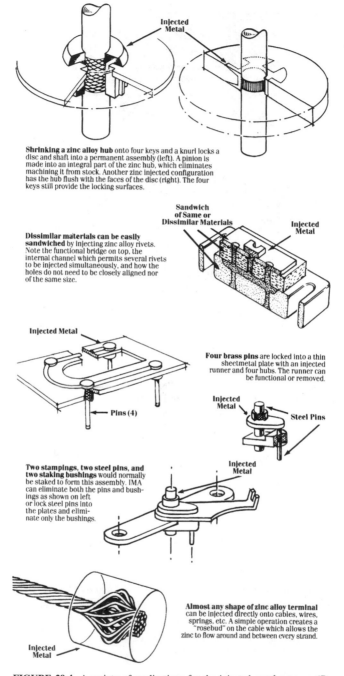

Shrinking a zinc alloy hub onto four keys and a knurl locks a disc and shaft into a permanent assembly (left). A pinion is made into an integral part of the zinc hub, which eliminates machining it from stock. Another zinc injected configuration has the hub flush with the faces of the disc (right). The four keys still provide the locking surfaces.

Dissimilar materials can be easily sandwiched by injecting zinc alloy rivets. Note the functional bridge on top, the internal channel which permits several rivets to be injected simultaneously, and how the holes do not need to be closely aligned nor of the same size.

Four brass pins are locked into a thin sheetmetal plate with an injected runner and four hubs. The runner can be functional or removed.

Two stampings, two steel pins, and two staking bushings would normally be staked to form this assembly. IMA can eliminate both the pins and bushings as shown on left or lock steel pins into the plates and eliminate only the bushings.

Almost any shape of zinc alloy terminal can be injected directly onto cables, wires, springs, etc. A simple operation creates a "rosebud" on the cable which allows the zinc to flow around and between every strand.

FIGURE 29.4 A variety of applications for the injected-metal process. *(By permission of Fishertech Div., Fisher Gauge Ltd., Peterborough, ON*

SAVINGS

If a component could be eliminated and replaced by the IMA process, money would obviously be saved. Also, it will save stock if a smaller shaft diameter can be used because you are eliminating a hub for a gear to be pressed against. In fact, that would save machining and the shaft could be a stock size. The elimination of any secondary operation saves expense. That would include machining, straightening, cleaning and deburring, and using the IMA tool for an inspection station.

STRENGTH

Injected-metal assemblies are permanent because the joints are formed of solid material. It is difficult to obtain meaningful test data because of the variety of materials joined and the many sizes and shapes used. However, reasonable tests have indicated that the joint strength is generally greater than the components joined. Zinc alloy is used for most IMA work, although lead is sometimes used when the assemblies are lightly loaded.

The zinc alloy used in this process develops excellent metallurgical properties. The average tensile strength is 41,000 psi, compression strength is 60,000 psi, Brinell hardness is 82, and Charpy impact strength is 43 lb.-ft. for ¼" × ¼" bar. Pull tests of terminations produced on an IMA system indicate exceptional results. The cable generally breaks under tension, but the terminations show little evidence of strain. The following assemblies were tested to fracture:

Maximum Torque for IMA Gear/Shaft Assembly

Steel shaft diam., in.	Length of knurl on shaft, in.	Length in zinc hub, in.	Result
0.062" (1.59 mm)	0.062" (1.59 mm)	0.120" (3.05 mm)	Fractured shaft
0.125" (3.18 mm)	0.125" (3.18 mm)	0.250" (6.35 mm)	Fractured shaft
0.250" (6.35 mm)	0.250" (6.35 mm)	0.500" (12.7 mm)	Fractured shaft
0.312" (7.94 mm)	0.250" (6.35 mm)	0.250" (12.7 mm)	Fractured shaft
0.625" (15.88 mm)	Diamond knurl	0.800" (20.32 mm)	Fractured shaft

RELIABILITY

Both the process and equipment of injection-metal assemblies have been improved to a point of high reliability and consistency. The results are predictable as soon as initial samples have been made and checked to the required specifications. The zinc alloy is very consistent. Its fluidity, strength, ductility, and shrinkage are accurately predictable. All variables of this process are precisely controlled. Assemblies are produced consistently to requirements. Pressures and temperatures in both the injected-metal and the parts-handling system are individually controlled. And although the timing of each phase of the assembly can be adjusted, it is regulated automatically.

PARTS FEEDING

Parts can be fed to the IMA machine either manually or automatically. The significant factors in this decision are: part configuration, their tolerances and symmetry, their orientation, sharp edges or burrs, ruggedness, part-to-part consistency, and the presence of foreign matter. The size of the equipment and labor costs also play a role in this decision. Manual loading and unloading should be considered if the parts involved are delicate, valuable, or difficult to feed. Sometimes a combination of hand loading and automatic feeding can be utilized to advantage. Assemblies that are not too sensitive can be produced on fast-running, automated machines.

C.

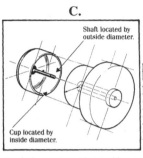

A.

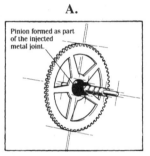

B.

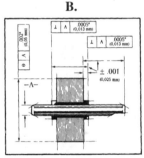

One of the main features of the IMA process is the ability to incorporate useful shapes into the final assembly without adding other components or operations. You eliminate components that are now manufactured separately, by producing them as an integral part of a common injected hub.

Another advantage of the IMA process is that dimensional variations in the injected metal joint are virtually non-existent. You can be sure of consistency throughout very long production runs.

Fishertech engineers can predict with high accuracy the final dimensions of injected metal sections because the alloys used are consistent in their flow and shrinkage characteristics.

The high accuracy of assemblies is primarily the result of the precision with which the tooling holds the components in their correct relationship until the injected metal solidifies.

There is usually no need for a special mechanical feature to be provided for location – virtually any functional feature can be used, including inside or outside diameters, flats and holes.

D.

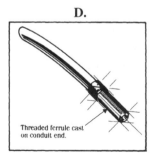

It is possible to form external threads as part of the injected metal joint.

As a parting line must run the full length of the thread, diametrically opposed flats to root diameter depth should be allowed whenever possible.

FIGURE 29.5 More applications for the injected-metal process: A. Combining functions. B. Tolerances. C. Locating by functional features. D. Threads. *(By permission of Fishertech Div., Fisher Gauge Ltd., Peterborough, ON*

ASSEMBLY: Several individual COMPONENTS joined together by the Injected Metal Assembly* process.

CAVITY: The hollow portion of the TOOL that gives the assembly JOINT its shape. The CAVITY fills with injected metal during the assembly process. The areas of the COMPONENTS where the assembly is to take place are located within the CAVITY.

COMPONENT: A discrete part to be joined together with other COMPONENTS during the Injected Metal Assembly process.

DIE: see TOOL.

FIXED TOOL: see TOOL.

GOOSENECK: The portion of the INJECTION UNIT which is immersed in the molten metal in a HOT CHAMBER DIE CASTING machine.

HEAD: see OPERATING HEAD

HOT CHAMBER DIE CASTING: A die-casting process in which the GOOSENECK is immersed in the molten metal.

HUB: see JOINT

IMA*: see INJECTED METAL ASSEMBLY*.

INJECTED METAL ASSEMBLY*: The Fishertech* process of using a die-casting method to assemble COMPONENTS.

INJECTION UNIT: The mechanism within the Injected Metal Assembly MACHINE which pumps molten metal from the MELTING POT into the CAVITY.

JOINT: The portion of the ASSEMBLY formed by the solidification of the injected metal. Also referred to as the HUB.

MACHINE: The portion of an Injected Metal Assembly SYSTEM that holds and controls the motion of the OPERATING HEAD and the TOOL during the assembly process as well as providing a means for holding, heating and injecting the molten metal.

MAIN SLIDE: The portion of the OPERATING HEAD on which the MOVEABLE TOOL is mounted. The motion of the MAIN SLIDE closes and opens as well as locks and unlocks the TOOL.

MELTING POT: The portion of the Injected Metal Assembly MACHINE that holds and heats the metal to be injected.

MOVEABLE TOOL: see TOOL.

NOZZLE: The part of the Injected Metal Assembly MACHINE that provides a passage through which the molten metal flows from the INJECTION UNIT, through the SPRUE HOLE, into the CAVITY.

OPERATING HEAD: The portion of an Injected Metal Assembly SYSTEM that holds and controls the motion of the TOOL during the assembly process.

PART: see COMPONENT

PARTING LINE: The place on the JOINT where the mating surfaces of the FIXED TOOL and the MOVEABLE TOOL meet. Also, the mating surfaces of the COMPONENTS and TOOL.

PLUNGER: The portion of the INJECTION UNIT used to pump molten metal out of the GOOSENECK into the CAVITY.

SHOT: The volume of metal pumped by the INJECTION UNIT into the CAVITY during a single cycle.

SHRINKAGE: The contraction of the injected metal in the CAVITY during cooling and solidification.

SLEEVE: The cylindrical portion of the GOOSENECK in which the PLUNGER moves during the injection of molten metal.

SOLIDIFICATION: The change of the injected metal from a liquid to a solid.

SPRUE: The portion of injected metal that solidifies in the SPRUE HOLE.

SPRUE HOLE: The tapered passage in the FIXED TOOL through which the injected metal flows from the NOZZLE into the CAVITY.

SYSTEM: The combination of MACHINE, OPERATING HEAD and TOOL arranged to assemble COMPONENTS using the Injected Metal Assembly process.

TOOL: The portion of an Injected Metal Assembly SYSTEM that holds the COMPONENTS in precise relationship to each other during the assembly process. The TOOL is comprised of two halves; the FIXED TOOL and the MOVEABLE TOOL. The FIXED TOOL is firmly fastened in the OPERATING HEAD. The MOVEABLE TOOL is mounted on the MAIN SLIDE and moves into and out of the FIXED TOOL during the process. The TOOL contains the CAVITY into which the injected metal is pumped.

*Trademarks

FISHERTECH®
DIVISION OF FISHER GAUGE LIMITED

P.O. Box 179, 194 Sophia Street, Peterborough, Ontario, Canada K9J 6Y9
Phone: (705) 748-9522 Telex: 06-962838
Facsimile (705) 748-6312

FIGURE 29.6 Injected Metal Assembly glossary. *(By permission of Fishertech Div., Fisher Gauge Ltd., Peterborough, ON*

THE IMA MACHINE

Whenever possible, IMA machines are installed right in the assembly line where they can help reduce material-handling expenses. All types of IMA equipment are easy to run and maintain. Operators need only a minimum of skill and can be trained in a brief period to achieve optimum performance. An installation requires no unusual services or costs for such things as environmental or pollution safeguards. It does require 220-volt electrical outlets, air for pneumatics, and cooling water. The machine stands 79" high and occupies 35 square feet. It does not need a special foundation or ventilation equipment. These machines can be used to perform one continuous assembly procedure or they can be quickly changed to accommodate other assemblies.

The IMA machine has a tool and an operating head. The head holds the tool, opens and closes it, and carries it to and from the injection nozzle. Operating heads can be changed in about 15 minutes. Add setup and testing time and you could be running the new job in 30 to 60 minutes. The tool itself could also be changed, if necessary, which takes just a little longer.

In many cases, parts that a company buys or makes become redundant because they can be formed in place with the IMA process. If the examination of an unfinished assembly shows parts that become integral when the assembly is complete, the IMA process should be considered. This process is ideal for integrating two or more parts in an assembly. Parts can be located and positioned by their functional surfaces. Injecting molten metal as a hub or as rivets accommodates inaccuracies, poor finishes, and burrs. In fact, using these functional surfaces as locating features could eliminate some inspection requirements because the assembly tool becomes the checking fixture. Figures 29.4 and 29.5 show a variety of uses for the IMA process. Figure 29.6 is a glossary of terms used in IMA.

P · A · R · T · 6

SURFACE-TREATMENT PROCESSES

CHAPTER 30
TITANIUM-NITRIDE (TiN) COATING

HISTORY

Physical vapor-deposition (PVD) thin-film coatings were introduced to manufacturing engineers in charge of cutting tools in the early 1980s and quickly produced dramatic and revolutionary results. The use of PVD-coated cutting tools spread rapidly throughout the metalworking industry because, in a short time, performance improvements could be definitely noticed. In most cases, the proof was visible in just hours.

Now, we are beginning to realize the vast array of precision metal parts that can benefit from the properties of PVD coatings. The hardness and lubricity of this process can eliminate both wear and release problems. And, of course, the thinness of the coatings preclude the necessity of reworking parts after coating (Figure 30.1).

In these days of intense industrial competition, the cost savings and productivity improvements associated with the titanium-nitride coating of cutting tools are too important to ignore. If tool engineers want to act aggressively, they would use this coating in every situation possible.

A TiN coating has a hardness of about 80 Rc and a coefficient of friction about one third that of the steel substrate. The coating protects tools from abrasive wear, and the lubricity reduces friction, heat, and the built-up edge, which forms on tool edges during use. These tools generally pay for themselves the first time used.

In the years since the introduction of coatings, the companies involved have performed much research. They now have several different products—each with an area of performance where they outshine the original coating.

MATERIALS, PROPERTIES, AND BENEFITS

The information in this section has been furnished by the Balzer Company. Other companies have much the same products, which should be compared by the astute engineers. The

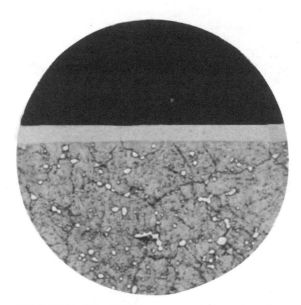

FIGURE 30.1 This 1000X photomicrograph shows the uniform thickness of Balzers TiN coating on M-2 high-speed steel substrate material. *(Photograph courtesy of Balzers Tool Coating, Inc., Agawam, MA)*

TiN coating is gold/yellow and 1 to 5 microns thick with a microhardness of 2300 Vickers (approximately 80 to 85 Rockwell C). It is thermally stable up to 1100°F and has a friction coefficient of 0.4-TiN/steel. This coating is useful on a wide range of cutting tools. It provides 3 to 8 times longer tool life along with higher speeds and feeds. It is particularly effective on simple molds and wear parts.

The titanium-carbonitride (TiCN) coating is blue/gray, 1 to 5 microns thick with a microhardness of 3000 Vickers (approximately 90 Rockwell C). It has a thermal stability of up to 750°F and a friction coefficient of 0.4-TiCN/steel. It is particularly useful when machining abrasive materials like cast iron and hard metals. It is excellent when machining copper and copper-based alloys at high speeds and feeds. Its wear resistance and toughness is superior to TiN.

The chromium nitride (CrN) coating is silver/gray, 1 to 5 microns thick with a microhardness of 1750 Vickers (approximately 78 Rockwell C). It has a thermal stability up to 1300°F and a friction coefficient of 0.5-CrN/steel. It is especially effective for cutting and forming copper, nickel, monel, and titanium metals and alloys. It has an enhanced thermal stability and oxidation and corrosion resistance. It has an improved performance with molds for diecasting aluminum and zinc alloys. Figure 30.1 shows the uniform thickness of TiN coatings and Figure 30.2 indicates an ideal situation for CrN coating.

The CrN coating will outperform titanium-based coatings in applications, such as resisting adhesive wear that occurs in higher-temperature diecasting, semi-hot forging of brass and in glass fabrication. It is harder than conventional chrome plating and requires no finishing operation, as does plain chrome plating in mold applications (Figures 30.3 and 30.4).

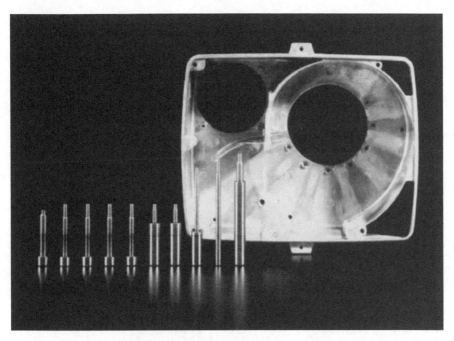

FIGURE 30.2 Chromium nitride (CrN) coated cores for aluminum die casting of oil burner housing. *(Photograph courtesy of Balzers Tool Coating, Inc., Agawam, MA)*

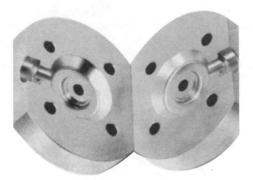

FIGURE 30.3 Plastic injection mold cavities. *(Photograph courtesy of Balzers Tool Coating, Inc., Agawam, MA)*

MAINTENANCE

Even after a TiN-coated cutting tool is resharpened, the tool life is still twice that of an uncoated tool. Coated tools are sharpened the same way as uncoated and they should be

FIGURE 30.4 Coated die casting cavity. *(Photograph courtesy of Balzers Tool Coating, Inc., Agawam, MA)*

taken out of service and resharpened before there is too much damage to the coating. It is very beneficial to use the coatings on molds not just for the increase in productivity, but also because of the increased lubricity. Wear points with little or no lubricity, such as shafts, cores, and especially bottle cap molds (where each part must literally be unscrewed from the mold) present problems. Therefore, wear warnings are extremely significant in these situations (Figures 30.5 and 30.6).

If a change in color is seen, the operator should call maintenance for a decision on re-coating the shaft, core, or the interior area of the mold. This color change is a most useful virtue of the coating process. Experience bears out the claim that recoated tools work better than they did when first coated. There are several speculations for this phenomenon. One reason might be the slight increase in coating thickness—especially where minor high spots wore more quickly than the rest of the area.

The TiN coating reduces cutting forces by about 17% in most drilling applications, and increases tool life 4 to 9 times machining cast iron, stainless steel, nonferrous metals, and carbon steels. TiN-coated taps reduce cold welding when tapping mild steel and reduce friction when tapping stainless.

TiN coating on thread rolling and similar tools, which move metal, rather than cut it, have the same saving potential. That is also true for countersinks, counterbores, reamers, end mills, milling cutters, hobs, broaches, turning tools, and saws. It might be necessary, on some occasions, to change grinding angles. But coating vendors can advise when this is required.

When the depth of cut is slight, it might occasionally require experimentation. The cutting edge might not bite into the work the same way as an uncoated tool would.

Longer tool life is reported when punching galvanized steel, carbon steels, bronze and soft iron. The punches have less material build-up on their sides and less galling. Stripping action is also improved. There is no doubt that TiN coating increases tool life, but this is only a minor feature in running presses. Downtime is the large, invisible cost that should be factored into the tool cost. Increasing the run life of a punch, significantly reduces the rate of tool change. In addition, load tests on presses indicate that TiN-coated tools do the same work with less strain on machine components. Reduced strain permits a smaller press to do the work of a larger one.

In semi-warm operations, excellent results can be expected. But coating forging dies that are used at high temperatures is not recommended. The thermal stability range of the coating should not be exceeded. Punches and forming tools made of D2 tool steel should be heat treated with 3 tempers at 950°F. This ensures dimensional stability and no loss of hardness when the tool is subjected to the coating-process temperature. Applying TiN to molds is very beneficial. Material build-up, like rubber or plastic, is easier to remove.

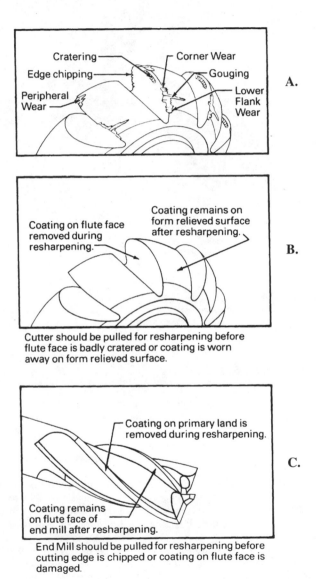

FIGURE 30.5 Different types of tool wear and titanium nitride coating.

FIGURE 30.6 Many types of cutting tools can be coated or recoated with TiN. Coating and recoating special-form tools is worth considering because the tools last longer between regrinds and will last through more regrinds, making these tools more economical. *(Photograph courtesy of Balzers Tool Coating, Inc., Agawam, MA)*

PREPARATION FOR PROCESSING

Certain steps have to be taken in the manufacture of tools. Cutting, forming, and punching tools generally are machined, heat treated, ground, and sometimes polished as integral steps in their preparation. Improper material, as well as inconsistent procedures, can adversely affect surface coating. To take full advantage of the process, tools must be properly prepared before coating. The following steps are not necessary in all cases, but they are strongly recommended.

Hardening should be carried out in a vacuum furnace with positive quenching capability. Tempering should be done in either a vacuum or inert gas atmosphere to preclude the formation of oxide layers. If tools are heat treated in air, salt baths, or quenched in oil, sufficient stock must be allowed so that all heat-affected material, such as decarburization or oxidation, can be removed.

Both the machining and grinding operations must be performed with sufficient and proper coolant to prevent overheating of the tool surfaces. Leaching must be prevented by adding inhibitors and surfactants to the coolant. Free-cutting wheels, preferably CBN or diamond, are best for this situation. Of course, the wheels should be dressed often enough to prevent overloading and smearing of metal.

Stress relieving is best done in a vacuum or an inert atmosphere. Even a slightly discolored surface because of overheating might cause poor adherence of the coating to the substrate.

Polishing is generally done to punches, dies, and molds to improve the surface smoothness. This step often smears metal to obtain the desired smoothness. Since smearing will trap dirt and create contamination, both of which cause poor adherence, it must not be permitted. Polishing should be done with water-soluble, silica-free compounds. Normal progression from coarse- to fine-grit polishing compound should be used to secure the required microfinish.

PROCESS DESCRIPTION

The best tool surface for coating is finely ground, clean, and with no prior treatment. That means no black oxide or any other surface treatment. A coating vendor will have to advise if you have doubts. Before polishing molds or any other tools, check with the coating vendor because some polishing materials are difficult to remove and the vendor will probably recommend some water-based polishing solutions. Figure 30.7 actually shows the deposition of titanium.

In these coating processes, the tools to be coated must be physically and chemically clean before they are positioned in fixtures, which are the cathode of a high-voltage system in a reaction chamber. The chamber is evacuated and charged with argon for a process called *sputter cleaning*. The positive argon ions are propelled by a high-voltage field to blast the tools so that they become atomically clean (Figure 30.8).

The first step in the PVD process involves placing physically and chemically clean tools in fixtures, which become the cathode of a high-voltage circuit in a reaction chamber. The chamber is evacuated and charged with argon. In a process called *sputter*

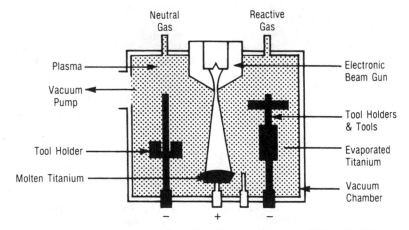

FIGURE 30.7 How titanium nitride is applied. *(Photograph courtesy of Balzers Tool Coating, Inc., Agawam, MA)*

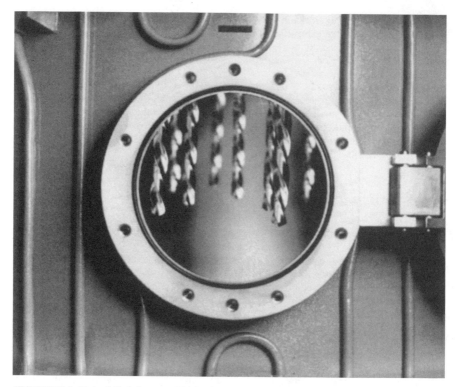

FIGURE 30.8 Twist drills being coated in the plasma of a coating vessel. *(Photograph courtesy of Balzers Tool Coating, Inc., Agawam, MA)*

cleaning, the positive argon ions are propelled by a high-voltage field and blast the tool so that it becomes atomically clean.

Solid titanium is heated by electron beam until it starts evaporating. At this time, nitrogen is introduced into the chamber and the titanium ions are electrically accelerated toward the tools. The titanium ion bombardment combines with the nitrogen gas to form a coating of titanium nitride about 0.0001" thick on the exposed surfaces of the tools. The adhesive bond between the coating and the matrix is so tight that the coating does not separate because of deflection when the tools are used.

The two complementary coating processes are PVD and CVD. The coating process, called "Physical Vapor Deposition" (PVD), operates at about 900°F, which is a little below the tempering temperature of high-speed steel. That means hardened tools will not soften during the coating process. As a directed ion-beam process, PVD coatings are deposited on line-of-sight surfaces. Consequently, the PVD will not penetrate a deep hole or an internal surface. The PVD procedure mixes gases in a correct proportion in the reactor (Figure 30.9).

The second coating process called *Chemical Vapor Deposition (CVD)*, which operates at a temperature of 1750 to 1950°F. It was originated to coat carbide inserts. The CVD bond is a chemical process; therefore, it will deposit more readily on interior surfaces. If tool steel is coated by this process, it would soften and require rehardening. For this reason, the PVD process is normally used for coating cutting tools. Another disadvantage of

CVD is that it tends to round over sharp edges and thicken corners. The performance of carbide tools coated by PVD is superior in these situations because the process minimizes edge embrittlement.

BREAK-IN PERIOD

When cutting tools are first used, they develop a wear land during the break-in period. Then, the wear of the cutting edge proceeds at a much slower pace. This is called the *low-wear period*. The useful life of the tool continues until rapid edge breakdown begins to occur. At this time, the tool should be sharpened. When a tool is TiN-coated, the low-wear

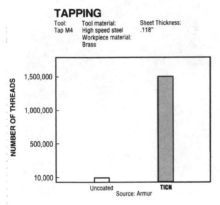

TAPPING

Tool: Tap M4 | Tool material: High speed steel | Sheet Thickness: .118"
Workpiece material: Brass

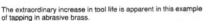

The extraordinary increase in tool life is apparent in this example of tapping in abrasive brass.

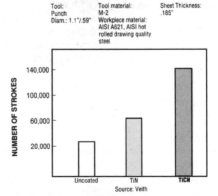

DRILLING

Tool: Spade drill Diam.: 1" | Tool material: CPM-M4 | Cutting speed: 200 SFPM
Workpiece material: 4150 Rc 30 | Feed: .012 in./rev.

This chart demonstrates how a TiCN coated drill performs 4-1/2 times better than the same tool with TiN coating.

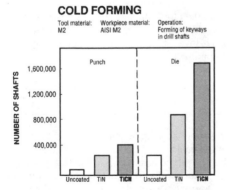

COLD FORMING

Tool material: M2 | Workpiece material: AISI M2 | Operation: Forming of keyways in drill shafts

In cold forming applications, Balzers **TiCN's** extra hardness contributes to increased performance, especially with the use of dies.

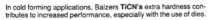

BLANKING

Tool: Punch Diam.: 1.1"/.59" | Tool material: M-2 | Sheet Thickness: .185"
Workpiece material: AISI A621, AISI hot rolled drawing quality steel

In this blanking operation, **TiCN's** greater lubricity contributes to longer tool life.

FIGURE 30.9   Differences in tool life for workpieces coated with titanium nitride. *(Photograph courtesy of Balzers Tool Coating, Inc., Agawam, MA)*

FIGURE 30.10   Coating technicians carefully prepare tools and parts for the PVD coating process.

period is extended far beyond that of uncoated tools. Forming tools, plastic molds, and mold components coated with titanium nitride show similar reduced-wear characteristics.

Numerous reports detail dramatic improvements in tool life and productivity because of the use of TiN-coated drills, end mills, milling cutters, and other high-speed steel tools. TiN-coated tools can now be used on all metal-removal operations to extend tool life, increase feeds and speeds and reduce machine downtime (Figure 30.10).

TOOL PREPARATION

Three areas should be considered by those who would like to understand the advantages of coated tools: the tool quality, the consistency of the quality, and the tool performance. Any rational evaluation of TiN-coated tools must begin with consideration of the tool prior to coating. It must be of good quality to achieve superior results after coating. The condition of the tool's cutting edge is a prime example. Often, tool edges should be honed before coating to minimize the eta phase's effects. A smooth cutting edge is less apt to chip. A ragged cutting edge and heavy burrs on any tool would waste the cost of coating. Coated tools with poor-quality surface finishes, would probably not show worthwhile improvement after coating.

Rough grinding will leave edges that break off easily when cutting starts, which would leave uncoated surfaces exposed to wear rapidly. Tools with grinding burns, rust and surface oxidation, or previous surface treatment, would have inferior adhesion of the TiN to the tool. Premium high-speed steel, when coated, will show better results than general-purpose HSS.

It is essential that all procedures in the coating process be carefully monitored if consistent quality is to be maintained. All tools should be evaluated at incoming to establish their suitability for coating. At this time, each tool should also be evaluated for the correct cleaning procedure. Chemical, electrical, or mechanical cleaning steps are possible.

All of the operating procedures, from arrival of the tool at incoming to packaging the tool for return, are significant. All the coating and inspection machinery should be well maintained. The final quality of the coating job is liable to be affected by any step in the procedure—even housekeeping. This type of work shows better results when "cleanroom" conditions are maintained.

Managers should do everything necessary to ensure that the coated tool is used in a manner that will derive the highest return. This means that proper machining operations should be followed. The workpiece should be fixtured rigidly. Tool holders, arbors, and any other aids used should be in good condition for running tools at high speeds and feeds. The full value of the coated tool will not be realized unless feeds and speeds are gradually increased to their maximum efficiency. The tool should be resharpened before the coating is seriously damaged. This step will allow less stock removal and increase the number of times that resharpening can occur. At some point, the tool should be recoated.

MATERIALS GUIDE FOR THIN-FILM PVD COATINGS

These materials can be coated:

1. Most high-speed steels that have been tempered at 950°F or higher; preferably T series, M series, ASP grades, CPM grades, and cobalt grades.
2. All grades of solid carbides: C-2, C-5, C-6, Micrograin, and more.
3. Tool and die steels that can be heat treated at 950°F or above, such as A-2, D-2, S-7, CPM10V, and more.
4. Carbide-tipped tools, provided that the brazing fillers are free of zinc or cadmium.
5. Pre-hardened and age-hardened mold steels: NAK55, NAK80, CSM21, P-20, and more.
6. Stainless steels, the 300 series (austenitic), the 400 series (martensitic), and some age-hardenable and maraging steels.
7. Some miscellaneous materials; Ampcoloy 940, 945, beryllium coppers, titanium and Ti alloys, nickel and Ni alloys, Inconels, monels, and more.
8. Other materials can be coated with the loss of some hardness points. Even carbide-tipped cutting tools that have softened during the coating process stand up well because their task is only to act as a support for the carbide. But all materials being considered for coating must be electrically conductive.

The following materials cannot be TiN coated:

1. Assemblies that are bonded, pinned, or screwed together permanently.
2. Leaded alloys and most aluminum, zinc, and magnesium alloys.
3. Any material containing cadmium or other low vapor-pressure alloys.

Surfaces that are best for TiN coating are:

1. Finely ground surfaces are best for maximum coating adhesion.
2. Surfaces should be free of burns or grinding wheel glazing.
3. Use free-cutting grinding wheels to grind with lower temperatures.

Tool surfaces that can be coated after special treatment:

1. Milled or machined surfaces.
2. Black oxide surfaces.
3. Nitrided surfaces.
4. EDM-machined surfaces.
5. Polished or lapped surfaces.

Painted or marked surfaces and any covered with plastic will be processed after special cleaning. This is also true for rusty surfaces.

CASES

Each vendor of surface-coating services has many case histories to quote that would parallel many, if not most, of the situations that clients wish to improve. Stamping houses are taking advantage of this procedure to be more efficient. One testimonial states that a mandrel was picking up metal fragments after about 5000 parts. Now, the same job has run 1,000,000 parts without pickup. The Transmatic company in Holland, MI, specializes in high-speed transfer-press production of deep-drawn parts, primarily for the automotive industry. About 150,000 parts were previously made with an untreated punch. They have currently produced 8,000,000 components without requiring tool maintenance. There are many similar situations, which is why you are urged to discuss your situation with the coating company.

The only negative about this process is that once a surface is coated, it cannot be repaired. It couldn't be welded, drilled, or altered to accommodate design changes.

The thing to remember is that this thin film, about 0.0001" thick, must adhere to the substrate—even while the forces that are generated by the cutting operation try to rip it away. Figure 30.9 shows a comparison of the two most popular coatings, TiN and TiCN, and Figure 30.10 shows a PVD operation in progress.

CHAPTER 31
HARDCOATING

INTRODUCTION

Hardcoating, in a way, is like deep anodizing. Anodizing is a process of forming an oxide coating on the surface of an aluminum component. This oxide film is created by combining atoms of aluminum with oxygen. The oxygen is amply supplied from an oxygen-containing electrolyte when a current flows from an anodically connected piece of aluminum through the electrolyte to the cathode. The properties of an anodic coating depends largely on the type of electrolyte used, as well as its temperature and concentration strength.

ADVANTAGES OF THE PROCESS

In addition to hardness and durability, hardcoating aluminum parts results in other advantages. In an application for which lubricity is required, the porous surface of a hardcoated piece will soak up a lubricant like a sponge. If a colored part is desired, the same porosity will absorb a dye. If corrosion resistance is required, the surface could be sealed. If an unusually fine finish is desired, the surface could be honed or lapped. The surface will protect the aluminum part in both high- and low-temperature environments. For this reason, hardcoated aluminum is often used as heatsinks in electronic circuitry.

DRY-FILM LUBRICANTS

These lubricants are solid materials that have the property of shearing within themselves. They function to separate rough spots. When applied over hard surfaces, they exhibit a lower coefficient of friction than when the base is unhardened. Therefore, their properties are enhanced when they are used over hardcoated aluminum. The most commonly used dry lubricants are teflon, molybdenum disulphide, and graphite.

Several companies that do hardcoating also have the capability of infusing teflon or some other dry lubricant into the surface. You could have a continuous teflon film that penetrates the hardcoated surface lower friction by as much as fivefold. The low wettability of teflon prevents contaminants from permanently adhering to the coated surface. Almost all substances release easily. Vendors using the Sanford (Duralectra) process generally handle dry film impregnation also.

Hardcoating provides an aluminum surface with the hardness of case-hardened steel (Rc 50-60) while maintaining the light weight of aluminum. Hardcoated surfaces are used under the sea, in hospitals, schools, factories, in the home, and on the moon. When a permanent dry lubrication is added to the hardcoated surface, you can find many more uses for it. A hardcoated, aluminum prosthetic arm can have its joints permanently lubricated by this process.

TESTING FOR QUALITY CONTROL

It is difficult to define the hardness of surface coatings in a meaningful, universal manner. Normally, hardness is defined as the ability to resist indentations. Measuring the depth of penetration of a standard indenter under a standard load (such as Brinell, Rockwell, Vickers, or Knoop) is unsatisfactory for thin-wall hardness testing. Consequently, a different approach is utilized here: the resistance of a surface to abrasion by another material being drawn across it. See Figure 31.1 for a comparison of wear resistance of hardcoat to other coatings and materials are also two different wear tests that are used sometimes for testing the quality of a surface coating.

We use abrasion resistance on a numerical ranking scale by means of Mohs' principle. Mohs, a nineteenth century minearalogist observed that brittle minerals can only be scratched by other minerals that are harder. His hardness scale goes from 1 for soft talc to 10 for diamond.

SEALING

Sealing the hardcoated surface against the introduction of undesired contamination and also to retain a color when that is required has been the subject of much research and development. Experiments have indicated that the most common problem leading to sealing defects has been a lack of adequate oxidation thickness. This is much more prevalent in anodized, rather than hardcoated, surfaces, but it does show the desirability for consistency and for the vendor's reliability.

Sometimes, when a hardcoated part is submerged in a hot solution (water or chemical) to seal the oxide covering the part, the large difference in heat conductivity between the aluminum matrix and the oxide will cause crazing. That is one important reason for creating a room-temperature hardcoating method.

Coloring hardcoated surfaces is similar to coloring anodized surfaces. It is done with the same dyes and procedures. The main difference is in the depth of colors. The oxide of anodized parts is much more porous, so it absorbs more color.

Hardcoated surfaces do not have to be sealed. In fact, a military specification covers non-sealed hardcoating. In one such case, the military is using an aluminum fin section on a missile. Without hardcoated surfaces, the intense heat of ignition would distort the fins, thereby altering the rocket's trajectory. So, the hardcoating is definitely required, but the sealing is not.

The conventional method of sealing either anodized or hardcoated aluminum is immersion in distilled or deionized water at or near boiling. In the case of dyed films, a bath

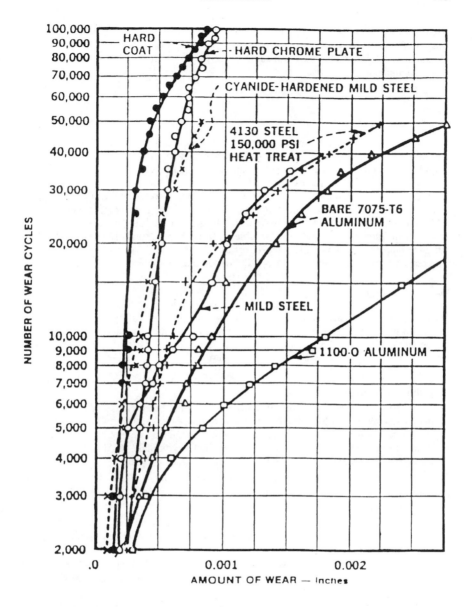

FIGURE 31.1 Wear resistance of hardcoat compared to other coatings and materials. *(Permission of Sanford Process Corp., Natick, MA)*

of nickel acetate, at the same temperature range, is used. In addition to the possibility of causing crazing, sealing can also degrade the abrasion resistance of hardcoating.

Some recent sealing products work at or near room temperature and should prevent crazing problems. The subject of sealing and coloring should be evaluated whenever hardcoating is considered because vendors, who have been on the scene for as long as 20 years, can give the best advice. Generally, vendors will not seal hardcoats unless it is specified on the purchase document, but they will automatically seal the anodized surfaces unless documentation dictates otherwise.

FINISHING

Sometimes it might be necessary to lightly grind a hardcoated surface to make it flat or parallel or for another reason. Many abrasive wheels and compounds are suitable for finishing hardcoated surfaces to achieve critical dimensional tolerances or very fine finishes.

For surface grinding, silicon carbide grit size 80 to 120, will provide a finish better than 10 microinch. For cylindrical grinding, a finer-grit wheel, Norton 39C120-J8VK, will be free cutting and yet produce a fine finish. For internal grinding, a fine-grit wheel, such as a Norton 39C100-J8VK, produces best results. In general, grinding should be performed wet, using a water coolant and a good soluble oil (mixed approximately 100 to 1). For polishing or lapping, a boron-carbide abrasive grain mixed with oil will produce good results. The range of grit size should be 400 to 1200, depending on the finish required.

Wrought Alloys

Hardcoating is recommended for use with virtually all aluminum alloys. Companies using the Sanford Hardcoat Process claim that their proprietary method will hardcoat any type of aluminum, even the 2000 series, which is troublesome because of its copper content. Perhaps other companies have solved this problem also.

1. *1100 series*   1100 is very common. Bronze-gray in color at 0.002" thickness. The alloy is difficult to machine and is soft.
2. *2000 series*   Very common are 2014, 2017, 2024, and 2618 forgings. Avoid sharp corners. Gray-black in color at 0.002" thickness to blue-gray at 0.004". Good machinability. Thickness to 0.006" for salvage work, but that thickness is not as hard as thinner coats. In fact, the coating becomes softer the thicker you make it. At 0.010" thick, the outside layer is actually dusty and the powder machines off easily.
3. *3000 series*   Most common is 3003. Gray-black in color at 0.002" thickness. It machines easily and is good for dye work
4. *4000 series*   Not commonly used.
5. *5000 series*   Most common are 5005 and 5052. 5005 is better for dye work. 5052 accepts only the black color. Both machine well. 5052 gets excellent dielectric value when coated 0.004" thick.
6. *6000 series*   6061 and 6063 are most common. Almost black at 0.002" thickness. 6061 forms an excellent hardcoat for grinding or lapping. Good dimensional stability. 6063 is used for extrusions.
7. *7000 series*   Most common is 7075. Blue-gray at 0.002" thickness. High-strength alloy. Not good for grinding or lapping. Maximum for salvage work is 0.008" thick.

INGOTS

Sandcast Alloys

Most common are 319, 355, 356, also 40E, Ternalloy, Tenzalloy, and many other proprietary alloys. 356-T6 is the most popular. Good for grinding and lapping. Hardcoat will not fill in the exposed surface porosity, which is common with sandcastings. Vacuum impregnation (plastic) will improve the hardcoated finish of a sandcasting.

Diecast Alloys

Most common are 218, 360, and 380. Only 218 produces a hardcoat that is comparable to that on wrought or sandcast material. However, 218 is difficult to diecast. Maximum thickness is 0.0025". Maximum thickness for 360 and 380 is about 0.001". Vacuum impregnation should be used especially where diecastings have thin walls with interior passages that cannot tolerate leakage.

The reason for the difference in hardcoating quality is the alloys. Silicon and copper are detrimental to a good hardcoat.

HOW TO ORDER HARDCOATING

To save time and trouble when ordering hardcoating, information on the following four items must be known:

1. *Alloy* Because the hardcoat builds up at different rates on each alloy, it is important to specify the alloy. Also, some alloys require procedures that vary slightly from one another.

2. *Coating Thickness* Hardcoating can be provided in a range of thicknesses from 0.0002 to 0.009", depending on the alloy and the application. It is important to remember that hardcoat thicknesses are 50% penetration and 50% buildup. Because half the coating is penetration and only half adds to the thickness, this change in dimension must be taken into account on the blueprint.

3. *Masking* If it is necessary to exclude (mask) certain areas from hardcoating, they should be clearly indicated on the blueprint. Generally, the need to mask adds cost to the job. If it is possible to hardcoat all over, this is the preferred way to proceed, unless not masking would require using special taps.

4. *Racking* Each part being hardcoated must have a good mechanical and electrical contact. This is called *racking*. Each racking point leaves a small void in the coating. These voids should be made in a noncritical area.

FACTS ABOUT HARDCOATING

Two types of hardcoats are currently available. The older method is formed at low temperature (32 to 45°F) in sulphuric acid. The newer type is formed at room temperature (60 to 70°F), also in sulphuric acid. Although hardcoating is quite similar to anodizing, it is about 10 times as thick and is much harder. Both anodizing and hardcoating processes form oxides on the surface of the treated parts. Very often, these parts have the oxide covering sealed to prevent the absorption of undesirable liquids (such as oil, undesired colors, or even perspiration from fingers).

Hardcoat is not plating, it is different. Hardcoating a shaft 0.002" thick, will increase the diameter by only 0.002". Plating the shaft 0.002" thick will increase the diameter 0.004". It is possible to save much money by salvaging parts in assemblies. If a shaft is worn 0.002", hardcoat it about 0.004" oversize and grind it to specifications

When hardcoating is called out on a drawing, the added phrase, "build up per surface" would prevent a misunderstanding.

Standard commercial tolerance is 0.0005" on a coating thickness of 0.002".

To allow for hardcoating a "V" thread, the formula is, "build up per surface multiplied by four."

Hardcoating is not compatible with anodizing. Parts can be damaged if they are anodized after hardcoating or vice-versa. When there is a requirement for hardcoat and any other type of chemical processing (such as anodizing, alodining, or irriditing), contact the hardcoat vendor for advice.

TYPICAL APPLICATIONS FOR HARDCOATING

1. Pump and turbine parts for resistance to erosion from high-velocity liquids.
2. Pistons for automotive brake cylinders.
3. Jet engine hydraulic controls, including valves, sleeves, gears, and cams.
4. Orthopedic braces.
5. Bearing race housings.
6. Heat exchangers.
7. Textile yarn wheels.
8. Gears and pinions.
9. Surveying instrument parts.
10. Hand tools.
11. Swivel joints.
12. Friction locks.
13. Leading edges of high-speed airfoils.
14. Helicopter rotor-blade tips.
15. Screw threads of hydraulic jacks.
16. Railway car buffers.
17. Aircraft landing-gear components.
18. High-speed air impellers.
19. Medical equipment components.
20. Die-cast metal spray guns.
21. Wrapping machinery parts.
22. Rollers for cardboard box forming machines.
23. Gears for ticket machines.
24. Feed plates in sorting and coating machines.
25. Cast electric motor brush boxes.
26. Fuel and oil pump housings.
27. Clutch and brake discs.
28. Fan blades.
29. Fuel nozzles.
30. Timing belt pulleys.

31. Boat and architectural hardware.

32. Truck beds and loading ramps.

33. Rocket tubes.

34. Gun scopes that are dyed black to minimize reflection.

35. Air-compressor pistons.

36. Gun components.

37. Cylinders for feathering propellers.

38. Guide tracks and guide rails.

39. Drill collar lifting components.

40. Fire-fighting equipment, valves, and nozzles.

41. Textile rolls.

42. Ammunition components.

43. Missile components.

44. Printed circuit boards.

45. Shoe molds.

46. Washers for electrical insulation.

And many more uses where hardness and lubricity, together with light weight, are required.

CHAPTER 32
ELECTROLIZING

DESCRIPTION

The electrolizing process uniformly deposits a very dense, high-chromium, nonmagnetic, hard (70-72 Rc) proprietary alloy onto the part being treated. This process provides an unusual combination of properties: wear resistance, low coefficient of friction, good anti-seizure characteristics, excellent corrosion protection, and a good sealant to the covered surfaces.

Electrolizing involves cleaning the part by removing a surface layer of atoms from the base metal and following immediately with an electroplate, which positively bonds the chromium alloy to the porosity of the base metal.

ADVANTAGES

Some desirable features of this process are:

1. Precision thickness from 0.000050" to 0.001".

2. Absolute adhesion of plating to matrix. There is no separation from the matrix—even when the part is bent 180 degrees on a diameter equal to the thickness of the part. The electrolizing coating will contract and expand at the same rate as the base metal being coated and will crack or fracture only when the base metal cracks or fractures. Thus, electrolizing is ideal in thermal shock applications.

3. Superior corrosion protection (Figure 32.1). Most mechanical properties are not adversely affected by this process. In fact, most are slightly improved.

4. Coating is conductive. This eliminates electrostatic buildup, which can be a problem with anodized or hardcoated aluminum.

5. No outgassing.

6. Stain resistance. The color will be satin-gray, smooth, and fine-grained. Most matrices will finish with the same color.

Fatigue Test Comparison

**All fatigue tests were conducted
on an R.R. Moore rotating beam fatigue testing machine.**

CONTROL GROUP SAE 4130 Steel		ELECTROLIZED (.0002'') SAE 4130 Steel		CHROME PLATED (.0005'') SAE 4130 Steel	
Stress psi	Cycles to Failure	Stress psi	Cycles to Failure	Stress psi	Cycles to Failure
100,000	142,000	105,000	110,000	90,000	66,000
98,000	152,000	100,000	280,000	80,000	118,000
96,000	341,000	98,000	609,000	70,000	169,000
94,000	357,000	95,000	981,000	65,000	269,000
93,000	469,000	93,000	633,000	62,000	596,000
92,000	445,000	92,000	768,000	60,000	15,255,000*
92,000	27,092,000*	91,000	1,449,000*	60,000	396,000
91,000	63,247,000*	90,000	24,755,000*	60,000	513,000
91,000	30,600,000*	90,000	40,484,000*	58,000	11,079,000
Fatigue Limit 91,000 psi		Fatigue Limit 90,000 psi		Fatigue Limit 59,000 psi	

* Indicates no failure.

Tensile Test Comparison

**Tension tests were conducted to determine ultimate tensile strength,
yield strength, percent elongation and percent reduction in area.**

CONTROL GROUP - SAE 4130 STEEL					ELECTROLIZED GROUP - SAE 4130 STEEL				
Sample No.	Ultimate Strength	Yield Strength	Elongation (2")%	Reduction of Area %	Sample No.	Ultimate Strength	Yield Strength	Elongation (2")%	Reduction of Area %
1	216,780	203,860	12.0	28.6	4	222,750	210,500	9.5	24.6
2	217,870	203,770	10.5	25.6	5	225,000	211,250	10.5	27.8
3	212,245	199,790	11.0	27.1	6	212,500	199,500	10.0	23.2
Average	215,632	202,473	11.2	27.1	Average	220,083	207,083	10.0	25.2

Salt Spray Testing (per ASTM-B-117)

**Sample with a plating thickness of .0001"-.00015" was subjected to a salt
spray test in accordance with the above specification. The conditions of the test
were as follows:**

> Time: 120 Hours
> Salt Solution: 5%
> Chamber Temperature: 97 ± 2° F

> The sample was monitored at 8, 12, 24, 36, 50, 72, 96, and 120 hours with no
> visible apparent changes to the plated surface.

FIGURE 32.1 Test results. *(Charts furnished courtesy Electrolizing, Inc., Providence, RI*

7. The coating can be applied to both heat-treated and nonheat-treated tool steels, all
   stainless steels, all nickel-alloy materials, most aluminum alloys, brass, copper, bronze,
   titanium, and all commonly machined ferrous and nonferrous metals.

8. There is no buildup of nodules at edges or corners.

GENERAL INFORMATION

Electrolizing should be performed after all metal processing is finished. That includes stress relieving or straightening. For best results, significant areas should have a finish of 32 RMS or better, and all surfaces must be completely free from other plated coatings or treatments.

Electrolizing is not intended as a substitute for heat treatment. Electrolizing will generally not be affected by thermal shock. In fact, it will adhere at temperatures ranging from −500°F to 1600°F, depending, of course, on the composition of the base or matrix. Because of the size of the facility's equipment, the size of the workpieces is limited to 10 feet in length, 20 inches in diameter, and 1000 pounds in weight.

The coating is not recommended for beryllium, magnesium, columbium, lead, and their respective alloys.

The cost of this process is competitive with others commonly used. The features that make up the price are:

1. Square inches to be covered.
2. Thickness of coating requested.
3. Tolerances required.
4. Areas can be masked, if required, but that operation adds to the cost.
5. The geometry of the part.

MOLDING PROBLEMS

The plastic and rubber-molding industries are plagued with difficulties related to platings or coatings. Electrolizing contributes to the efficiency and profitability of the following moldings: thermoplastic, thermosetting, injection, compression, extrusion, and blow molding. Electrolizing improves release, which means the separation of the plastic part from the cavity without particles sticking and causing a waste of time to clean tool parts.

There is a growing use of tougher resins, filled with glass, minerals, flame retardants and other abrasive and corrosive additives. That means such processing equipment components as barrels, screws, nozzles, mold cavities, pins, runners, and mold plates will wear rapidly, unless they are protected with a coating like electrolizing. This coating increases the rate of flow, which means shorter cycles and increased production.

ATTRIBUTES OF PROCESS

Galling is caused by a pile-up or gouge-out of metal, which can lead to seizure. *Seizure* is the welding of one metal to another. *Scoring* occurs when small particles scratch the metal during movement. Electrolizing will either prevent these harmful occurrences or retard them.

Surface corrosion of metal is a common problem that always must be considered in manufacturing. It is caused when unprotected metal is exposed to chemical environments, water, or even moisture in the atmosphere. If metal surfaces are left unprotected, surface erosion and pitting of the surface will result. Then, the integrity of the metal must be restored by remachining or repolishing. Electrolizing protects against attacks of corrosion.

Friction and heat caused by metals rubbing together cause wear. Electrolizing, with its high surface hardness (Rc 72) and low coefficient of friction, increases wear life.

It is interesting that most geometries can be uniformly coated. However, when slots, grooves, and threads are in your part, the vendor should be consulted.

The processing temperature should not exceed 180°F. This precludes heat damage and hydrogen embrittlement.

Aluminum alloys are generally anodized or hard coated because a conversion coating is normally preferred by engineers for aluminum. That is because it is difficult to electroplate aluminum, which has a high affinity for oxygen. Also, most metals used to plate aluminum are cathodic to aluminum. This means that any voids in the coating, no matter how small, will lead to localized galvanic corrosion of the aluminum. And finally, the thermal expansion of aluminum is so different from most metals electroplated that it could result in flaking and nonadherence.

Electrolizing, on the other hand, can be applied directly to aluminum surfaces without intermediate coatings of copper or nickel. Electrolizing can be applied to most aluminum alloys including 2014, 2017, 2024, 6061, 7075, and most cast alloys. Electrolizing is an ideal coating for aluminum. This has been proven through countless applications over the years.

CASES

The following case histories might introduce possibilities:

1. The Pershing Missile Program changed the hard chrome coating of monel bearings to electrolizing to reduce the coefficient of friction.

2. A metal-stamping house working on hard brass, increased tool life from 100,000 hits to just over 1,000,000. Again, the change was from hard chrome to electrolizing.

3. In coil winding, a change from hard chrome to electrolizing gained a 200% improvement in tool life.

4. In wood forming, a change from a noncoated mold to an electrolized surface created a 300% increase in the tools forming maple, beech, and oak.

5. A web-handling company changed the nickel coating of their printing rolls to electrolizing to eliminate a flaking problem.

6. Terminal crimping tools, which had been plated with 0.0007" of chrome and then polished, were electrolized, eliminating polishing and gaining a 700% improvement.

7. An electronic chip manufacturer, plagued with static-electricity problems, changed from hardcoating to electrolizing to save the assembly line from serious quality failures.

8. Finally, during the early 1950s, I was using four, 1" diameter stainless-steel rods to control the guidance system of the Atlas Missile. When the missile was fired, the rapid sliding of the guidance equipment over the four rods caused galling, then seizure. Electrolizing the rods prevented further catastrophes.

The following seven photographs (Figure 32.2) have been furnished by the courtesy of Electrolizing, Inc., of Providence, RI.

Valve control parts: 0.0004"/0.0007" onto I.D.
high temperature: high pressure.

Cutting tools:

Jet engine
wheel: Repair of
0.0001" to 0.006"

Injection mold parts: 0.0002"/0.0003" for release

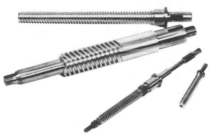

Jackscrews: 17-4PH with 0.0004"/0.0006": wear
and lubricity.

Injection mold parts: 0.0002"/0.0003" for wear
and release.

Aircraft gear and safety clutch:
Each with 0.00015"/0.00025" for
wear and lubricity.

FIGURE 32.2 Electrolized parts. *(Charts furnished courtesy Electrolizing, Inc., Providence, RI*

CHAPTER 33
POLY-OND

A NEW TECHNOLOGY FOR PLATING METAL

Poly-Ond is a liquid-bath process that chemically deposits nickel phosphorus, impregnated with polymers, on the surface of metal parts. It is a proprietary formulation developed in 1976. This process permits the use of less-expensive metals when anti-corrosive materials are called for. But more significant in the economics arena, this process is responsible for the success of many disparate applications.

ENGINEERING SPECIFICATIONS

The hardness, as deposited, is Rc 50. However, this is a highly controlled process that can achieve any Rockwell surface hardness within the range of Rc 50 to 70. The process will duplicate the hardness of various cycles with an accuracy of ±1 Rockwell point. The hardness after a standard one hour bake at 750°F will be Rc 68 to 70. If the part material will not permit the standard processing temperature, discuss the situation with the vendor because some gains in hardness are possible—even at 250°F.

The coefficient of friction under a 200-pound kinetic load is 0.06. If you want to run two Poly-Onded surfaces together, it would be best to mismatch their hardness by 10 Rc points. This type of mismatch should be used no matter what hardness process is on two or more parts that slide against each other.

Salt-spray resistance per ASTM-B-117 test is 300 hours. This is a difficult, accelerated-time "salt fog" test, used by the military to weed out equipment that will not survive conditions that had been encountered in the South Pacific during World War II. In the Poly-Ond facility, many components of the processing machinery and equipment have had constant exposure to water and chemicals. Still, they show absolutely no corrosion after four years of service. This process provides uncommon resistance to corrosion that lasts as long as the coating itself.

The thickness range per coating is 0.0002" to 0.003" per surface and it can be controlled within a tolerance of ±0.0001". The standard plating thickness is 0.0005" ±0.0001". This coating will stand up to operating temperatures from freezing to 550°F, and an intermittent temperature of 700°F.

LUBRICITY

The process creates an infusion of polymers throughout the thickness of the coating. As the coating wears, there is a continuation of lubricity. Poly-Ond plating has had the approval of the U.S. Department of Agriculture for applications on food-processing equipment since 1983. The coating is good for this purpose because the dry lubricity precludes product contamination that can occur with other types of lubrication (Figure 33.1).

The lubricity achieves an exceptional release for all types of molds and is commonly used in applications where a low coefficient of friction is required, in addition to hardness of material. In fact, as a mold release, it is so exceptional that many compression- and injection-mold operators are able to drastically cut (or entirely eliminate) the use of liquid or spray releases. The high lubricity action of Poly-Onded molds also lowers rejection rates.

The size for processing ranges from watch parts to 5 tons in weight and up to 30" diameter and 12 feet in length. Parts are generally racked for achievements of economy.

COST INFORMATION

In the early 1990s, the approximate cost of this process varied from 35 to 50 cents per square inch, depending on the requirements of stripping and/or masking, and the plating thickness and tolerance. Of course, there is a minimum charge of about $50 for any order. Any type of plating would ordinarily have a similar charge. In fact, to be thorough when considering the cost question, a comparison should be made with the cost of competitive finishes (Figure 33.2).

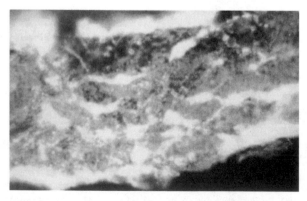

FIGURE 33.1 A photo enlargement cross section of the coating, showing the infusion of high-lubricity polymers throughout the thickness of the Poly-Ond coating. *(Poly-plating, Inc., Chicopee, MA)*

FIGURE 33.2 The process permits the handling of many identical parts at one time, thus achieving pricing economies. *(Poly-plating, Inc., Chicopee, MA)*

ADDITIONAL TECHNICAL INFORMATION

Avoid masking, if possible. It is generally less expensive to coat an entire surface, rather than mask areas.

Most standard manufacturing materials can be coated, including tool steels, stainless steels, aluminum, copper, and bronze (Figure 33.3).

This is an electroless process of chemical reduction and as such will deposit very evenly. The dogbone buildup, common to electroplating, does not occur with this process. Normally, surface geometry does not affect uniformity of thickness. That is especially important for bends, grooves, and cutouts.

The coating appears the same in color, regardless of the matrix to which it is applied.

For repair work or simply to make design modifications, Poly-Ond can be chemically stripped from the matrix without harming dimensions or degrading existing surface characteristics.

USES

The designer should be aware of the different types of surface hardening available. There are cases when one type did not work well very long and another substituted quite successfully. In some situations, the degree of hardness is crucial; in other cases, lubricity is the significant factor. The material (matrix) is often critical in making a process selection. And sometimes the function of the part is most important.

FIGURE 33.3 A sampling of the many different types of products that can be Poly-Onded. *(Poly-plating, Inc., Chicopee, MA)*

Recently, a heavy assembly fixture was being moved back and forth across a granite surface plate. It moved on three standard fixture feet, each ¾" in diameter. After a short period of time, the feet scratched marks in the granite. This, of course, required expensive repair work. But before the fixture was used again, the three feet were Poly-Onded, which precluded a repetition of the same wear problem.

During repair of a high-pressure steam valve, steel was substituted for expensive bronze. After four years, there was no evidence of rust, pitting, or corrosion on the steel surfaces that had been Poly-Onded. In many cases, Poly-Onded steel has replaced stainless steel.

Another report has the process being used on cutting tools. It reduced the amount of metal removed during sharpening operations. More importantly, tool life had been extended and downtime, because of tool sharpening, had been reduced.

Mr. Edward Bodine, Chairman of the Board of the Bodine Corp. of Bridgeport, CT, a manufacturer of automation, has spoken on the subject of robotics on countless occasions. He has stated, "Poly-Ond has made a substantial contribution to our need for high lubricity of certain components on our assembly machines, especially when conventional lubrication could contaminate products being assembled."

A plastic extruder was able to successfully extrude six different plastics on a Poly-Ond coated screw and found an output increase of 17.5%.

In the manufacture of ultrasonic dental scalers, Syntex Dental of Valley Forge, PA, found that the hardness and lubricity of the Poly-Ond applied to oscillator motor parts met their goal of zero wear for the life of the product.

Parker Hannifin uses Poly-Ond on spur gears in their metering pumps.

Husky Injection Molding Systems, Ltd. of Ontario, Canada, uses the process to build up worn surfaces in molds to make them usable, rather than scrap them. The exceptional adhesion of the coating and dry lubricity makes the process ideal for this use.

CHAPTER 34
MAGNAPLATE

QUESTIONS

Designers should consider the finish their product should have well before the design is completed. By the time they have read this far in this book, they should know that there are several alternatives to that selection. There are many questions to answer first:

1. What is the function of the part?
2. What is the purpose of the coating?
3. Is the entire sample to be coated? Which portions? Is masking critical?
4. Is color important?
5. What is the anticipated annual volume?
6. What are the coating tolerances? Is there a specific thickness of coating that must not be exceeded?
7. How are the samples currently being finished?
8. Has a cost parameter been set on the finishing process? If so, what is it?

MAGNAPLATE SYNERGISTIC COATINGS

Magnaplate coatings are a series of multi-step, proprietary processes that become an integral part of the top layer of the base metal. Each coating series serves a particular purpose by protecting a specific metal or group of metals. Most coating series can be further adapted to solve specific problems. For instance, these are some benefits:

1. To protect mechanical wear by covering the sample with a harder-than-steel, dry lubricating surface.
2. To resist abrasion and galling by providing an extremely low coefficient of friction.

FIGURE 34.1 A variety of parts treated with Magnaplate "synergistic" coatings. *(General Magnaplate Corp., Linden, NJ)*

3. To protect against attack and deterioration from chemicals and corrosive agents.

4. To eliminate problems of chipping, peeling, or flaking.

5. To act as a superior mold release.

6. To aid in material flow by eliminating sticking and product hangups.

7. To putting a smooth surface on castings thereby reducing the need for some machining. This is an unusual attribute because most finishes cannot accomplish this, unless, of course, some intermediate step (such as plastic impregnation) is taken. Incidentally, if castings have internal passages where leakage cannot be tolerated, a plastic impregnation should be done as a matter of course.

8. To improve sanitary conditions on food- and drug-processing equipment.

9. Many coatings are USDA approved and FDA compliant.

10. To permit the substitution of less-expensive metals for substrates.

11. To contribute dielectric strength where required.

The process begins with a thorough cleaning of the parts, followed by a deposition of an intermediate high-density film or by thermal spraying, depending on the coating. The process continues with a controlled infusion of selected polymers or other dry lubricating particles. These particles are then mechanically cross-linked and locked in by a proprietary process to become a permanent, integral part of the newly enhanced surface layer.

This company divulges nothing it doesn't have to. Although they will not disclose specifics of any of their processes, they will guarantee to solve your coating problems, whatever they are. They claim to be problem solvers, rather than platers. About eight years ago, I was aware of 10 series of coatings that the company offered. Soon, it became 12. Recently, their vice-president claimed two additional series would soon be available. A variety of parts treated with Magnaplate are shown in Figure 34.1.

FRICTION

Before progressing into a description of each different Magnaplate coating, it is important to understand just a little about friction. Starting friction is called *static friction* and friction in motion is called *dynamic friction*. Friction data guides have been prepared to assist engineers select combinations of materials that can improve the service life of mating components. However, the data guides should only be considered as guides. Other factors can alter the figures derived from laboratory constants. For example:

- Applied loads
- Temperature conditions
- Point loading
- Environmental variations
- Loading stresses
- Micro finish of components
- Substrate hardness

Because of the complex nature of the wear and friction phenomena, all of these conditions must be considered when designing any system.

THE MAGNAPLATE SERIES OF COATINGS

Tufram

This coating imparts hardness, lubricity, and unusually high wear and corrosion resistance to aluminum. It is basically hardcoating, as in Chapter 2, with the addition of an infusion of dry lubricants. These particles mechanically bond to the anodized layer and create a hard, dry, lubricated surface that won't chip, peel, or delaminate. Other characteristics are protection against abrasion and chemical attack. It has a high dielectric strength and an attractive appearance. Some of these coatings are USDA approved and FDA compliant. The end use of the aluminum part, together with its engineering specifications, determines which type of Tufram coating to select.

Nedox

A Nedox coating protects most base metals, including aluminum and titanium against abrasion and corrosion. In this process, a part is first cleaned and prepared.

Then a coat of nickel alloy is deposited on the surface. Next, micro pores of the nickel are enlarged by a proprietary step. After that, the surface is infused with a polymeric or other dry lubricant. Finally, the part is subjected to another exclusive step to ensure a thorough mechanical bond of the system. The process creates a hard, smooth, dry lubricated surface that is exhibited, in addition to its corrosion and abrasion protection, and permanent nonstick and anti-static properties. Some of this series meets USDA and FDA codes.

Plasmadize

This thermal spray composite coating features wear and corrosion resistance, dry lubricity, and mold-release qualities. Plasmadize exceeds the performance of conventional thermal sprays because it is composed of a combination of ceramic particles that are infused with polymers to create structural integrity and lubricity. It is completely nonporous and more ductile than chrome plate. It can be applied in coatings up to 0.010" thick, then remachined to save worn parts. Several types of Plasmadize meet USDA and FDA codes.

Hi-T-Lube

This dry lubricant prevents wear, galling, and fretting at temperature extremes. This space-age dry film is used on steel, stainless steel, and copper alloys at both high- and low-temperature extremes and under heavy load. A matrix of metallic layers becomes a permanent, integral part of the base metal and is effective in rolling or sliding actions. The coating is compatible with most chemicals, metals, gases, and oils. After burnishing, it offers the lowest coefficient of friction of any Magnaplate.

Lectrofluor

Lectrofluor provides the maximum corrosion resistance on parts made of any metal or combination of metals. This series of elastomeric, polymer-based, corrosion-resistant coatings provides maximum protection against a wide range of corrosive solvents, acids, and alkalies—even in hostile environments. It can operate at any temperature between −400 and 550°F. It can be applied to an assembly of more than one type of metal, such as steel inserts in an aluminum part.

Their rigidity and creep resistance qualify them for many load-bearing applications. Many of this series meet USDA/FDA codes for use in food and pharmaceutical processing and packaging equipment. These coatings are excellent for mold releasing.

Canadize

This is a hydrogen-free, super-hard, fracture-free coating for titanium. It was developed specifically to prevent hydrogen absorption, a major problem in the surface treatment of titanium. The titanium drill that took core samples out of the moon was protected by a Canadize treatment. In this treatment, the surface is first hardcoated, then the hard layer is infused with selected engineered polymers (molybdenum disulphide, graphite, or other dry lubricants).

The process creates a steel-hard, high-fatigue, hydrogen-free, corrosion-resistant, low-friction surface. The Canadize treatment significantly increases wear and abrasion resistance and eliminates galling, binding, and seizing. A typical thickness is 0.0001 to 0.0005". The hardness is around Rc 55.

Magnaplate HMF

This coating imparts optimum wear performance to copper, steel, and other alloys. It is characterized by a Rc 70 hard, smooth, micro finish with a low coefficient of friction—even when unburnished. It precludes breakaway vibration and slip stick. It is USDA approved and FDA compliant. It maintains anti-wear, abrasion, and chemical resistance even up to a temperature of 950°F.

Magnaplate HCR

By introducing a series of bimetallic particles in an aluminum-oxide formation, this coating provides corrosion resistance and hardness to aluminum. This synergistic coating provides corrosion resistance of 15,000 hours in an ASTM B-117 salt-spray atmosphere, and a Rc 68 hardness in 6061 T6 aluminum. Other alloys, such as 2024 or 7075, are limited to 5000 hours and a hardness of Rc 65. Magnaplate HCR permits the use of 6061 T6 in corrosive and abrasive environments.

Magnadize

Magnesium has been a material that the military would like to use more often. Unfortunately, its surface protection in unfriendly environments (such as the oceans) has always presented problems. A pinhole or micro crack will initiate rapid oxidation and destruction of integrity. The Magnadize system allows a greater utilization of magnesium because it prevents oxidation and galling of this lightweight material. In this process, magnesium parts are treated in an electrolytic bath to create a hard coat with a smooth, porous, crystalline surface structure. This becomes an excellent base for impregnation with dry lubricants. The four standard thicknesses range from 0.0003 to 0.0020".

Magnagold

The Magnagold process is based on the physical vapor deposition of titanium nitride by reactive plasma ion bombardment in a vacuum. It extends service life by resisting wear and abrasion, and improving all types of cutting and forming tools. Magnagold is FDA compliant.

Magnaplate TCHC

This surface-enhancement, precision anodizing technology was developed to solve a problem with a geometrically complex electronic component. This part required thermal conductivity and it had to be electrically insulative.

Magnaplate SNS

This surface provides superior wetting resistance to any metal. A serious problem in manufacturing assemblies is splatter from solder and brazing rods and welds sticking to the fixture or other places, where it is not wanted. Magnaplate SNS creates an ultra-hard surface, close to Rc 80, from 0.0003 to 0.0010" thick, which eliminates most tedious and costly cleaning.

CASE HISTORIES

An aluminum air-compressor impeller made by the Turbonetics Company had its life shortened by corrosive and erosive attack of CPI gas streams. The application of Tufram coating gave the impellers hard, permanently dry-lubricated, smooth surfaces that had less air drag and protected the surfaces from hostile atmospheres (Figure 34.2).

FIGURE 34.2   An air compressor impeller. *(General Magnaplate Corp., Linden, NJ)*

FIGURE 34.3 Noncontaminating NEDOX protects food-processing equipment parts, such as this Nabisco cookie mold, against product residue clinging to surfaces. General Magnaplate Corp., Linden, NJ)

See Figure 34.3 for a Nedox application, Figure 34.4 for a sample of Plasmadize, Figure 34.5 for a sample of Magnaplate HCR application, Figure 34.6 for a sample use of Hi-T-Lube, Figure 34.7 for a sample of Lectrofluor use, Figure 34.8 for a Canadize application, and Figure 34.9 for a Magnaplate HMF application.

The last case to be mentioned concerns the Ekofisk drilling platforms in the Norway sector of the North Sea. The platforms had sunk about 12 feet and, for safety's sake, had to be raised. Because revenues from this operation came to $4.7 million per day, it was deemed unfeasible to shut down for repairs. Phillips Petroleum of Norway determined that the five platforms of the complex would be jacked up together by a computer-controlled hydraulic lifting system.

It was quite an undertaking to cut the legs, weld flanges to them, and lift against the flanges. Then, large-diameter piping was inserted above each cut and welded to raise the platforms to their original height above water. The machines used to cut the legs were split-frame "Clamshell" pipe lathes, which could sever in-line piping and simultaneously bevel the pipes to prepare them for welding the flanges and rejoining. The big complication was the protection of the equipment, both steel and aluminum components, from the destructive attack of sea water and salt spray. All steel parts were Nedoxed and all aluminum parts were coated with Magnaplate HCR. None of these parts required much maintenance during the entire lifting operation. However, one lathe, ordered and delivered late, was seriously damaged by its exposure. Unfortunately, there had been no time to protect its metal surfaces as the other tools had been protected.

PLASMADIZE®

For superior abrasion resistance.

APPLICATION:
Molds for forming closely dimensioned ceramic parts

CUSTOMER:
Producer of precision ceramic components

Problem: Aluminum molds used in the production of slip cast ceramics were experiencing excessive wear as well as a sticking problem which was causing damage to the product.

Solution: All surfaces contacting the slurry were coated with PLASMADIZE W11F. This not only solved the wear and the mold release problems but also facilitated clean-up of the molds.

FIGURE 34.4   The Plasmadize coating. General Magnaplate Corp., Linden, NJ)

FIGURE 34.5   The Magnaplate HCR "Synergistic" coating protects aluminum parts against abrasion and salt-water corrosion. Pictured is an underwater sonar device used for oil exploration.

FIGURE 34.6 A sample of HI-T-LUBE. *(General Mag-naplate Corp., Linden, NJ)*

FIGURE 34.7 Electrofluor coating on a Tollhurst centrifuge. *(General Magnaplate Corp., Linden, NJ)*

For titanium.

APPLICATION:
Aerospace

CUSTOMER:
NASA

Problem: Prevent galling at the joints and in the drive shaft of the titanium core sample drill tubes. Also to prevent contamination of moon surface sample by titanium particles or other foreign materials.

Solution: A CANADIZE coating on both the inside and outside of the entire tube eliminated any danger of galling or contamination. All surfaces were made abrasion-resistant and permanently dry-lubricated.

FIGURE 34.8 The Canadize coating. *(General Magnaplate Corp., Linden, NJ)*

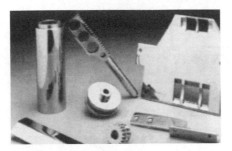

FIGURE 34.9 Magnaplate HMF coating on paper packaging equipment. *(General Magnaplate Corp., Linden, NJ)*

From this text, it should become apparent that this company thrives on difficult surfacing problems that are encountered under severe operating conditions. They can readily customize special jobs, in doing that, they have developed several processes that now find common acceptance. Also, several of these coatings are similar to those described in Part VI Surface Treatment Processes.

CHAPTER 35
POWDER COATING

HISTORY

World War II created a European shortage in the type of solvents required for conventional painting operations. Dr. Irwin Gemmer, a German engineer, discovered a way around this shortage. He found that when heated metal parts were dipped into a cloud of powdered plastic, suspended by air turbulence, the particles would melt and cling to the part. He then discovered that when these coated parts were heated and cooled, the plastic particles fused and solidified into a tough, durable, totally encapsulated finish with no drip marks. This was the birth of the powder-coating industry (Figure 35.1).

After 10 years of coating parts and continually experimenting, it was learned that charging the plastic particles electrostatically would make them cling to the work pieces better. The electrostatic method, Gemmer's (now called *fluidized bed*) method, or a combination of the two are the main methods still being used today. The techniques have matured until they represent a superior finishing process.

USES

Electroplating, anodizing, bonderizing, phosphating, and painting are good, useful processes. However, all of them have lost business to powder coating, which has its own area of efficiency. Powder coating doesn't affect the rigidity or strength of the matrix material. When powder coated, the part will resist weathering, abrasion, ultraviolet radiation, thermal energy, and electrical charge. The coating can also be used for decorative purposes. Figure 35.2 shows some of these parts.

These coatings are nontoxic, low-friction, resist steam sterilization and a wide range of chemicals, and have the ability to cover sharp edges, weld joints, and corners. This makes them ideal for hospital and other sanitary applications. In addition, they resist corrosion, deaden sound, and are pleasant to the touch (Figure 35.3).

FIGURE 35.1 Products removed from a fluidized-bed coating process. *(Plastonics, Inc., Hartford, CT)*

Rough castings of all metals can be clad in plastics to improve salability. The castings are heated to a temperature just above the melting point of the plastic being applied. Then, the casting is immersed in finely divided powder, which melts on contact with the hot metal. By controlling the temperature and time of exposure, a very uniform, smooth coating encapsulates the article.

These coatings protect both manufacturing personnel and customers from sharp edges and rough surfaces. In many cases, they reduce polishing, grinding, and deburring requirements. Wire fabrications, castings, forgings, stampings, springs, and subassemblies have all been powder coated successfully, saving considerable funds. Size, shape, or complexity are not limiting factors (Figure 35.4).

MATERIALS

The range of plastic powders available for coatings is constantly growing because new formulations are continually found that possess a needed, special characteristic. At present, five families of powder are commonly used: nylon, vinyl, epoxy, polyethylene, and polyester.

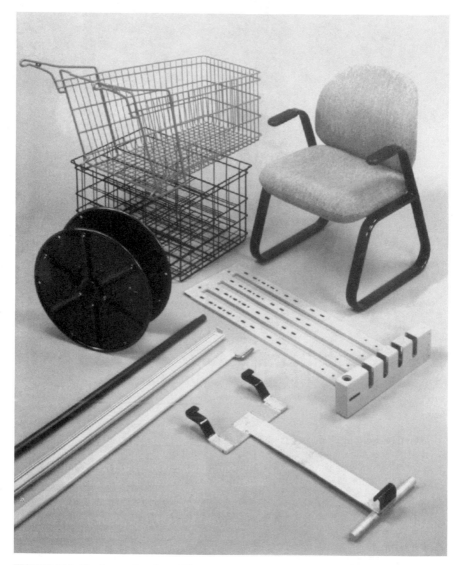

FIGURE 35.2   Plastic-coated products. *(Plastonics, Inc., Hartford, CT)*

Nylon

Nylon is the conventional choice when a low coefficient of friction or a tough plastic is desired. Nylon is also excellent when abrasion or resistance to solvents and alkalis is a problem. Nylon is autoclavable. Metal shelving is coated with nylon to protect it against food stains, rust, dirt accumulation, and fungus formation. Available in a wide range of colors,

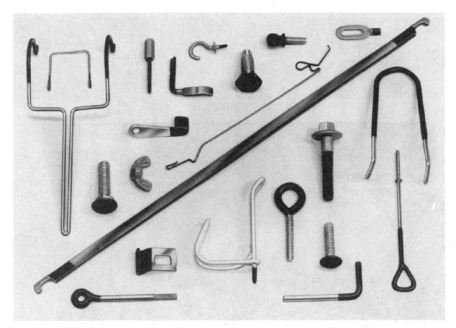

FIGURE 35.3 Plastic-coated products.

nylon has excellent impact resistance and is also good for weathering. It is FDA approved and is machinable.

Vinyl

Vinyl performs very well when superior resilience and waterproofing are needed. The vinyl surface cleans easily, is nonslip, and is resistant to alcohols, and most acids, alkalis, oils, inks and foods. Vinyls are tasteless and nontoxic. Vinyls are soft and attractive, and they have good impact resistance. Vinyl is used on outdoor furniture—especially at the seaside because it withstands salt spray and weathering.

Epoxy

Epoxies provide a super-hard coat with excellent temperature resistance. They have the capability of unusual adhesion to most surfaces—even without primer. They possess excellent toughness, strength, thermal shock resistance, and maximum electrical insulation. They have a full range of colors and are resistant to salt spray and most chemicals.

Polyethylene

Polyethylene exhibits superior chemical resistance, inertness, and high impact resistance. It is tough even though it is soft and warm to the touch. It's durable, flexible, and

FIGURE 35.4 Hinges entering the electrostatic spray bath. *(Plastonics, Inc., Hartford, CT)*

has good electrical insulation. Compared to most other powder coatings, this is a low-cost material.

Polyester

Polyesters are hard, tough, and have excellent adhesion and impact strength. This is probably the best choice for weather resistance and protection against ultraviolet discoloration. Polyesters resist most chemicals, and have a full range of colors, textures and glosses.

The following is a partial list of plastics in general use, but not used as commonly as the five aforementioned:

- Fluorocarbons
- Polyurethanes
- Polypropylenes
- Acrylics
- Cellulose acetate butyrate

COST

The cost of powder coating varies widely, depending on the plastic used, the surface area covered, and the thickness desired. A high-production rate would permit the use of automation. If powder coating cost is investigated in a manner similar to that used when considering plating or painting, the price would be competitive.

Plastic-coating companies installing new types of equipment are able to hold down rising costs while maintaining high quality. One such installation consists of an automatic four-gun system in the coating chamber, a conveyor, an oven with combined infrared and convection curing, and a powder-recovery system. When necessary for better coverage of certain parts, the spray guns can be programmed to oscillate automatically. As a result, the conveyor, which is operating at speeds up to 25 feet per minute, need not be slowed down or stopped for manual touch-up spraying of complex-shaped parts.

Cost-efficient production rates are maintained because of the efficiency of the curing oven. Extremely fast, high-intensity infrared panels melt the powder coating to begin the curing process. Then, the gas-fired convection portion of the oven completes the curing process, in which the final cross-linking occurs and the ultimate physical and chemical properties are achieved. Rounding out the overall cost effectiveness of the system is its powder-recovery capability, which is close to 98% efficient.

This new coating system is an electrostatic spray. It charges the plastic powder while it is directed at a grounded workpiece. The charged powder is attracted to the part, which is then conveyed into the oven. Plastic powder-coating companies, which are installing the latest equipment available, remain on the forefront of the industry. Some have mastered the "dip to a line" process. This is to attract the business of products that require a definite separation line, above which they do not want any coating.

METHODS OF COATING

Earlier, mention was made of the fluidized bed and the electrostatic methods of applying plastic coats. These are the two most common methods and most coating vendors use them. However, some companies create their own methods. Plastonics, Inc., of Hartford, CT, has experimented at length with, and become very successful using, the following five powder-coating methods.

Electrostatic Coating

In this process, cold parts are coated with unheated, electrostatically charged plastic powders. The powder is sprayed on. Unlike the fluidized bed process, this method is well-suited

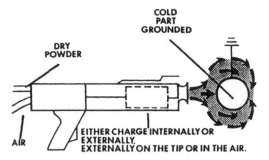

FIGURE 35.5 The electrostatic coating process. *(Plastonics, Inc., Hartford, CT)*

to the application of powder to localized areas, as well as complete parts. Thin coatings can be applied. Many plastics in many colors are available (Figure 35.5).

Electrostatic Fluid Bed

This process passes unheated parts over an electrostatically charged fluidized bed of dry plastic powder. The method is excellent for small parts and where thin coats are wanted. It is recommended where masking of areas is required (Figure 35.6).

Flocking

Flocking is used when small quantities are needed. The process is similar to the electrostatic, except that the parts must be preheated before applying powder by spray, and no electrical charge is involved. Flocking should be considered as a suitable alternative, only where part size or weight precludes using the fluid bed process (Figure 35.7).

Fluidized Bed

Fluidized bed involves the preheating of metal parts and their immersion into a fluidized bed tank filled with dry, plastic powder. Components or assemblies coated by this process can be handled in small or large quantities. The process is very economical. Heavy coats of a wide variety of materials are used (Figure 35.8).

Flo-Clad

This method is generally limited to very small parts not exceeding three ounces in weight, three inches in length, and required in large quantities. It is very economical (Figure 35.9).

FACTORS TO CONSIDER

A multitude of problems have been solved by plastic coating and many functions have been improved by introducing the process. Parts that have unsightly welds exposed can

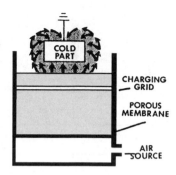

FIGURE 35.6 The electrostatic fluid bed process. *(Plastonics, Inc., Hartford, CT)*

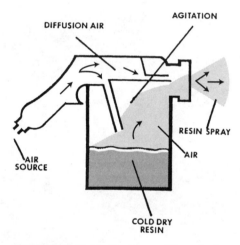

FIGURE 35.7 The flocking process. *(Plastonics, Inc., Hartford, CT)*

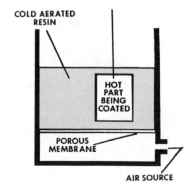

FIGURE 35.8 The fluidized bed process. *(Plastonics, Inc., Hartford, CT)*

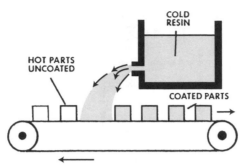

FIGURE 35.9 The flo clad process. *(Plastonics, Inc., Hartford, CT)*

FIGURE 35.10 The manual masking and demasking of small quantities. *(Plastonics, Inc., Hartford, CT)*

have the surface imperfections hidden under plastic coating. Masking of parts should be done only when absolutely necessary. It adds cost (Figure 35.10).

A manufacturer of nickel-cadmium batteries had a service-life problem because acid leakage ate through the painted finish, causing the batteries to short out. They found that epoxy coating was several times more resistant to acid than paint.

Many pieces of hardware are improved functionally and aesthetically by plastic powder coatings. Any furniture or hardware, in or around swimming pools, will survive much longer if they are coated with vinyl. Wine display racks are required to be attractive, withstand abrasion, resist impact, and remain corrosion-free. Grocery-bagging systems are now being powder coated for the same reasons.

SURFACE-COATING RECOMMENDATIONS

This section of the book covers: titanium nitriding, plastic coating, hardcoating, electrolizing, Poly-Onding, and Magnaplating. All these coatings are advantageous for specific applications. They all pass military specifications in their categories. If you contemplate using any of them, it is recommended that you discuss the situation with vendors. Then, you can establish your own criteria and inspect the results accordingly. By comparing the costs of the process with their relative results, you can establish your own values for each.

For instance, titanium nitriding (and other PVD and CVD coatings) stands alone. It is the only process that I can recommend for coating cutting and forming tools. I can recommend TiN by Balzer, Inc., because they have supplied us with high-quality coatings. The General Magnaplating Corp. has a process they call "Magnagold," which is their TiN coating for cutting tools. Probably other vendors are capable of applying some coating to do the same job just as well.

Plastic coating has a unique place in industry. It is used more as a decorative finish, rather than for lubricity or hardness. It also protects against corrosion, scratching, and cutting. Its main competition comes from painting.

The third process, hardcoating, is only for aluminum. It has been the traditional method for hard-surfacing aluminum. When followed by Teflon impregnation or some other dry lubricant, it is a very useful design process.

Now, here is where the serious competition begins: electrolizing, Poly-Onding, and several of the Magnaplating processes are all excellent for finishing aluminum. All are hard coats (Rc 50 to 72) and have good lubricity. They are outstanding for finishing ferrous materials, as well as aluminum. All are proprietary, so there is a limit to the information obtainable. These last three processes are reliable. You'll simply have to compare the results with costs.

Another method of surfacing metals is covered in another chapter. It has not been mentioned with this comparison because it is used only when lubricity is the goal. It is a method of spraying a lubricant on a matrix of any material that can stand the curing temperature.

FUTURE DEVELOPMENTS

The most active area of development in the future will probably be in new resins and harder coatings. Because of their excellent mechanical properties, epoxy resins and nylon are the most versatile of the powder coatings. New resin hardeners will enhance that versatility. Improvements of this type, especially to the epoxies, will permit radiation curing without sacrificing high performance standards.

Considerable development activity is occurring with polyesters for use in decorative coating systems. Called *polyester hybrids*, their manufacture is relatively simple. Also on the horizon are powder coatings that combine the attributes of acrylics and polyesters.

Other improvements will be in the smaller size of powder. If the particle size can be reduced sufficiently and without increasing cost too much, it would allow thinner coats and curing with radiation. The less heat energy required to melt and cure fine particles of powder coatings, will expand the application of powder coatings into the fields of wood and paper. When the finishes are smooth enough, they will find use in automotive interiors and exteriors.

CHAPTER 36
SPRAY COATING

GENERAL INFORMATION

Another method of treating metal surfaces is spray coating. This process has become significant because of the many product design uses discovered for it. Recent advances in this coating procedure involving new materials, application methods, surface preparation techniques, and curing technologies, have revolutionized traditional coating shops and the services that they offer.

Spray coating is performed on metals that are generally unhardened or surface treated in any other manner. Reducing friction is its most common purpose. Sometimes the part is masked to expose only the area that requires the almost frictionless surface. It is also used as a release agent for molds, household cookware, and for anti-corrosion and anti-icing purposes.

Industrial applicators licensed by DuPont and other independent applicators can now handle jobs that they considered impossible five years ago. Working with various industries (such as the automotive industry), they have improved part quality and now are much in demand. The ability of these vendors has been driven to a large extent by the increasing availability of high-performance coatings, which are capable of broader uses and more-demanding requirements. Fluoropolymers, nylons, epoxies, and urethanes have all undergone a steady improvement in performance characteristics. Potential customers have approached these applicator jobbers trying to solve problems to meet competition by using a spray listed in Figure 36.1.

SURFACE PREPARATION

The durability and performance of any coating is only as good as the preparation of the surface to which it is applied. The first step is cleaning. The parts, large and small, are grit blasted using quartz sand, silicon carbide, aluminum oxide, or some other abrasive. Attention must be paid to the depth of the blast profile, which is critical to adhesion.

Non-stick release

Very few substances permanently adhere to fluorocarbon coatings. Release is dependent on both the coating system as well as the tackiness of the contacting material.

Electrical properties

Coatings are available with dielectric strengths of 500 to 1000 v/mil. There are also some coatings which are semi-conductive... useful for draining static charges.

Non-wetting release

Coatings are available that are both oleophobic and hydrophobic, therefore, offering both non-wetting and ease of cleaning.

Permanent lubrication

Coatings offer permanent dry lubrication under a wide range temperatures and pressures. They also are used where convention lubricants would cause contaminatic.

Low friction

Coatings with either moderate or very low coefficient of friction may be used to reduce surface drag.

Chemical resistance

Should corrosion be a problem an organic coating can usually provide the answer. Careful analysis will dictate the correct coating to be used.

Temperature stability

A broad spectrum of coatings provide operating temperatures ranging from as low as -450°F to as high as 500°F without effecting their mechanical properties.

A licensed industrial applicator for DuPont's "Teflon" finishes

FIGURE 36.1 Some coatings that can be sprayed on metals. *(Precision Coating Co., Dedham, MA)*

The workpiece must be cleaned and degreased before and after blasting using special solvents.

Some coatings adhere better if a phosphate of zinc or manganese is applied. Phosphates passivate the surface and make a better anchor for coatings, such as Teflon-S, in applications that require corrosion resistance. In other applications, special primers must be used to provide adhesion to systems that cure at high temperatures. It is wise to make your vendor certify that all these steps have been taken because some workers might try to skip a step or two, not realizing their importance (Figure 36.2).

A.   **B.**

C.

FIGURE 36.2 Production capabilities: A. Surface preparation, consists of degreasing, preheating, and grit blasting. B. Coatings are predominantly sprayed. Using the latest equipment enables the application of a wide range of coatings. C. Batch ovens with capacity up to 14 feet allows the processing of a wide range of parts. *(Precision Coating Co., Dedham, MA)*

APPLICATION AND CURING TECHNIQUES

Both water and solvent-based coatings can be sprayed from a suction gun or a pressure pot. In jobs where large quantities of parts are coated, a bulk-handling system, including at least a conveyor, should be utilized. This would create a cost advantage for small hardware parts or anything like that. It should be apparent by now that there are alternative approaches for most coating jobs. It is the duty of the design and procurement team to compare the costs vs. the advantages of each.

Most curing is done in gas-fired or electrically heated furnaces in either batch or continuous modes. Substantial advances have been made in recent years in controlling the heat within the furnaces to ensure an even, overall cure. New technologies for curing have emerged, such as infrared lamps and induction heating. Knowledge of automated controls for curing is becoming more significant because the curing process is ideal for computer control.

MATERIALS

The most difficult assignment for the spray applicator is to determine the best material to meet the customer's unique combination of performance requirements without exceeding cost parameters. In response to design demands, coating materials have proliferated over the last few years. DuPont now offers a complete line of Teflon coatings in five basic categories:

1. *Teflon PTFE*   This primer/topcoat system can be used in high operating temperatures.
2. *Teflon FEP*   This nonstick, nonporous finish flows during cure.
3. *Teflon-S*   This one-coat, nonstick finish combines with resins or modifiers to improve abrasion resistance.
4. *Teflon dry lubricants*   This one-coat system that provides excellent lubrication.
5. *Teflon-P PFA*   A powder coating for tough, nonstick abrasion resistance and chemical corrosion resistance.

In addition, DuPont has recently introduced Tefzel ETFE copolymer, another powder coating, which produces durable, corrosion-resistant finishes up to 30 mils thick. Teflon-P PFA and Tefzel ETFE are applied either by electrostatic coating or the fluidized bed process; consequently, it belongs in Chapter 9. However, the intent here is to list the DuPont products together. Each of these series of Teflons can be altered by the DuPont licensed industrial applicator (LIA) to tailor-make a specific coating that will accomplish the most for your product. Some coatings can be safely used at operating temperatures of 600°F, but others will not survive above 325°F. These LIA's are kept informed by DuPont of the latest advances in their laboratories.

Some coatings have a normal thickness of 0.001", but others can easily go as thick as 0.006" and Tefzel can be applied up to 0.030" thick. Most coatings are excellent insulators although, when necessary, they can be made conductive, as perhaps to drain static charge. Most coatings require a curing temperature of 300 to 700°F. The chart, Figure 36.3, is a quick-reference guide. It tells what the coating is made of, its outstanding properties, its continuous operating temperature, normal thickness range, curing temperature, colors, and whether it's a thermoplastic or thermoset.

This is a good spot to remind you that Teflon is a registered trademark of the DuPont Company. Other large companies produce their own version of the generic fluoropolymers and they offer their types of similar coatings.

COATING MATERIAL	CHIEF USES	OPERATING TEMP. MAX.	NORMAL COATING THICKNESS	COATINGS TOLERANCES	NORMAL DIELECTRIC STRENGTH	BAKING TEMP.	OTHER USEFUL CHARACTERISTICS	COLORS	TYPE
TEFLON* (TFE)	ANTI-STICK	600 F	1 mil (.001") (.5 mil/coat)	+ −.0005"	450v/mil	750 F	LUBRICITY INERTNESS	VARIOUS COLORS	SINTERED
TEFLON* (FEP)	RELEASE INSULATION ANTI-CORROSION ANTI-ICING	425 F	3-4 mils (.3 mil/coat)	+ .0005"	1000v/mil	700 F	LOW TEMP. OPERATION −325 F FDA/CHEM RESISTANT	BLUE GREEN GRAY	THERMO-PLASTIC
TEFLON* (PFA)	LOW FRICTION INSULATION ANTI-CORROSION CHEM RESISTANT	500 F	4-6 mils (1.5 mil/coat)	+ −.0005"	2000v/mil	700 F	DURABLE AND ABRASION RESISTANT	GREEN BLACK	THERMO-PLASTIC
CONDUCTIVE TEFLON (TFE)*	TEFLON TO HELP DRAIN STATIC CHARGE	600 F	1 mil (.5 mil/coat)	+ .0005"	CONDUCTOR	750 F	CONDUCTOR	BLACK	—
TEFLON-S* #400	LOW FRICTION ANTI-STICK	325 F	2 mil (1 mil/coat)	+ −.0005"	1500v/mil	450 F	ICE RELEASE LOW TEMP. CURE	GREEN BLACK	THERMO-SET
TEFLON-S* #550	HARD TEFLON FOR ABRASION RESISTANCE	450 F	1.5 mils (.7 mil/coat)	+ −.0005"	1500v/mil	650 F	FOR DIFFICULT POTTING MOLDS SAW BLADES	BLACK	THERMO-SET
DURAFILM TEFLON*	DURABLE WEAR RESISTANCE RELEASE	425 F	1 mil (.5 mil/coat)	+ .0002"	1000v/mil	700 F	FOR DIFFICULT EPOXY OR URETHANE MOLDING	BLACK GREEN	THERMO-PLASTIC
TEFLON (UF)*	RUBBER "O" RING LUBRICATION	AS ALLOWED BY RUBBER	3 mil (1 coat)	+ −.002"	—	AS ALLOW BY NUMBER	INSULATION ON FERRITE CORES	GREEN	THERMO-SET
TEFLON* SFDA	FOOD CONTACT COOKWARE COVERS	425 F	1.5 mils (.7 mils/coat)	+ .0005"	1500v/mil	650 F	PROCESSING EQUIPMENT WEAR RESISTANT	BLACK	THERMO-SET
DURAFILM FLUORO-CARBON "A"	LOW TEMP. CURE	450 F	1.5 mil (.7 mil/coat)	+ −.0005"	—	300 F - 1 Hr 400 F - ½ Hr		BLACK	THERMO-SET
KYNAR*	ANTI-CORROSION CHEM RESISTANT	450 F	12-15 mils (3 mils/coat)	+ −.001"	1280v/mil	500 F	DURABLE AND ABRASION RESISTANT	YELLOW BLUE GOLDEN BROWN BROWN	THERMO-PLASTIC
DURAFILM "K" (COMPOUNDED EPOXY)	ANTI-CORROSION	400 F	5 mils (1 mil/coat)	+ .001"		385 F	SALT WATER RESISTANT	GREEN	THERMO-SET

FIGURE 36.3 Coatings data chart. *(American Durafilm Co., Holliston, MA)*

Another factor that requires expertise on the part of the applicator as well as the design engineer is matching the coating with the substrate material. Teflon coatings, for example, can be used on such diverse substrates as carbon steel, stainless steel, aluminum, titanium, brass, tin, magnesium, ceramics, glass, fiberglass, and plastics. Most applicators would be glad to assist engineers select and try out the proper coating for the job.

VENDORS: LICENSED INDUSTRIAL APPLICATORS

Most of these copolymer and dry lubricants are applied by very expert companies that have been in the business 25 to 45 years. If you select one of these, you will not have bad experiences. They have seen all the problems and know how to avoid them. The proper finish can be used to decrease production costs in a vast number of areas. Once the parts are coated, they have such a low coefficient of friction that they slide, glide, roll, and turn punch or pull without lubrication. The fluorocarbons are self-sealing, nonsticking, non-wetting, heat and chemical resistant, and have cryogenic capabilities and excellent dielectric properties. They will save wear and tear on your machinery.

These LIAs can show you how to increase sale potential. Your products will look better, work better, and last longer. You can advertise that your products have been surface treated. Pots and pans that are Teflon-treated sell easily. Snow shovels that are coated sell themselves. These coatings can be used on very small components, like a ball bearing or on large conveyors or on everything from zippers to boat propellers to hand tools to the inside of a tank car. See the 20 examples of commercial uses in Figures 36.4 and 36.5. Most products can be improved in some way by utilizing this process.

When you think of a possibility, call a nearby LIA and discuss your idea. He will inform you of the military specifications he can meet, including quality control. He will describe his electronic check for pinholes, the large batch furnace, and the medical

Product:
Ripsaw blade

Benefit:
Dry lubrication and ease of release provide longer blade life with sharper and more efficient cutting

Product:
Parts feeder

Benefit:
More efficient processing of sticky and gummy parts

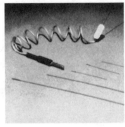

Product:
Step-in ski binding

Benefit:
Prevents ice build-up which would jam safety bindings

Product:
Needles for neuro-muscular examination*

Benefit:
Dry lubrication provides easier insertion while providing good insulation and durability

Product:
Glove lay-off board

Benefit:
Ease of removing molded glove, cut production time and rejects

Product:
Heart starter*

Benefit:
Good electrical insulation and ease of cleaning

Product:
Nylon stocking preform board

Benefit:
Eliminates weekly cleaning; provides easy removal of stocking

Product:
Filter screen

Benefit:
Being hydrophobic, "Teflon" **filters out free water from hydrocarbons and provides a permanent filter separator

Product:
Gravity flow volumetric feeder

Benefit:
Provides 100% release of low density fine powder

Product:
Slides for Freezer drawers

Benefit:
Excellent ice release prevents jamming of drawers by ice build-up

*The suitability of "Teflon" finishes for specific medical uses must be determined by competent medical authority.
**Reg. U.S. Pat. Off. for Du Pont's non-stick finishes.

FIGURE 36.4 Sample products and their benefits. *(E. I. DuPont Co., Wilmington, DE)*

Product:
Electrical switch parts

Benefit:
Dry lubrication and electrical insulation result in longer maintenace free operation

Product:
Paper guide for computer

Benefit:
Permits easier paper passage

Product:
Nuclear submarine ball valves

Benefit:
Lowers torque and corrosion resulting in longer maintenace free life

Product:
Water valve— "Teflon-S"* coated

Benefit:
Corrosion protection and longer life by reduction of salt deposits

Product:
Industrial blower wheel

Benefit:
Reduces dust and product build-up; does not go out of balance, provides longer life for bearings

Product:
Roller used in Xerox #2400 copier

Benefit:
Provides release properties at high operating temperatures

Product:
Exhaust fan blade

Benefit:
Used in spray booth exhaust system to prevent build-up of sticky solids

Product:
Carburetor

Benefit:
Dry lubrication eliminates jamming of throttle and choke shafts especially in cold and icy weather

Product:
Paper feed guide

Benefit:
Dry lubrication prevents paper from jamming sorting machine

Product:
Air compressor Connecting rods

Benefit:
Good lubrication, release and corrosion resistance

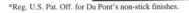

*Reg. U.S. Pat. Off. for Du Pont's non-stick finishes.

FIGURE 36.5 Sample products and their benefits. *(E. I. DuPont Co., Wilmington, DE)*

instrument-coating clean room. They will accept work from companies involved in aerospace, chemicals, marine products, food processing, medical, automotive, electronic, or textiles.

CHECKLIST FOR DESIGNERS

1. Closed vessels, such as metal rolls, must be vented, drained, and removed of any bearings.
2. Gaps, cracks, and other voids in the part can create problems in coating. During the coating process, the wet spray might bridge the imperfection and entrap air. When the part is cured, the air will expand and blow pinholes in the coating.
3. Sharp edges and corners should be radiused for more uniform coating coverage. Otherwise, a "dog bone" effect might result.
4. For best results, most coatings are cured at 700°F. You must be sure that the elevated temperature doesn't cause a material deterioration. If in doubt, consult a specialist.
5. Copper, bronze, and brass tend to oxidize at most curing temperatures, which could affect the adhesion of the coating. Aluminum, steel, and iron are preferred materials.
6. Thermoplastic coatings will melt flow during the cure. Thus, holding "points" must be selected to hang or support the part during the cure.

The following information for design personnel has been supplied by The Precision Coating Co. of Dedham, MA, concerning their subspecialty, coating-tank interiors:

1. Prior to sending tanks to have their interiors lined, all seams must be completely bead welded and smooth. In addition, all surfaces requiring coating must be free of weld splatter, scale, porosity, and sharp edges. For optimum corrosion resistance, flanged fittings should be used wherever possible.
2. Vendor surface pretreatment includes degreasing, initial 700°F burn-off to degrade residual contaminants, plus a final aluminum oxide grit blast to white metal for guaranteeing surface cleanliness and to help promote mechanical adhesion.
3. Bisonite Phenoflex 957, manufactured by Bisonite Corp. and Plasite 3055, manufactured by The Wisconsin Protective Coating Company are pure phenolic-type coatings that offer good solvent, alkali, and abrasion resistance. Approximate film buildup is 3 to 6 mils for a two-coat primer and a three-coat (or more) clear system, which is cured at 400°F.
4. Kel-F manufactured by Precision Coating is a fluorocarbon coating offering good acid, alkali, and abrasion resistance. Approx-imate film buildup is 6 to 12 mils in a two-coat primer and a five-coat (or more) clear system that is cured at 500°F.
5. F.E.P. Teflon, manufactured by DuPont is a fluorocarbon coating offering an excellent wide range of chemical resistance, but has limited abrasion resistance. Approximate film thickness is 2 to 4 mils in a system that incorporates a primer, two intermediate coats, plus multiple clear top coats cured at 700°F.
6. P.F.A. (perfluoralkoxyethylene), manufactured by Precision Coating is a fluorocarbon coating with an excellent wide range of chemical resistance, along with good release properties and a 500°F working temperature. The approximate film thickness is 2 to 5 mils in a system that requires a primer plus multiple clear top coats that are cured at 700°F.

CASE HISTORIES

Ten years ago, *Motor Magazine* printed a story about the use of Teflon coatings on race cars. The perennial winner, A. J. Foyt's team had been using the coating on bolts so that they could be more accurately torqued without friction-induced errors and also to minimize seizure—especially in castings. Then, they began putting it on wheel hub nuts so that wheel changing would be easier and faster. Now they are trying it on internal parts in the engine and the drive train. They feel that any place where friction exists should be considered. The company doing this type of application for them said that a smooth, clean surface is the first requirement. Cleaning and preparation seem to be the key to this endeavor.

When an automobile engine first starts, the oil pump can't get fresh oil up from the sump for a few revolutions. The pistons only have the oil trapped by the rings for that short duration, so it would be helpful for the pistons to be Teflon coated. Coating valve springs with Teflon cushions the interference between primary and secondary springs and dampens high-frequency vibrations. Both of these actions produce heat, which gradually takes the temper out of the springs.

Coatings applied to any gears, such as in the transmission and differential, permits the use of a lighter grade oil. Teflon coatings can also be effective on timing chains and wheel bearings. Several other car racers are now experimenting with coatings and automobile manufacturers also are seeking places to use coatings and replace greasing as much as possible.

The following is a list of parts where the automotive industry is either already using fluorocarbon coatings or is considering its use:

1. Valve springs, as in the racing cars.
2. They have improved gas mileage by coating valve stems.
3. Experiments have demonstrated that an aluminum piston can be used in an aluminum housing without galling, provided that the piston is coated.
4. Fuel pump cam rings make good use of dry lubrication.
5. Coatings have extended the life of throttle body shafts and levers.
6. Coating gaskets when they are used between different materials prevents expansion leaks.
7. Coating a clutch plate's inner spline reduces maintenance costs.
8. On solenoids, coatings prevent stickups.
9. On hinges, locks, and springs, coating reduces friction and noise.
10. Coating brake parts reduces corrosion and helps deliver uniform pressure.
11. On MacPherson strut bushings, coatings provide dry lubrication and corrosion resistance.
12. Coatings lubricate and seal pores of powdered metal bearings.

The chemical industry is using coated carbon steel mixers, instead of the stainless-steel material. Oil-recovery systems are applying a 1.5-mil coating (a primer and two top coats) to their oil-filter scavengers. The list of possibilities is endless, but the following might stimulate creative thinking:

1. agitator paddles
2. aluminum hosiery forms
3. bakery dough-handling equipment
4. baffle plates
5. business machine parts
6. glue dispensers and glue pots
7. candy equipment
8. linings for kitchen ovens
9. snow plow blades
10. carton guide rails
11. cookie pattern rolls
12. solid propellant pouring mandrels
13. pumps
14. conveyor chutes
15. paint mixers and drums
16. cheese molds
17. dough divider hoppers
18. valve stems and seats
19. dry-ice platens
20. extrusion dies
21. ice-making equipment
22. pails
23. packaging equipment
24. paper-making machinery
25. wax pans
26. storage bins
27. valve stems and seats

CHAPTER 37

INDUSTRIAL FINISHING SYSTEMS

WHAT IS A BURR?

Many metal-working manufacturing personnel have encountered circumstances when their interpretation of edge quality differed considerably from that of other workers. Frequently, this difference caused delays, extra costs, and frayed tempers. Accidents, damage, and even loss of contracts have resulted. Litigation has also been caused by this lack of proper communication and understanding.

There is no standard definition for a burr. One company might examine for burrs at a magnification of 10, and another might use a magnification of 50 (or even higher). A company might declare their product burr-free if nothing falls off when the assembly or part is turned over. Each industry treats this situation in their own way. Furthermore, each company might have its own interpretation of what is a burr. With some products, safety is jeopardized by the presence of burrs. One application might consider a raised or rough edge as part of the parent metal and no concern whatsoever. Another company might judge such an imperfection as an undesirable burr. One company defined a burr as any material that could be dislodged by an 80-psi blast of air. Consequently, edge conditions should be specified clearly in a note on the drawing. The note should be quantitative, but not over specified.

This book contains several chapters on deburring and surface finishing, but those subjects would be incomplete without a few words on industrial finishing systems (such as dryblast, impact treatment and vibratory finishing machines). I have sought the advice of three well-known, reliable manufacturers of this type of equipment and you will read their contributions in this chapter. The vibratory tubs and sand-blasting equipment of the World War II era have been superseded by technologically superior equipment.

The Vibrodyne Company of Dayton, OH, produces precision finishing vibration machines. Almco, Inc. of Albert Lea, MN, also produces vibratory tubs and a few more sophisticated specialty machines. Guyson Corporation of Saratoga Springs, NY, manufactures dryblast, impact-treatment machines. Each company also markets abrasive and nonabrasive media for any of these machines, plus many types of ancillary equipment. If you are interested in industrial finishing equipment, it is suggested that you investigate other sources in the field and satisfy yourself that you are dealing with the proper company. I felt secure that the information furnished by these three companies is accurate and indicates a profound depth of knowledge.

VIBRATING FINISHING MACHINES

The vibrating tub type of finishing machine is simple, sturdy, and effective. The only moving part in the patented Vibrodyne Finishing machine is the tub, lined and molded in place with a heavy thickness of polyurethane and suspended on rugged, alloy steel I-beam springs. Epoxy-encapsulated electromagnets provide the energy to vibrate the tub at 3600 cycles per minute. Parts and media roll full circle within the tub providing complete contact and continuous scrubbing action on all parts. The components will not pack or tangle when properly processed. Complex parts with hard-to-reach areas will receive positive scrubbing all over (Figure 37.1).

There is nothing to lubricate—no expensive, complicated mechanical drives, such as motors, bearings, sprockets, chains, belts, or cranks. In addition, each unit is dynamically balanced, making it unnecessary to bolt it to the floor. A separate unit contains the variable-voltage transformer, which controls the power and range of vibration. No other adjustment is needed. Fragile parts can be processed at a lower amplitude, but rugged, heavily burred parts receive a more aggressive cutting action with a higher amplitude.

This machine comes in several sizes, from a scaled-down table-top version used in laboratories and by jewelers, to a large model with a six cubic foot capacity. The capacity of the small unit is only ⅛ cubic foot. The accessories available include:

1. A compound-metering device that ensures accurate proportioning of chemical concentrate to water and controls the flow to the machine.
2. A compound recirculation system, which maintains fresh compound mixture while removing debris, burrs, chips, and shavings.
3. Tub dividers, which clamp on and permit separate processing of parts.
4. Two sizes of screen separators, which facilitate the separation of media from components.
5. The usual assortment of tote pans, hopper pans, and media bins.

The following is a general list of uses for the various media that are readily available:

1. Ceramic media:
   -Steel deburring
   -Aluminum castings
   -Matte finish
   -General parts cleaner

2. Plastic and synthetic media:
   -Nonferrous metals deburring
   -Preplate finisher
   -Frost finishes

3. Porcelain media:
   -Polishing and burnishing
   -Ferrous and nonferrous metals
   -Combines with abrasives to clean and polish

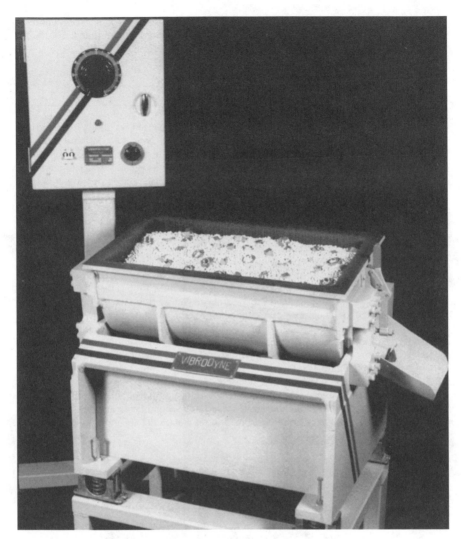

FIGURE 37.1 A Vibrodyne finishing machine. *(Vibrodyne, Inc., Dayton, OH)*

4. Steel media:
   -High-quality polish
   -Fast cycle times
   -Specialty deburring

5. Wood and miscellaneous media:
   -Dry process polishing
   -Glass and plastic applications

6. Sizes and shapes available:
   -Individual design required for special components

7. Compounds:
   -Selected to suit requirements

One thought must be foremost in your mind. The three companies that are mentioned in this chapter offer competing equipment, as do others that aren't mentioned. Your job is to select the best for your application.

GENERAL CAPABILITIES OF DRYBLASTING

Dryblast treatment is one of the most widely used (and misused) methods for deburring parts. When used properly, it can be efficient and save costs. New materials and processes in engineered finishes and coatings have introduced challenges and new opportunities for surface preparation. The simple sandblaster of two generations ago has been supplanted by an automatic dryblaster. This machine can be used to clean, condition, texture, and strip. For rapid removal of rust, scale, and other surface deposits, dryblast cleaning can offer advantages over other mechanical cleaning processes or chemical treatments, such as acid pickling and caustic baths. Even when such treatments are required for downstream processes, blast precleaning can yield major savings in time, materials, and the reduction of wastes, such as chemical sludge (Figure 37.2).

The two most common types of dryblasting processes are airblasting—using compressed air to propel the blast media against the workpiece, and airless, turbine-wheel blast systems that use centrifugal force to propel the media. Within airblast technology there are two types of equipment: suction and pressure blast. The suction system uses a blast gun, which is connected to two hoses—one for the blast media and the other for incoming compressed air. The incoming air to the gun creates suction, which siphons the blast media from a feed hopper and forces it out through the gun's nozzle. An automated system would have several guns with individual controls so that different forces could direct media where it is needed (Figure 37.3).

In cabinet-type airblast systems, the operator stands or sits outside of the machine and, placing his hands into large gloves, which are part of the equipment, handles the blast nozzle and/or the component. In an automatic system, the components and nozzles are manipulated in an effective manner and the operation is standardized for production.

FIGURE 37.2 A complete dryblast system. *(Guyson Corp., Saratoga Springs, NY)*

Batch deburring and cosmetic finishing of metal components on custom built magnetised component holder

Removal of encapsulation deposits from electronic components

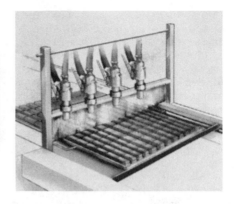

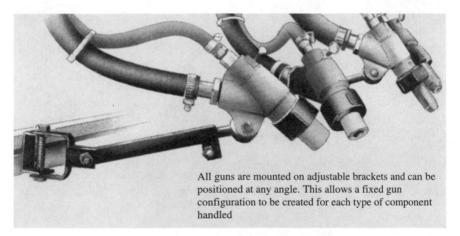

All guns are mounted on adjustable brackets and can be positioned at any angle. This allows a fixed gun configuration to be created for each type of component handled

FIGURE 37.3   Suction guns in action. *(Guyson Corp., Saratoga Springs, NY)*

In the direct-pressure system, the blast media is propelled under a regulated pressure from the media storage container at a higher pressure than is attainable in the suction system. This is generally used when processing large parts, when high production rates are required, or when the larger suction-type hose is too big to effectively hit the desired areas.

With manual blast equipment, the particle delivery is, of course, by hand. Constant attention is required and the quality can vary from operator to operator. The most common blast cabinets are the suction feed, in which, as described before, blast particles are drawn into the blast gun by a vacuum and accelerated within the gun by a metered stream of compressed air. Suction guns always have two hoses: one to supply regulated air and the other to feed abrasive to the gun.

In contrast to the suction-feed systems, direct-pressure systems provide faster coverage of larger surfaces and more energy efficient media delivery. In pressure-blast equipment, the media is held in a pressurized vessel and metered into a single hose. At a given pressure level, nozzle particle velocity is higher with direct pressure than with suction-feed delivery. Thus, more work can be done in a shorter time (typically, about four times faster).

Direct-pressure systems make more efficient use of expensive compressed air than the suction-feed systems because all of the air is used to accelerate particles; in the suction gun, some of the air is used to generate vacuum for media pickup.

Blast treatment is often the fastest and least costly way to achieve a uniform surface condition that improves, speeds, and simplifies subsequent processes. Surface irregularities, scratches, tool marks, etc., which interfere with coating or cosmetic requirements, can be blended to an overall even condition, without removal of material or generation of wet wastes. Dryblast methods are commonly used to remove unsatisfactory organic coatings, leaving the surface in excellent condition for recoating.

Many applications of versatile dryblast technology exploit the properties of nonabrasive or mildly abrasive blast media in such processes as cleaning, deburring, and deflashing. This same technology is used as a cost-effective alternative for the controlled modification of surfaces, both as a final cosmetic finish and as a preparation for coating or bonding.

Both automatic airblast and turbine-blast (wheelblast) systems are excellent for preparing surfaces for coating. New engineered materials that have been specifically developed for use as blast media have increased the scope of this technology. Satisfactory applications of automatic dryblast techniques depends on the successful combination of numerous factors including:

1. Selection of the blast medium. The use of an inappropriate medium wastes time and money.
2. A reclamation and delivery system adjusted for optimum effectiveness.
3. Component handling facilities that effectively present components to the blast.
4. Adequate monitoring and control systems.
5. Equipment designed for the harsh abrasive blast environment.

All types of blast equipment incorporate a media reclaimer and a dust collector. The media reclaimer automatically separates the reusable media, recycles it, and removes the fractured particles and debris to the dust collector.

TURBINE BLAST-MEDIA DELIVERY

In situations where general cleaning or overall surface treatment is required, the airless wheelblast equipment performs without the added expense of compressed air. In wheelblast (airless) systems, media is metered to the center of a high-speed turbine wheel and fed onto blades that sling particles at the component. The turbine is generally positioned about 18" from the part being treated to permit the blast stream to fan out. Because the media must retain sufficient kinetic energy to do the work, it is normally restricted to heavier metallic shot.

Turbine blast equipment is much more energy efficient than the airblast delivery. A small-diameter turbine driven by a 5-horsepower electric motor can throw 150 pounds of blast material per minute, creating a blast pattern about 4" wide and 24" long. For comparison, a 5-horsepower compressor will power a single-suction gun that delivers about 5 pounds per minute of the same medium, creating a blast pattern about the size of a half dollar. Because of the volume of media thrown and its weight, media used in airless systems is generally conveyed to the top of the machine by a bucket elevator.

Primary separation of dust and fines occurs when particles are dumped from the elevator bucket, and they cascade through an airwash zone where the lighter particles are pulled out of circulation. When extremely light particles are used, a secondary separation might be required.

Selection of Blast Media

The surface specification should state quantitatively the roughness measurement by profilometer or other verifiable means. Unless the choice of the medium is already known, the selection process should include a demonstration of the alternatives. Because more than 100 blast media of natural and man-made materials are available in many sizes and screenings, the test results should be thoroughly evaluated. Experience and/or experimentation should determine the media selection.

Consider the shape, size, mass, hardness, and fracture characteristics of media for a given application. This is significant because most nonmetallic media fracture quite easily under harsh mechanical acceleration. A general rule is to use the smallest particle size that will do the work because this yields the most impacts per second of blast exposure. It is also a good idea to use the lowest blast velocity to minimize consumption of blast medium. Based on the surface specifications, most media can be quickly eliminated from consideration. Sample testing should then proceed on the alternatives (Figure 37.4).

Suction-feed equipment is good for precise work, where delicate parts are being handled. The machine does very well using nonabrasive plastics, such as nylon or polycarbonate. Yet, these same machines can deliver finished components more quickly if the proper metal beads are used. Again, experience is vital in selecting media.

A. Plastic media

B. Plastic media

C. Ceramic media

D. Compounds

FIGURE 37.4 Material used in finishing machines: A and B. Plastic media. C. Ceramic media. D. Compounds. *(Almco Corp., Albert Lea, MN)*

The experience of the machine manufacturers who have been in this business for many years, cannot be excluded from consideration. That experience helps determine the variables such as:

1. Distance from nozzle to component.
2. The angle of impingement.
3. Should fixtures be used?
4. What is the best way to present the component to the blast?

MACHINE TYPES

The goal of efficiency has led to construction of various equipment designed for specific applications. In addition to manual systems (operator holding component and/or blast nozzle), three basic processing concepts are available in automatic blast systems, including batch-tumbling rotary tables, and in-line conveyors.

The batch-tumbling machine has a gentle tumbling action to preclude component damage. In Figure 37.5, twin guns blast a basket full of components. A rotary table system is shown in Figure 37.2 and an in-line operation is shown in Figure 37.3.

Model	Basket Capacities		Cabinet dimensions (H" × W" × D")	Approximate maximum component size (length")
	Volume (cu. ft.)	Weight (lbs.)		
T-86	1.0	150	74 × 44 × 60	6
T-50	1.0	75	74 × 48 × 50	6
T-40	0.5	50	74 × 40 × 40	4

Unique basket design and gentle tumble action eliminate part-on- part impingement.

1. Tumble basket rotation variable at control panel.
2. Blast gun(s) and airwash nozzle(s) mounted on multi-adjustable brackets.
3. Advanced abrasion protection available for harsh media applications.
4. Cyclone reclaimer separates fractured media, dust and debris; recirculates reusable media.
5. Dust collector is totally automatic–no shut-down for cleaning.
6. Easy-to-use controls regulate all cycle parameters–even loading/unloading on fully automatic Model T-86.

FIGURE 37.5   A typical tumbleblast system. *(Guyson Corp., Saratoga Springs, NY)*

Before   After

FIGURE 37.6 Molds cleaned in a mold-and-die cleaning, turbine-blast system. *(Guyson Corp., Saratoga Springs, NY)*

Normally, a rotary table system is used when parts cannot be tumbled or when selective treatment is required. The machines can operate in three distinctive modes: continuous rotary, rotary index, and rotary index with satellite work spindles. Operating in continuous rotary mode, the components are placed on the table, rotated into the machine under the blast nozzles, passed through an airwash, then exited from the machine.

The rotary index mode operates in much the same way, except that the components to be finished are placed in designated areas of the table. Each part is then indexed into place, where it is blasted for a preprogrammed time before the next index. The third mode operates much the same as the rotary index, except that the components are fixtured onto individual rotary spindles. The spindle is rotated when in the blast work station, which allows a single nozzle to cover the entire part.

For large components, a conveyor is a natural choice. It can be integrated into the current material-handling system. The components can simply be positioned on the conveyor, passed into and through the machine under the blast nozzles, through an airwash, and out to be unloaded or turned over for a trip to do the second side.

Most models allow for single-operator loading and unloading. All cabinets are designed to be free standing, floor mounted, and requiring no special foundations. Totally enclosed to house blast guns, air-wash facility, and handling system, each cabinet has wide access doors, full-view inspection windows, and interior lighting, which provides excellent visibility during operations. Safety switches installed in cabinet doors ensure that all operations will cease if they are opened when blast cleaning is in progress.

Guyson makes a popular automatic mold and die-cleaning turbine-blast system. They also manufacture multiple-gun airblast systems, which normally use their flexbead or metalbead media to deliver nondestructive cleaning. Molds are automatically cleaned while rotating on a heavily reinforced table inside a completely enclosed cabinet. A full table of molds can be cleaned in 10 to 15 minutes per side (Figure 37.6).

The purity of the cleaning compound is maintained by a closed-loop reclamation system with rotating sieve, air and centrifugal separators, and a dust collector. This equipment cleans fast without damaging the tooling. The operation reduces labor, and eliminates operator error and the need to handle chemicals. It is an easily maintained airless operation.

The standard features of most machines include:

1. Safety interlocked doors.
2. Heavy-duty construction with continuous, welded steel plates.
3. An airwash with manual or automatic blow-off nozzles on all models.
4. Direct-pressure or suction-feed media delivery available individually or in combination.
5. Reclaim systems to clean media and maintain visibility in the cabinet.

Optional features normally available include:

1. Transfer carts to handle heavy components.
2. Turntables for convenience and assistance.
3. Custom configurations are available from pre-engineered designs: stretched versions, custom access, etc.
4. Advanced abrasion-protection package consists of durable rubber cabinet linings, wear-resistant urethane ductwork, and inner-abrasion window in quick-change frame for harsh media applications.
5. Enhanced reclaim options include vibrating screens, magnetic particle separators, etc.

MEDIA RECLAMATION

Deburring and deflashing are dryblast applications that require attention to the "back end" of the process: specifically, the reclamation system that removes contaminants from the blast medium and maintains effective blast particle size.

The blast cabinet is a negative-pressure enclosure from which volumes of air are constantly extracted and filtered in a dust collector. Fractured blast media particles and contaminants removed from components are not released back into the plant. A properly adjusted separator is key to the optimum operation of airblast systems. In the case of Almco and Guyson equipment, a cyclone separator is placed between the blast cabinet and the dust collector; it separates dust, fines, and fractured media particles from the recirculating media mixture. This helps maintain effective sizing of blast media.

As reusable media, fractured media, and dust, are carried into the separator (cyclone) by a controlled flow of air, the heavier particles of "good" media travel to the outside and drop down, to be returned to the blast system. Lighter particles are suspended in the central extraction zone of the cyclone and drawn off to the dust collector.

Adjustment of the airflow through the separator (cyclone) is critical to effectively remove lighter, undersized media particles. Too little airflow will result in slower, dustier blast operations, and too much flow will carry good media particles to the dust collector. Inspection of the contents of the dust collector will indicate the performance of the separator. Five to ten percent of the material in the dust bin should be "good" media particles.

If the contents are purely dust, it means that fractured, ineffective particles are abundantly present in the blast medium.

ENHANCED MEDIA RECLAMATION

The presence of undersized, fractured particles in the blast mixture can result in an unacceptable variation in the surface finish. The separator (cyclone) alone is not sufficient for the degree of separation required by some specifications. This problem might be caused by the similarity in mass between the fractured particles and the "good" particles. In such cases, adding a mechanical screening process to the reclaim stack-up will permit tighter control of particle size. A two-screen vibrating classifier is used to sift the incoming media to the feed hopper. The first screen captures particles that are larger than specified and the second screen allows only particles smaller than specified to pass. Together, they effectively remove flakes, chips, burrs, and flash contaminants, which can cause clogs and finish defects. With the screens in operation, the medium is continuously sieved to precise size.

GRITBLAST SURFACE PREPARATION

Impact surface modification has the effect of increasing surface area. Angular particles will penetrate deeper than rounded particles. This is a critical consideration in preparing surfaces for coating applications. Hard grit media, such as aluminum oxide, generates a nonreflective, matte finish. Soft plastic or agricultural grit will alter only the surfaces of soft substrates. Mineral, ceramic, or metallic media are used in airblast equipment, and iron or steel grit is normally used in turbine blast systems. Blast-machine maintenance costs are higher with more aggressive media, which requires more frequent inspection and replacement of worn parts necessary. Dryblast impact surface treatment is a dynamic technology that is undergoing continuous development in response to the needs of manufacturers.

Each of the three companies listed on the first page of this chapter have their own particular approach to the process of deburring and finishing. Vibrodyne presents a vibrating tub. Guyson has the dryblast approach. Almco offers the round-bowl vibratory machine. That is, of course, a simplistic explanation because all three companies market products similar to each other's besides the ones mentioned. For instance, Almco manufactures vibratory finishing systems, spindle finishing machines, horizontal and through-feed machines, material-handling equipment, media and compounds, filtration systems, and washers and dryers.

For any company to recommend equipment, certain conditions must first be established. The burr condition should be determined and made part of the drawing radii required, and cleaning that is required. It is also necessary to know production (pieces per hour) required in terms of time (such as next year and the following year). The Almco Company, for instance, would run tests on sample components and establish time cycles, recommended media, and the proper equipment to accomplish the job.

Sample processing means extensive laboratory testing to acquire the information that verifies the process. That would include media selection, along with proper finishing compounds. The report should list all trials and their results, such as what media was tried and what happened to the media, as well as to the samples. Steel media burnishes the surface

FIGURE 37.7 The popular round-bowl vibratory machine. *(Almco Corp., Albert Lea, MN)*

FIGURE 37.8 A special vibratory-finishing machine. *(Almco Corp., Albert Lea, MN)*

and leaves a bright finish. In some applications, you would want the workpiece peened without removing material. This type of finish might be desired for oil retention, as on crankshaft journals.

Star-shaped plastic or ceramic media is used to enter holes and massage the side walls. These media must be the correct size and the operator must know when they have worn enough to be replaced. The right media is significant in most jobs. Compound selection is also important. This finishing compound cleans the components and the media. The compound selection can also affect optimum speed setting and cycle time. Some pressworked parts need a compound that will clean heavy drawing lubricants. An alkaline compound that will attack aluminum might be just right for steel parts. The percentage of compound added to water affects cleaning. Again, operator knowledge is important.

The industry started with basic tumbling barrels, but it has evolved over the years to meet the needs of a growing market. Those tumbling barrels have given way to vibratory machines, spindle equipment with robot loading and unloading, filtration systems, and closed loop washer/dryers. Almco specializes in the round-bowl continuous vibrator (Figure 37.7). Now, these machines are being used by the automotive, aircraft, machine tool, and sporting goods industries. Many companies are using fully automated systems that incorporate programmable controllers and computerized robots. Figure 37.8 shows one of their special machines used by the Boeing Company. As mentioned, these machine manufacturers will design and build special equipment to suit specific needs.

It is your job as an engineer, purchasing agent, or manager, to sift through this information, compare it with your own investigation, and select a reliable source to supply your needs. You require an engineering, process-oriented team to function as your problem solver. Most any vibratory finishing machine manufacturer would be glad to cooperate with you, finish some of your samples, and outline a system that would do everything that is required. Your job is to select the company that you believe will supply the experience and equipment most compatible with your needs.

P · A · R · T · 7

ELECTROCHEMISTRY PROCESSES

CHAPTER 38

ELECTROCHEMICAL MACHINING

PROCESS DESCRIPTION

Electrochemical machining is one of those nontraditional machining processes that is finding more applications as equipment improves and manufacturing engineers learn more about it. First the aerospace industry found it very useful. Then, applications were found in the following products: automobiles, appliances, machinery, military hardware, and even medical implants.

Electrochemical machining (ECM) is a controlled, rapid metal-removal process with virtually no tool wear. It leaves no burrs and does not generate high thermal stresses. Unlike conventional processes, ECM removes metal, atom by atom. It can be regarded as a solution to a variety of metal-removal problems, such as cavity sinking, radiusing, contour machining, and machining helices (e.g., rifle barrels).

There are some standard ECM machines. However, much equipment for production is built to suit a specific machining task. Either way, they have the same type of components: The machine itself, the frame, work head, table, and the work enclosure, which prevents the electrolyte from spilling over the floor. Then, there is the electrical system, which includes the machine control for electrodes and electrolyte, and the dc power supply. Third is the electrolyte system, which is a closed system, including tank, filter, sludge-removal circuit and treatment units, regulating components, and supply system and finally the electrodes.

The ECM system is controlled from the operator's panel. Power supplies for ECM equipment are currently built with outputs ranging from 200 to 40,000 amperes. The voltage is usually variable between 7 and 20 volts. Gap-protection devices for ECM equipment consist of five different detectors that respond to short circuit, turbulence, passivation, contact, and overcurrent.

Sometimes a standard machine tool will be modified and transformed into an ECM. I visited a company in Connecticut that had a dozen large machine tools that had been converted. Large drills and milling machines seem to be the favorites for conversion. For information concerning the suitability of the ECM process for your particular task, telephone

Anocut, Inc., near Elk Grove Village, IL, at (708) 427-1199 or the Robert Bosch Corporation Surftran Division in Madison Heights, MI, at (810) 547-8103. These two companies sell machinery to do ECM work and will also act as vendors to accommodate companies that want to try out the process.

The type of machine depends on the workpieces. Should it be a single-station or twin-station machine? Should it be vertical or horizontal? Should the machine be automated, have an indexing table, or have some other special equipment? There are many questions for the uninitiated to have answered before embarking on a large investment.

Electrochemical machining (ECM) uses components of mechanics, electricity, and chemistry, usually arranged as depicted schematically in Figure 38.1. The cathode tool (Figure 38.1A) is shaped to provide the form desired in the workpiece (Figure 38.1B) with appropriate compensation for overcut. The Anocut power unit supplies 5 to 20 volts dc, with a negative charge through cables to the tool and positive charge to the workpiece. An electrolyte (electrically conductive solution) is pumped under high pressure between the tool and the work. In the meantime, a mechanically driven ram feeds the tool at a constant rate, preset by the operator, into the workpiece to machine the desired shape.

This is a good spot to explain some differences and similarities. Figure 38.1 details the ECM process, showing all necessary components. Figure 38.2 shows how electroplating works. The electrodes are immersed in stagnant electrolyte and positioned away from each other. The current density is very low and the anode is attacked very slowly. In ECM, the close proximity of the shaped tool to the work provides high current density for fast material removal and accuracy.

Figure 38.3 shows electric-discharge machining (EDM) in which a pulsating dc current sends small sparks through a nonconductive oil. The energy of the sparks vaporizes small portions of the metal from the workpiece. The big differences are:

1. ECM is much faster than EDM.
2. ECM has no tool wear; EDM can cause tool wear.

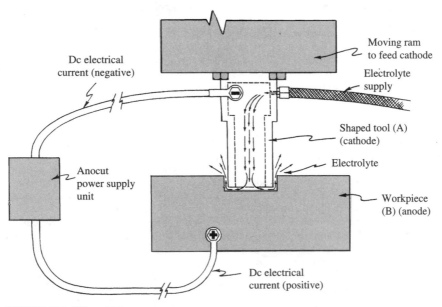

FIGURE 38.1 EC machining.

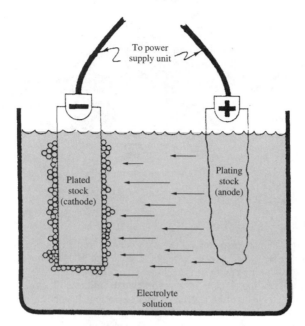

FIGURE 38.2 Electroplating.

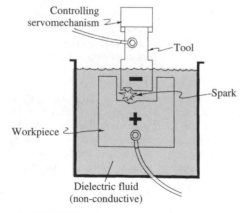

FIGURE 38.3 EDM process.

3. ECM has no heat damage; EDM can cause heat damage.

4. ECM provides better finish and accuracy.

Figure 38.4 shows chemical milling. The part is immersed in a tank of chemical solution (acid or alkali) and the chemical etches or dissolves the metal uniformly over its entire surface. The part is masked around areas where no etching is desired.

Electroplating (Figure 38.2) is the commonly used method of finishing metal (sometimes nonmetals) parts. It generally adds a thin coating of a stainless, exotic, or hard material to

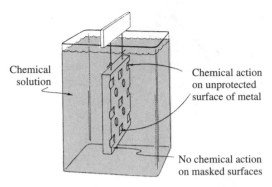

FIGURE 38.4   EC milling.

a part to protect it against corrosion. Figure 38.3 is an example of electric-discharge machining, which is covered in detail as a chapter in this book. EDM is mainly used for such tooling purposes as sinking cavities or machining hardened steel.

Electrochemical milling (Figure 38.4) is somewhat similar to electrochemical machining in that both processes are the reverse of electroplating. Instead of adding metal to a part, these two processes remove metal. The milling is done to very thin conductive metals. It is generally used to remove just a few thousands of an inch of metal. A photoetching step prepares the surface very much like making a printed circuit board by depleting the unprotected (unmasked) areas. Stock is removed, atom by atom, from both sides simultaneously toward the center of the material.

Tooling for electrochemical machining (ECN) must be designed so that all areas of the tool are properly irrigated and the flow across these areas is smooth and even. Improper flow will cause poor surface finish, striations, and unreliable operation. Tooling that is not designed correctly to supply sufficient flow can damage the electrode or workpiece as a result of sparking.

The most familiar use of Faraday's Law is electroplating. ECM is an application of Faraday's Law in reverse of electroplating: controlled metal removal. Faraday's law governs the rate of metal removal; in ECM, the law is assisted by controlling the space between the anode and cathode and the rate at which the ram is fed. All common metals are removed at about the same rate, although the exact rate is determined by the electrochemical equivalents. For estimating purposes, it is convenient to assume that 10,000 amperes will remove 1 cubic inch of metal per minute. The tool never touches the workpiece, so there is never any tool wear or damage from heat or sparking.

STRAIGHT-FLOW TOOLING

Previously, the common method of providing for electrolyte flow was to design electrodes with holes or slots in them, through which electrolyte, generally a conductive salt solution (such as sodium chloride) was pumped. It normally passed across the face of the tool and exited along the outside of the tool between it and the wall of the cavity being machined. For example, to machine a round hole, a hollow tube becomes the electrode. Electrolyte is pumped down the inside of the tube and up the walls of the machined hole (Figure 38.5).

This general method has some disadvantages:

1. The freely exiting electrolyte created striations and a poor finish.
2. The flow from inside out thins and sometimes is insufficient to provide a smooth finish.
3. Complex electrodes can end up with areas starved for electrolyte.
4. The electrolyte tends to spatter in this type of flow.

REVERSE-FLOW TOOLING

Some of these problems can be overcome by reversing the flow of electrolyte. Using what is called *reverse flow tooling*, the electrolyte is introduced between the electrode and the work by means of a dam or supply chamber. The electrolyte enters the work zone around the periphery of the tool. The exit passage could be holes or slots in the electrode face. Each exit passage is connected by a hose to the electrolyte tank (Figure 38.6).

This procedure leads to:

1. A better finish.
2. A uniform and predictable overcut.
3. Freedom from sparking.
4. Cleaner operation, no spatter.
5. This direction of flow prevents undesirable erosion.

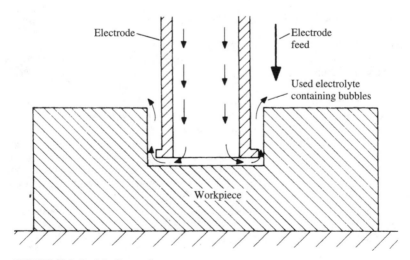

FIGURE 38.5 Straight-flow tooling.

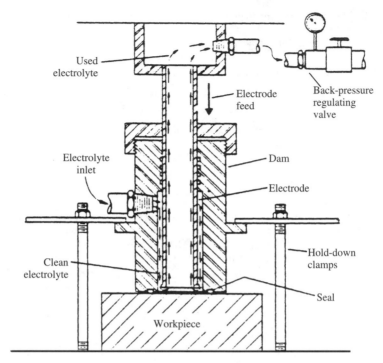

FIGURE 38.6   Simple reverse-flow tooling. *(Anocut Corp., Elk Grove Village, IL)*

Design Characteristics of Reverse-flow Tooling

Figure 38.6 shows a simple form of reverse-flow tooling. Clamps are used to hold the dam down against the workpiece. It is customary to provide more than one inlet connection to ensure equal distribution of electrolyte. The dam can be made of brass, stainless steel, or green glass. If the dam is made of metal, a plate of green glass must be secured to the side of the work to insulate it.

Figures 38.7 and Figure 38.8 show improved reverse-flow tools. Clamps are eliminated and the dam is constructed so that the pressurized electrolyte seats it against the workpiece.

If an electrode has an irregular outline, its shank can still be made straight or round. This type is easier to build. To prevent stray etching of the workpiece, a plate, having a cutout in the shape of the electrode, can be mounted on the end of the dam. The cutout should be about 0.030" larger than the electrode that fits through it. This is to ensure a uniform electrolyte flow around the electrode.

The green glass plate, which has been cut to conform to the electrode's shape doesn't have to be very thick because it transmits the hydrostatic force directly to the workpiece. However, the side walls of the dam must be strong enough to withstand substantial hydrostatic forces.

ECM ELECTROLYTE SYSTEM

Electrolyte is pumped through the working gap. It removes metal depleted from the workpiece and heat generated during the electrolytic action. The electrolyte is stored in a tank and filtered regularly. The pressure is regulated and controlled. To keep the electrolyte machining characteristics constant, the pH value, the concentration, and liquid temperature, are regulated to preset values by means of three individual regulating units.

The dissolved metal precipitates in the electrolyte as metal hydroxide and can be removed by gravitational forces, either by means of a centrifuge or a settling tank. Normally, the wet sludge should be treated to a neutral pH, then pressed into a solid block, 50% solid. The pressed blocks should be sent to a suitable landfill to avoid environmental pollution.

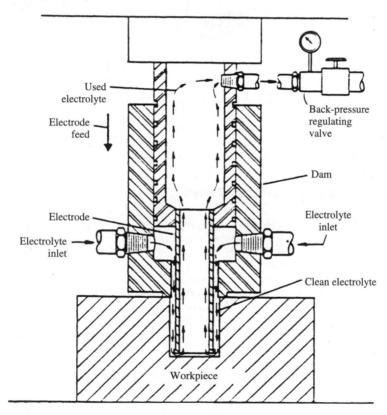

FIGURE 38.7 Improved reverse-flow tooling. *(Anocut Corp., Elk Grove Village, IL)*

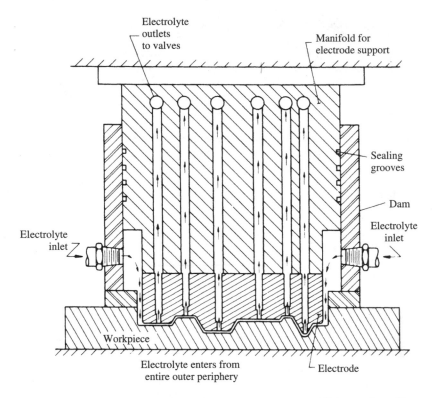

FIGURE 38.8 A second version of reverse-flow tooling. *(Anocut Corp., Elk Grove Village, IL)*

ECM PRODUCTION APPLICATIONS

Turbine wheels are electrochemically machined, blade by blade. For this job, the electrode is a metal plate with a hole of similar cross section to that of the blade to be machined. Each surface of the blade is machined simultaneously under CNC control. ECMing speed for this assignment is between 0.2 and 0.3″ per minute (Figure 38.9).

Constant-twist or gain-twist rifling can be ECM broached into barrels of weapons (such as pistols, rifles, shotguns, or even cannons), at rates up to 20″ per minute. Figure 38.10 shows a simplified, cutaway illustration of vertical rifling. The barrel is slid over a stationary tool that is mounted on a manifold fixture. The ram feeds down delivering a positive charge to the barrel. Electrolyte is introduced through the manifold under pressure and it flows upward, spiraling through the tool's grooves. The positive-charged barrel (anode) allows material to be removed when electrolyte is flushed between it and a negative (cathode) electrode. After removing the barrel, complete electrolytic action is stopped by washing the barrel in water.

The following are advantages of electrolytic rifling:

1. Stress-free grooves.
2. Full selection of groove geometries.
3. Any conductive metal or exotic alloy.

A. Large stationary gas turbine blades showing tip cavity contoured by ECM.

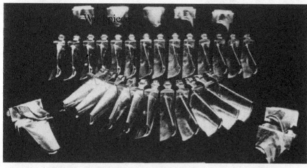

B. Small stationary gas turbine blades showing tip cavity contoured by ECM.

Large aircraft gas turbine compound blade. The midspan pocket contouring was the first production application of electrochemical cavity work in this country. These pockets were first run in 1960 and are still in production.

Aircraft gas turbine blades mass produced by ECM from oversized forgings. The airfoil and the shroud and root shelf and fillets are finished simultaneously, first on one side and then the other.

FIGURE 38.9 Electrochemically machined turbine wheels. *(Anocut Corp., Elk Grove Village, IL)*

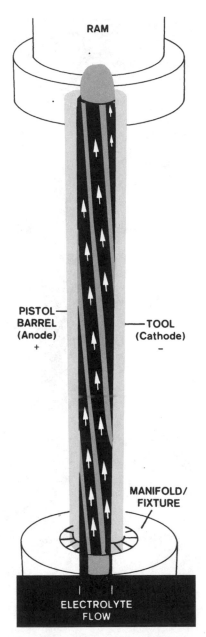

FIGURE 38.10   Rifling a pistol barrel. *(Bosch Corp., Surftran Div., Madison Heights, MI)*

4. Burr-free rifling.

5. Any number of grooves.

6. Nonwearing tool.

7. Constant or gain twist.

8. Repeatable accuracy of ±0.0005".

9. Surface finish of 32 microinches.

10. Fast cycle times.

Profiles of irregular shapes in high volume can be done economically. Six or more parts can be processed simultaneously, using a cassette to hold them in position. Irregular cavities and shapes that are difficult for conventional machining can be easily processed by ECM.

Machines equipped with integral rotary tables are used to make a variety of cavity configurations in engine casings (Figure 38.11). The rotary table is the positive connection,

FIGURE 38.11 An eight-component tool used to machine ratchets. *(Extrude Hone Corp., Irwin, PA)*

A. A 20,000-ampere fixed bed horizontal Anocut machine. This machine is similar to the 10,000-ampere fixed-bed machines, but it has a larger, more rigid ram (17" square) and a much heavier frame.

C. The roll-open enclosure in the open position shows the large fixed-bed table of the horizontal machine, allowing for overhead loading, not possible with vertical machines. Having the tank at the end of the machine conserves floor space.

B. The large, rugged ram of this horizontal heavy-duty ECM machine is required to withstand the forces exerted by the electrolyte when machining large areas with 20,000 amperes.

FIGURE 38.12   ECM equipment produced by the Anocut Corp. *(Anocut Corp., Elk Grove Village, IL)*

the workpiece is fastened to the table, and the cathode is quick-connected to the horizontal ram via hydraulic clamping. As many as six different cathodes can be used for multiple machining around the casing OD.

Relatively large cavities can be economically machined by ECM. Some shapes can be processed in 1/10 the time that the job would require by EDM.

Anocut has developed a machine tool that is capable of cutting deep splines into long tubes with an accuracy and speed not previously possible. The technique uses the principle of reverse-flow ECM using a unique cutting-element design.

Electrochemical machining is not a well-known or popular process. But it is better for manufacturing engineers to consider its cost-saving possibilities when preparing for new production or reviewing old.

Figures 38.11, 38.12, 38.13, and 38.14 will serve to better acquaint the reader with the ECM process.

These shafts and cam plates have been formed as onepiece forgings. ECM was used to finish the contour of the cam plates.

A fully automatic ECM process was used to machine the slots in these adjusting screws at a rate of five seconds per workpiece.

FIGURE 38.13 Parts machined by the ECM process. *(Extrude Hone Corp., Irwin, PA)*

It is more economical to machine the cavities in these cold forming dies by ECM when quantities exceed 20 to 60 (depending on shape).

FIGURE 38.14 Dies finished by ECM. It is more economical to machine the cavities in these cold-forming dies by ECM when quantities exceed 20 to 6 (depending on shape). *(Extrude Hone Corp., Irwin, PA)*

CHAPTER 39
ELECTROCHEMICAL GRINDING

THE PROCESS

As can be perceived from the title, the mechanism of electrochemical grinding (ECG) is similar to that of electrochemical machining (ECM). It uses the deplating power of electrolytes, driven under pressure and guided to the workpiece, to remove stock much faster than conventional grinding, and it creates no heat or stress as is caused by conventional grinding.

A negative potential is applied to the conductive grinding wheel while the workpiece is grounded to the reciprocating table and acquires the positive polarity of an anode. An electrolytic solution is pumped onto the grinding wheel just before the wheel contacts the workpiece and the liquid is drawn into the tiny gap maintained between the wheel and the part.

The grit on the wheel performs mechanical agitation for the deplating process. As the grinding progresses, ions form a thin, microscopic coating on the surface being ground. The force of the pressurized electrolyte sweeps the part's surface clean for the release of the next layer of ions. The grit on the wheel, just barely touching the part, leaving microscopic lines in the metal, and increasing the surface available for ion release. The effect of the grit increases the metal-removal rate by as much as 50 to 1 over electrolytic action alone. This violent scrubbing action, created at the workpiece by the high-speed grinding wheel, changes a process (the electrolytic action) that is slower than traditional grinding into one that is much faster.

The deplating power of pressurized electrolytes is magnified by the vigorous agitation of the whirling grinding wheel. The deplating occurring in this process is similar to that of the process which reduces the size of the workpiece (anode) in ECM and also the size of the anode in electroplating. Once again, it is a case of Faraday's Law in reverse.

Traditional grinding removes metal by abrasion. It can consistently hold tolerances of ±0.0001". But this process creates heat and stress. It can be difficult to grind thin stock. In ECG, a production tolerance of ±0.001" is fairly easy to maintain and it is possible to hold a tolerance of ±0.0003" by special means. Stock removal by ECG is about four times faster than conventional grinding and it always results in burr-free parts that are unstressed. This process can grind material as thin as 0.040", whereas the heat and pressure of traditional grinding would cause such a thin part to warp, making tight tolerances difficult to control. Warping of even 0.002" will place many different thin parts out of tolerance and require either a straightening step or relegating them to scrap.

PROCESS CHARACTERISTICS

Electrolyte flow has a great effect on surface smoothness and flatness. As in the other electrochemical processes, the solution's pH, its cleanliness, and its point of impact, all have an effect on the quality and speed of performance. It is difficult to machine a sharp inside-corner radius. Only by repeated wheel dressing can a 0.010" radius be achieved. Shallow cuts make this easier to attain. The reason for this problem is that the point of highest pressure of the electrolyte is the wheel corner. Outside-corner radius formation is a different story. It can be maintained consistently at 0.002". High-speed grinding benefits both inside- and outside-corner quality.

Historically, the role of ECG has been misunderstood. There really never has been a reason to relegate ECG to the machining of hard metals only. Certainly, that chore is easily handled. But most grinding operations on soft steel could be performed by ECG simply because it is a faster process and consequently, cheaper. Because there is little contact between the wheel and the workpiece, there is no tendency to warp as there definitely is in conventional grinding. In the manufacture of computers, there are many small, thin parts that continually require straightening after grinding. It would be wise to change this process to a stress-free action.

As an example with ECG, most workpieces can be rough ground removing 0.003 to 0.005" per pass at a surface speed of 40" per minute (ipm). The final pass should remove 0.0005 to 0.002" at a speed of about 60 ipm. When thin parts are ECG'd and only one or two passes used for stock removal, no more than 0.001 to 0.002" of stock are taken per pass at a slow table speed of 5 ipm.

Electrochemical grinding alone is seldom used. It would be used only when the metal depth for removal is very shallow or the wheel speed is too slow. As the need for removal of larger quantities of material become necessary, the proportion of metal removed by abrasive grit actually touching the workpiece becomes greater.

At the other extreme is a situation where a heavy removal rate is required, but a close-tolerance, fine finish is desired. This demands a different approach. The first cuts should have grinding parameters adjusted to removing only 10 percent of the stock mechanically. The final pass should remove ¾ of the stock mechanically. This would hold a tolerance of ±0.0001" and yet ensure a flat, stress-free surface.

Many types of thin parts in modern electronics require slotting by a gang cutter, or punched or drilled holes. All of these parts should be ECGed to leave a good finish and deburr the parts at the same time. Thin parts with two or more thicknesses are difficult to grind within tolerances the conventional way. These jobs are easy for the ECG process.

GRINDING WHEELS

The closest action to pure electrochemical grinding is with a graphite wheel. Graphite is very soft and has no abrasive action. The only agitation is provided by the spinning wheel.

The extreme softness of these wheels permits them to be easily dressed by a mild steel shoe to any contour needed. This material is ideal for sinking curved shapes. With the absence of agitation, the table movement must be slowed down to ¼ ipm.

Copper-base aluminum ceramic wheels have the ability to retain the ceramic grit through heavy roughing cuts and can also be used as very thin wheels, for instance, to grind keyways in shafts or making thin slots. Heavy mechanical grinding will wear down the grit and require redressing. If electrochemical action predominates, the current will roughen the wheel's surface beyond its original condition.

Carbon-base ceramic wheels do not require conditioning; their natural surface is very rough. This increases the wetting ability of the wheels, and maintains a very consistent and uniform flow of electrolyte, which, in turn, tends to provide a higher percentage of electrolytic action than mechanical grinding. However, the carbon wheel does not possess the bonding power for ceramic grit as the copper wheel. The grit breaks free relatively easily, resulting in more rapid wheel wear.

This section on grinding wheels is only to indicate the significance of selecting the correct wheel for the job. Many more types of grinding wheels are available; this becomes a separate issue that must be investigated thoroughly.

SPECIMEN GRINDING

Specimens for metal-fatigue testing and tensile testing can now be machined quickly and expertly on a special ECG specimen machine produced by both Anocut and Surftran (Figure 39.1).

Specimens have presented problems when machined the traditional way: by being turned or ground. The surface in the middle of each specimen must have a fine finish and it must be unstressed. Some tests have been invalidated because the break area had not been polished sufficiently to provide consistent, reliable data. And some specimens had

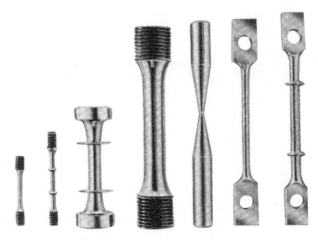

FIGURE 39.1 These specimens are tested to failure. *(Anocut Corp., Elk Grove Village, IL)*

been suspected of having been stressed in their machining process. ECG positively makes unstressed parts and the finish can be maintained consistently at 10 microinches.

Much of this material has been obtained from the Anocut Corporation as a report from the *Machine and Tool Blue Book*, a Hitchcock Publishing Co. Division.

CHAPTER 40

ELECTROPOLISHING

OVERVIEW

Electropolishing streamlines the microscopic surface of a metal object by removing metal from the object's surface through an electrochemical process similar to, but the reverse of, electroplating. In this process, the metal is removed, ion by ion, from the surface of the metal object being polished. Electrochemistry and the fundamental principles of electrolysis (Faraday's Law) replace traditional mechanical finishing techniques, including grinding, milling, blasting, and buffing, as the final finish.

In basic terms, the object to be electropolished is immersed in an electrolyte and subjected to a direct electrical current. The object is maintained anodic, with the cathodic connection being made to a nearby metal conductor. During electropolishing, the polarized surface film is subjected to the combined effects of gassing (oxygen), which occurs with electrochemical metal removal, saturation of the surface with dissolved metal, and the agitation and temperature of the electrolyte (Figure 40.1).

The smoothness of the metal surface is one of the primary and most advantageous effects of electropolishing. During the process, a film of varying thickness covers the surfaces of the metal. This film is thickest over microdepressions and thinnest over microprojections. The electrical resistance is at a minimum wherever the film is thinnest, resulting in the greatest rate of metallic dissolution. Electropolishing selectively removes microscopic high spots ("peaks") faster than the rate of attack on the corresponding microdepressions or "valleys." Stock is removed as metallic salt. Metal removal under certain conditions is controllable and can be held to 0.0001 to 0.0025", depending on the amount to be removed and the starting surface condition of the object.

In summary, electropolishing removes metal. It does not move it or wipe it. As a result, the surface of the metal is microscopically featureless with not even the smallest speck of a torn surface remaining. The basic metal surface is subsequently revealed—bright, clean, and microscopically smooth. By contrast, even a very fine mechanically finished surface will continue to show smears and other directionally oriented patterns or effects.

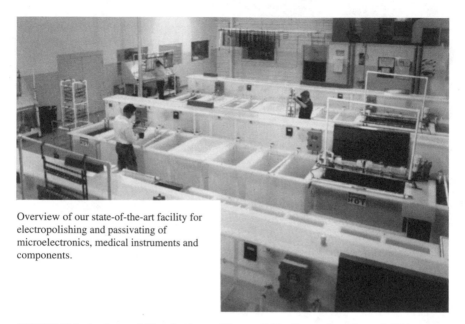

Overview of our state-of-the-art facility for electropolishing and passivating of microelectronics, medical instruments and components.

FIGURE 40.1   An electropolishing plant layout. *(Electropolishing Systems, Inc., Plymouth, MA)*

HISTORY OF ELECTROPOLISHING

In 1912, the German government issued a patent for the finishing of silver in a cyanide solution. This was history's first reference to electropolishing. Experimentation led to the successful electropolishing of copper in 1935. Within a few years, solutions for electropolishing stainless steels and other metals were developed.

During World War II, extensive research by both allied and axis scientists yielded a substantial number of new formulas and results. Dozens of new patents were issued between 1940 and 1955. Important applications were developed for the military during World War II and the Korean conflict.

Today, electropolishing is being rediscovered by a wide range of industries as a replacement for mechanical finishing. In addition to making a surface smoother, it is an excellent means of brightening, passivating, and stress relieving most metals and alloys.

BENEFITS OF ELECTROPOLISHING

This process presents a better physical appearance because of:

1. No fine directional lines from abrasive polishing. This would be ideal for finishing test specimens and providing uniformity for a base line.
2. Excellent light reflection and depth of clarity.
3. Bright, smooth polish and a uniform luster—even on contoured parts.

There are better mechanical properties because:

1. There is less friction and surface drag.

2. There is a longer duty cycle because the process reduces fouling, plugging, and product buildup.

3. The surface of stainless steel is super passivated. A chromium-rich surface, as measured by Auger Electron Spectroscopy, results from electropolishing stainless steel.

4. Galling of threads is reduced considerably.

5. This process yields maximum tarnish and corrosion resistance in many metals and alloys. Steels, including stainless steels, proceeding through the usual shop processes, pick up metallic and nonmetallic inclusions. In fact, some of these are present on the surface of steels, as delivered from the mills. Mechanical polishing not only fails to remove inclusions, but tends to push them deeper into the surface. These inclusions can eventually become starting points of corrosion.

Ease of cleaning any product after it has been electrolized provides:

1. Substantially reduced product contamination and adhesion of foreign matter because of the microscopic smoothness of an electropolished surface (much like glass).

2. Decreased cleaning time considerably.

3. Improved sterilization and maintenance of hygienically clean surfaces for drug-, food-, beverage-, and chemical-processing equipment.

Special effects:

1. Simultaneous deburrs as it polishes. See the chapter in this book on electrochemical deburring.

2. Produces a radius or a sharp edge, depending on the racking position.

3. Polishes areas inaccessible by conventional methods.

4. Reveals metal surface flaws undetectable by most other means. In fact, this process is an effective inspection tool for judging surface quality.

5. Permits micromachining of metals and alloys.

6. Processes large numbers of parts simultaneously.

WHAT CAN BE ELECTROPOLISHED

Most metals can be electropolished, but the best results can be obtained with metals that have fine grain boundaries, which are free of nonmetallic inclusions and seams. Metals with a high content of silicon, lead, or sulfur can be troublesome. Stainless steels are the most frequently electropolished alloys (Figure 40.2). Castings will polish to a bright finish, but not to the same brightness or smoothness as wrought metals. In addition, removing the surface of a casting might reveal subsurface porosity.

A. This 1000x photomicrograph represents the surface of 316L seamless stainless steel tubing as received from the mill. Note the elongated grain structure, cavities and surface distortions resulting from the mandrel used in the tubing's manufacture.

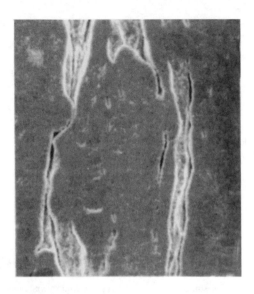

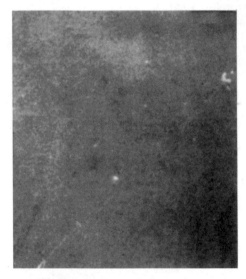

B. This 1000x photomicrograph shows the same surface after electropolishing. Note the disappearance of grain patterns and other imperfections which were in the pre-electropolished surface.

FIGURE 40.2   Result of electropolishing stainless (316L) steel. *(Delstar Electropolish, Princeton, NJ)*

Other commercially electropolished metals include:

High-temperature alloys; molybdenum, waspaloy, tungsten, and nimonic alloys.

Low- and high-carbon steels	Tool steels
Aluminum	Titanium
Copper	Kovar

Cupronickel	Inconel
Brass	Columbium
Leaded steel (low lead)	Bronze
Beryllium copper	Beryllium
Vanadium	Monel
Hastelloy	Tantalum
Silver	Gold
Nickel silver	

Many of these metals can only be electropolished in large production runs and in controlled environments. That is because of the sometimes costly setups, tooling, and special environmental and safety requirements associated with some of these materials.

CHARACTERISTICS OF THE ELECTROPOLISHED SURFACE

A smooth, highly reflective surface is determined to a large extent by the surface conditions of the metal prior to processing, and the process controls. If your work must be highly reflective, you should seek the most reliable metal suppliers. Some producers have more rigid controls than others. Base metal conditions that can result in less-than-optimum electropolished finishes include the presence of nonmetallic inclusions, improper annealing, overpickling, heat scale, large grain size, directional roll marks, insufficient cold reduction, or excessive cold reduction. These conditions can be inherent in the metal as it comes from the mill. During electropolishing, metal is removed, revealing these flaws.

Many products require a superior finish. Companies making these products would be well advised to have their own metallurgist survey the manufacturing source, the mill. She should satisfy herself as to the quality and maintenance of the basic equipment, the controls in place, and the honesty and reliability of the quality-control department.

The quality control of the company doing the electropolishing is the next place to survey. A lack of sufficient controls produces inconsistent and unpredictable quality. Although some variables are functions of technology, others fall under what can be addressed as the art of electropolishing. Most variables have been studied so long that they are now statistically controllable.

Start with using the proper electrolyte. Maintain its temperature and chemistry. Critical factors of the chemistry are specific gravity, which is a major consideration in hygroscopic electrolytes, the acid concentration, and the metals content. A supply of ripple-free dc power must be available to drive the process. Appropriately size the cables and connectors to the anodes and cathodes. The dc power must be applied at the correct voltage and current density (amperes per square foot).

Now the "art" of electropolishing enters the picture. Cathodes for optimum polishing in inaccessible areas, corners, and areas of low density must be properly configured. Equally significant is the knowledge of where, when, and how to agitate either the electrolyte or the part to prevent gassing streaks, flow marks, and other undesirable markings.

For many products, electropolishing's mirror-like luster is the goal. Other goals require additional advantages of electropolishing not attainable by mechanical means. The electropolished surface possesses the true structure and properties of the bulk metal. Mechanical polishing procedures leave a layer of disturbed structure and the surface will not have the properties reported for the bulk metal. The surface of the metal is said to be the place where the metal ceases to exist. Although this can be truly said for the electro-polished surface, it does not hold true for the surface mechanically finished by cutting, smearing, rolling, buffing, drilling, broaching, or grinding.

That mechanically polished surface yields an abundance of scratches, strains, metal debris, and embedded abrasives. In contrast, an electropolished surface is featureless. It reveals the true crystalline structure of the metal without the distortion produced by the cold-working process that always accompanies mechanically finishing methods.

The differences between electropolishing and mechanically finishing is often not readily obvious to the unaided eye, particularly if both are polished to the same microinch finish. However, when the metals are viewed under high magnification, the differences are clear. These differences are more than simply topographical. The damage associated with cold working penetrates the metal, and abrasives have a similar effect. The tensile strength of a mechanically finished part might be somewhat decreased.

Burnishing metal by lapping, buffing, or coloring processes decreases microinch roughness, but it never completely removes the debris and damaged metal caused by mechanical finishing. Surface smoothness is not simply an independent variable in surface definition. It is, rather, one factor in the subject of surface metallurgy.

APPLICATIONS

Some parts that have been successfully electropolished are:

- Pipe/tubing
- Valves
- Fittings
- Sheet metal
- Stampings
- Spinnings
- Weldments
- Castings
- Wire goods
- Forgings
- Fasteners
- Drawn metals

Electropolishing produces important improvements in reactor vessels, heat exchangers, blenders, storage tanks, clean-room equipment, nuclear applications, medical equipment, and various food- and chemical-processing equipment.

CLEAN-ROOM AND MEDICAL APPLICATIONS

Today's clean rooms operating at the class 10 level or better demand noncontaminating and nonparticulating metal surfaces. With the advent of the new sub 0.5-micron contamination control regimes, electropolishing metal surfaces will be mandatory in all clean rooms operating under this cleanliness level. Electropolishing is the ultimate finish for clean-room tables, chairs, waste containers, light fixtures, electrical conduits, and outlet boxes. Major users will be semiconductor manufacturers and pharmaceutical firms, as well as many other high-purity industry leaders.

For many years, the medical field has been a beneficiary of electropolishing. Medical and surgical equipment (such as scalpels, clamps, saws, bone and joint implants, prosthetic devices, burn beds, and rehabilitation whirlpools) are good subjects for the process. All metal articles exposed to radiation and requiring regular decontamination are prime candidates for the process. The process is used to polish metal parts, which are located in a nuclear environment. It improves the effectiveness of decontamination over the standard procedure. Any contamination located on or embedded in the surface can be removed by electropolishing.

The world of biotech and medical research is demanding ever more stringent contamination controls. Major applications in this industry include vacuum chambers, cryogenics, high-purity gas systems, process vessels, and other similar process-related equipment.

In food and beverage processing, the easy cleaning and cosmetically attractive surfaces coupled with the noncontamination and sanitary qualities demanded by these industries are finding that electropolishing does the job best.

MORE INDUSTRIAL APPLICATIONS

Electropolishing has been used on the following:

1. Filters, screens, and strainers.
2. Dry product delivery systems.
3. Product trays and dryers.
4. Pumps and valves.
5. Compressors, wheels and impellers, and turbine blades.
6. Condensers and cooling coils.
7. Vacuum chambers.
8. Paper mill equipment, such as slurry pipe and head boxes.
9. Electronic and communication equipment.
10. In passivation of stainless steels, it produces a chromium-enriched layer on the surface, which can be as deep as 50 angstroms.
11. Occasionally, parts are made to improper tolerances (oversized) and need a slight change. A small amount of metal can be removed by electropolishing to bring the part within tolerance.
12. Offshore oil-field equipment and anyplace where it is vital to reduce oxidation and corrosion, such as parts located in a chlorine environment.

HIGH-QUALITY ELECTROPOLISHING

High-quality electropolishing should exhibit brilliant luster and reflectivity. If the process is used correctly, the surface will be free of the following flaws: frosting, shadows, irregular patterns, streaks or stains, pitted areas, orange peel, erosion, and pebbly appearance.

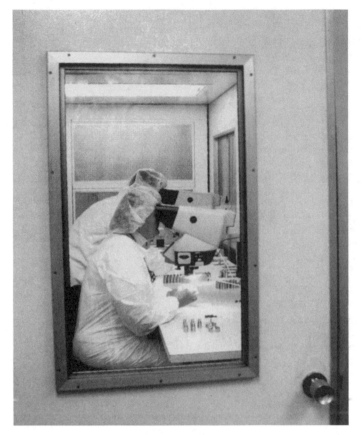

FIGURE 40.3   Final inspection in class 100 clean room. *(Electropolishing Systems, Inc., Plymouth, MA)*

Satisfactory results are obtainable by starting with good material and continuing with proper techniques. A uniform, homogeneous crystal structure produces the best electropolished finish. Any material imperfection will detract from the finish. Learning to recognize quality electropolishing is like learning to distinguish the difference between a real one dollar bill and a counterfeit. You know by looks, feel, and performance. Examination by SEM photomicroscopy or Auger electron spectroscopy will show the difference (Figure 40.3).

This chapter is only one of several in the book that describes a process using electrochemical means as a tool. Delstar, a company with five locations across the country, has been doing electropolishing since 1982. Electropolishing Systems, Inc., has a new facility in Plymouth, MA. Since 1965 they were known as Accurate Metal Finishers. A search through the Thomas Register might provide some additional competition.

CHAPTER 41
ELECTROFORMING

PROCESS DESCRIPTION

In its simplest terms, electroforming is an electrochemical process of metal fabrication. The technique uses an electrolyte, an anode to supply the metal, controls of electrical current, and special methods to monitor the deposit of metal on the electroformed part. Sound familiar? Of course, it's the same description that might be used for the electroplating process.

Electroforming is a process that, by definition, produces parts and is not to be confused with electroplating, which provides only a final finish on previously produced parts. The implication can immediately be drawn that the electroformer must have a far broader knowledge than an electroplater. The supervising electroformer should not only be an expert in the field of electrochemistry, but should also be proficient in mechanical engineering, metallurgy, and a host of ancillary fields, including microwave engineering. Electroforming is being called upon increasingly to produce the accurate and complex waveguides required in radar and microwave communication equipment.

Electroforming is drastically different from conventional forming, cutting, or casting methods. If you combine the light weight of sheetmetal fabrications, the possible complexities of castings, and the accuracy of machining methods, you would have all these characteristics in one process, electroforming.

This process has been available for decades, but it is still considered to be a space-age technology. In electroforming, metal is electroplated onto a part; the part is removed and thrown away, keeping the plating. Actually, the techniques of electroforming are not quite so simple. They differ considerably from ordinary electroplating, yet the family resemblance remains. The basic capabilities and chemistry of plating, Faraday's Law, governs electroforming.

In electroforming as in electroplating, metal ions are transferred electrochemically through an electrolyte from an anode to a surface (cathode), where they are deposited as atoms of plated metal. In electroforming, the surface that is to receive the plated metal, called a *mandrel*, is conditioned so that the plating does not adhere. Instead, the plated metal (*electroform*) is lifted away and retains its as-deposited shape as a discrete component.

A part formed by this process has several unusual characteristics:

1. It can have extremely thin walls, less than one mill. In fact, minimum thickness is generally limited only by the fact that a part requires a certain amount of sturdiness to avoid being bent or broken by normal handling.
2. Surface features of the mandrel are reproduced with extreme fidelity on the surface of the electroform. High surface finish and intricate detail are easily obtained.
3. Complex contours are easily produced.
4. Dimensional tolerances can be held to high accuracies: ±0.0001" are not unusual.
5. Maximum size is limited only by the size of the available plating tank. Parts over 7 feet long have been successfully electroformed.

The present increased emphasis on precise, thin-walled metal components, has placed a new light on this process. Waveguides, optical-quality reflective surfaces, and even aerospace components are now being manufactured on a production basis. Electroformed dies are lowering the cost of some intricate dies for both metal and plastics (Figure 41.1).

Formerly, this process was an art; accordingly, it provided erratic results. What is new is the application of advanced technology to the field. Now, a better understanding of elec-

A.

C.

B.

D.

FIGURE 41.1 Typical electroforming products: A and B. Large, complex, "grown-on," quad-type wave guides. C. Nickel-type wheel with master. D. Wave guide coupler, showing "grown-on" aluminum flanges. *(GAR Electroforming Corp., Danbury, CT)*

trochemistry and the role of chemical additives in plating baths permits much closer control of electroformed parts. Results are now reproducible the same as in casting, forging, and other conventional processes.

This process endows parts with unusual characteristics:

1. They can have very thin walls, down to 0.001" thick.
2. Surface features of the mandrel are reproduced with extreme fidelity and high surface finish.
3. Complex contours are no more difficult to reproduce than the simplest shapes.
4. Some dimensional tolerances can easily be held to 0.0005".
5. The available plating tanks are the only limit to the size of parts to be electroformed.

The thickness of electrodeposits tends to be excessive on protrusions and relatively thin in recesses. Therefore, wherever possible, rounded edges and generous fillets in internal corners should be incorporated. This will provide relative uniformity of thickness in the electroform.

MANDRELS

Perhaps the most interesting aspect of electroforming is the ease with which complex shapes are formed. The same geometric relationship exists between mandrel and electroform that exists between a mold and its casting. Internal features of the electroform are formed as negative images of the mandrel.

Two types of mandrels are used: permanent and disposable. When there is no undercut surfaces on the electroform and it can be lifted directly off the mandrel, a permanent mandrel is generally used. Also, if the electroform is open at one or both ends, it could simply be withdrawn from the permanent mandrel. If undercuts are necessary on the electroform, the mandrels must be destroyed by dissolving or some other device.

Stainless steel and aluminum are the materials most often used for mandrels. Stainless steel polishes easily and provides internal finishes to the electroform as fine as 2 microinches. External finishes are normally about the same as that of a diecasting. For mass production, 440C stainless steel, hardened to 58 Rc, would probably be the choice for the original mandrel. This would likely allow 30 to 40 nickel electroforms to be produced before the mandrel would require polishing.

Aluminum mandrels machine much easier, but do not have the service life of stainless steel. Principal among aluminums are 2024 and 6061. Both are removed from the electroform by a hot caustic etch. Solid aluminum mandrels up to one foot in length are readily dissolved out in hot sodium hydroxide.

Several other materials are used for mandrels: bismuth alloys, invar, and cast alloys of many different metals (such as brass or nickel). Plastics, wood, wax, and quartz can also be used. Low-temperature melting materials, such as cerrocast, are favored when usable. Nonconductors must first get painted with a conductive material before use as mandrels. Because even the smallest scratch will reproduce, mandrels must be carefully handled.

It is nearly always necessary to pretreat expendable mandrels chemically before starting the electrodeposit. These techniques, including pickling and zincate treatment, are well known. Several common methods of rendering expendable mandrels conductive include electroless copper, vacuum-deposited metal, conductive paints, and Aquadag dispersions.

When large numbers of disposable mandrels are required, castable or moldable units are used, if at all possible. These can then be made economically by casting them in reusable molds. Disposable mandrels generally cannot provide the accuracy or finish of permanent mandrels. A number of trade-offs must be considered when contemplating mandrels. Because of the quantity of mandrel materials available and the special features of many of them, this becomes a serious decision.

The necessity for undercuts should not be combined with a requirement for high finish or close dimensions. If such requirements are unavoidable, more than one material can be used for the mandrel by bonding. This technique is often used for waveguides and other electronic equipment, which have numerous cavities.

Although most mandrels are male, sometimes the need for a fine external finish or a tight external dimension requires the use of a female mandrel. In fact, there is such a broad choice of materials and methods of preparing mandrels for electroforming that the designer should consult with a qualified electroformer. Too often, the electroformer is confronted with a design that precludes the use of a more economical approach or makes it difficult and even impossible to produce a satisfactory electroform.

WALL THICKNESS

Because plated metal is deposited more or less uniformly, electroforms are usually parts of constant wall thickness. The thickness can vary from 0.005 to 0.500". Most electroforms are in the range of 0.010 to 0.050" thick. Uniformity of thickness is subject to the same vagaries as electroplating. Areas of high plating current will thicken the deposit locally, as will sharp edges or convex surfaces.

The simplest way to avoid these positions of thickness variations is to provide adequate radii at edges and corners, and to make holes and slots as wide as they are deep. Otherwise, the plater must use shields and "thieves" to reduce current density or conforming anodes can be used to boost current density.

INSERTS AND GROW-ONS

The desired part need not be made entirely from deposited metal. Other materials, even nonconductors, can be incorporated into the component by plating over or around separate pieces attached to the mandrel. Threaded inserts, shafts, or other "grow-ons" are often used in this manner.

Sometimes the nondeposited metal constitutes a larger portion than the electroformed material (Figure 41.2). The joining of two or more pieces is now a large percent of all electroformed output. An electroformed part can be mated to one or more parts, then all of the components can be reelectroformed to provide a seamless assembly. This method of joining multiple components involves a "grown-on" process patented by the Gar Electroforming Corporation. This process permits the electroforming of mounting flanges or supports on reflectors. Such a procedure vastly simplifies the mounting and dismounting steps when changing reflectors in the field. Very often, these multiple components involve dissimilar metals.

MATERIALS AND PROPERTIES

At this time, nickel is the material most often electroformed. It provides good strength and corrosion resistance, and it is an easily plated metal. Copper is the next most popular material. Gold, silver, and rhodium are used when extreme resistance to oxidation is the prime consideration.

Lockhead L-1011 components fabricated via electroforming. In the center is a nickel electroformed large tubing with welded flange. At the bottom center is an electroformed nickel fire wall WYE with welded stainless steel fittings. On each side of the electroformed WYE are two electroformed drain masts. One is called the *APU drain mast* and has welded stainless tubing. The other is called the *Keelson drain mast* with grown-in stainless flame arrestors. Both drain masts have brazed stainless flanges for attachment.

Rectangular to cross transition waveguide with grown-on Brass Flangles.

The center of the photo shows a cavity filter with grown-on tuning bosses and electroformed iris's. At left and right side of filter are electroformed nonseamless rectangular copper cavities with grown-on mounting brackets.

FIGURE 41.2   Electroformed "grow-ons." *(GAR Electroforming Corp., Danbury, CT)*

Before progressing too far in the design, the assistance of an electroformer should be sought because the actual plating conditions, such as the composition of the plating solution, has a lot to do with the results. Properties of deposited metal can vary considerably because of changes in the plating bath. Residual stresses in an electroform can occasionally cause difficulties for highly stressed parts. Proper heat treatment can make quite a difference in properties.

Sometimes, one metal will be deposited on top of another when the properties of the first are not good enough for the task. By manipulating with masks and shields, many combinations of shapes and functions can be achieved. Chromium, for instance, is too brittle for electroforming. But, if deposited with nickel, the combination is usable. See the engineering standards in Figures 41.3 and 41.4.

General Chemical Practice	**ENGINEERING STANDARDS**	GCP-116

GAR Pure Electroformed Nickel Metal Physical Properties*

1. Temperature tolerance without change of specific physical properties is up to 700° F.

2. Resists normal atmospheric corrosion and oxidizing atmospheres.

3. Possesses the same mechanical properties of an annealed low carbon steel.

4. Accepts welding, brazing and soldering.

5. Magnetic properties and curie point is 665° F.

6. Ductility: can be crimped around its own radius and restraightened without cracking.

UTS	75 to 85 KSi
Yield	45 to 55 KSi
Elong. (2″)	10 to 20%
Hardness	150 to 300 VHN (20 to 30$_c$)
Purity	99.8% Ni. Max.**
Density	.322 lb. / cu. in.
Specific Gravity, 20° / 40°	8.908 g / cc
Melting Point	2650° F (1455° C)
Specific Heat (32-212° F)	.11 (B.T.U. / lb. / ° F)
(0-100° C)	.11 (Cal. / g / ° C)
Linear Coef. of Expansion	
(32-212° F)	7.5 (10^{-6} in. / in. / ° F)
(0-100° C)	13.5 (10^{-6} cm / cm / ° C)
Thermal Conductivity (32-212° F)	470 (B.T.U. / sq. ft. / hr. / ° F / in.
(0-100° C)	.16 (Cal. / sq. cm. / sec. / ° C / cm.)
Specific Resistivity (20° C)	7.0 microhm-cm.
Temp. Coef. of Sp. Resistivity	
(0-100° C)	.006 per ° C.
Modulus of Elasticity in Tension	28 (10^6 lb. / sq. in.)
Crystal Structure	Columnar, interspersed with grain

    *Deposited from GAR Proprietary Sulfamate Electrolyte.
  **Also includes cobalt.

GAR Hard Nickel Electroforming Process

1. Resists normal atmospheric corrosion and oxidizing atmospheres.

2. Very high wear, resistance and erosion.

3. Adequate ductility. Machining performed by grinding only.

UTS	100 to 150 KSi
Yield	70 to 75 KSi
Elongation (2″)	3 to 6%
Hardness	450 to 600 VHN (45-55$_c$)*

4. Purity is 99.7%.**

FIGURE 41.3   Engineering standards. *(GAR Electroforming Corp., Danbury, CT)*

General Chemical Practice	**ENGINEERING STANDARDS**	GCP-117

GAR Pure Acid Sulfate Electrodeposited Coppers*

1. High electrical and thermal conductivity.

2. Good plating abilities.

3. Same joining properties as wrought copper i.e., soldering, brazing, etc.

4. Non-Magnetic.

UTS	25,000 to 30,000 psi
Yield .05% offset	7,500 to 9,000 psi
Elong. (% in 2")	15 to 30%
Hardness (VHN)	45 to 60 VHN
Electrical Resistivity	1.72 (Ohm - cm X 10^6)
Density	.32 lb./cu. in.
Specific Gravity, $20^o/4^o$	8.9
Melting Point	$1980^oF.$ (1080^oC)
Specific Heat (32-212°F.)	.092 B.t.u. / lb. / $^oF.$
(0-100°C.)	.092 Cal. / g. / $^oC.$
Linear Coef. of Expansion	
(32-212°F.)	$9.2 \, 10^{-6}$in. / in. / $^oF.$
(0-100°C.)	$16.5 \, 10^{-6}$ cm. / cm. / $^oC.$
Thermal Conductivity	
(32-212°F.)	2700 B.t.u. / sq. ft. / hr. / $^oF.$ / in.
(0-100°C.)	.93 Cal / sq. cm. / sec. / $^oC.$ / cm.
Specific Resistivity (20°C.)	1.7 microhm-cm.
Temp. Coef. of Sp. Resistivity	
(0-100°C.)	.004 (per ^oC)
Modulus of Elasticity in Tension	16 (10^6 lb. / sq. in.)
Crystal Structure	F. C. C.
Lattice Constant (A_o)	3.608 A. U. = 10^{-8} cm.
Purity	99.8%

Submitted by: Approved by:

Russell A. Richter   George A. Ray
Q. C. Manager   President

*Both the GAR Coppers are either the pure non-additive type, which can be used with high temperature silver soldering materials up to 1200° F, or the very smooth bright finish type copper can only be used with low temperature solders.

FIGURE 41.4 Engineering standards. (*GAR Electroforming Corp., Danbury, CT*)

Electroless nickel (92% nickel and 8% phosphorus) deposited on electroformed copper can be heat treated to a hardness of 70 Rc. Many such combinations are used. Research and development in electroforming has led to the addition of ceramics to certain plating baths. These additives become embedded in the electroform as the plated metal is deposited, substantially increasing strength and resistance to creep.

One feature of an electroform establishment is the unusual cleanliness in evidence. To ensure clean and flawless parts, free from imperfections caused by dust contamination, the interior air must be continuously filtered. All production electroforming baths have unitized filters for each tank. Physical characteristics of the desired electroforms are maintained by conformance to specialized formulations.

CASES AND USES

The demanding, ultra-precision criteria of space-age technology is met by this process, which can replace multiple operations. Different materials can be plated over one another to improve desired characteristics, such as strength, appearance, or electrical conductivity. When material purity is essential, this process will provide it. The following products have been produced by electroforming:

- Heat shields
- Wave guides
- Optical mirrors
- Electronic components
- Complex aircraft parts
- Injection molds for plastics
- Prosthetic appliances
- Floor tile molds
- Record stampers
- Printing plates
- Missile and rocket hardware
- Embossing dies
- Layup molds for aircraft body parts
- Foil
- Tubing
- Jewelry
- Plaques
- Nameplates
- Reflectors

Master sheet mandrels have been used to make phonograph record stampers and surface texture comparison scales. These two jobs require the utmost integrity in reproduction.

Bell Aerospace Textron of Buffalo, NY, has been developing the electroforming process for use in its airplanes. Their engineering management feels that certain delicate and intricate shapes, which are difficult or impossible to machine, achieving the precise dimensions required, can be economically produced by electroforming. Bell has specific requirements that many of these components and assemblies function properly at both cryogenic and elevated temperatures.

Bell is trying to develop processes for electrodepositing heretofore difficult metals (such as aluminum, titanium, columbium, beryllium, technicium, and tungsten). Each of

FIGURE 41.5 Electroformed reflectors are replacing fragile glass. *(GAR Electroforming Corp., Danbury, CT)*

these materials has unique properties for specific applications in aerospace, energy, nuclear, and commercial industries.

Certain parts cannot be made by any method, except electroforming. Some have problems attaining tight tolerances. Sometimes reflectivity is the problem or it could be exact duplication or complexity. But when there is a genuine need, and profits hang in the balance, a way will be found to accomplish what is currently only speculation.

P · A · R · T · 8

MACHINE
DEBURRING
PROCESSES

CHAPTER 42
ELECTROCHEMICAL DEBURRING

PROCESS DESCRIPTION

Electrochemical deburring (ECD) is a process that uses electrical energy to remove burrs in a very localized area, as opposed to thermal energy deburring (another chapter in this book), which provides general deburring. The part to be deburred is placed in a nonmetallic fixture, which positions an electrode in close proximity to the burrs. The workpiece is charged positively (anode). The electrode is charged negatively (cathode), and an electrolytic solution is directed under pressure to the gap between the electrode and the burr.

This flow of electrolyte should precede the application of the current to flush out any loose chips that probably would cause a short in the system that could damage the part, the tooling, or the equipment. As the burr dissolves, a very controlled radius is formed. The process is consistent from part to part.

Because burrs are projections, they are high current-density areas and are thus electrochemically removed preferentially. This is an advantage because no direct contact between the tool and the part being deburred is required. In contrast to processes involving metal removal via grinding or polishing with abrasives, the workpiece is not exposed to mechanical or thermal stress.

The process always requires fixturing to establish the anode-cathode relationship. A typical fixture consists of a plastic locator, which holds the part and insulates (masks) areas of the part that do not require ECD. The fixture also positions a highly conductive electrode, designed with a contour that conforms to the desired dimensions of the area to be deburred. The locator and electrode direct the flow of electrolyte. The variables of voltage, current, electrolyte flow, and cycle time provide precise control of the ECD process.

ECD will change the dimensions of a part only to the extent of removing burrs, leaving a controlled radius. Thus, the dimensional changes are desirable and are generally required to make parts to print. ECD will only affect areas of the parts in the vicinity of the electrode, if the fixture is properly designed and maintained. The plastic locator should protect such areas as threads. On the other hand, the electrode will remove burrs in threads that have occurred from milling and cross-drilling through the thread form.

APPLICATIONS OF ECD

ECD is effective on all electrically conductive materials. Copper alloys and stainless steel are both good materials for the process. The benefits of using ECD include:

1. Elimination of costly hand deburring.
2. Fixed manufacturing costs with no variances.
3. Ensurance that all burrs are consistently removed.
4. Increased quality and reliability.
5. Reduced personnel and labor costs.
6. The radius generated during ECD is controllable.

This latter advantage solves such functional problems as removing sharp edges from the ID of valve bodies, where cross-holes intersect. This is also applicable to ports in hydraulic components. In hydraulics, there are many occasions when seals must slide over sharp edges, both at assembly and sometimes during operation. The slight radius resulting from the ECD operation precludes the possibility of harming elastomer seals during this assembly.

In the instrument industry, burrs created on delicate gears during the hobbing operation must be removed without damage. ECD does an excellent job here. In fact, ECD will actually improve the surface of these gears. In both the ordnance and aeronautic industries, ECD and Thermal Energy Machining (TEM) are sometimes combined when specific finishes are required. Each process is used where it excels.

The ordnance industry has an exacting specification. The ordnance component must work once upon demand. There are no second chances. Many of the components function only because of the tremendous forces, accelerations and RPM's created by the power of an explosion. This same power can cause malfunctions if the sliding, indexing, or rotating surfaces and edges are not properly deburred. These same forces can dislodge a single, tiny burr and jam a critical movement. The manufacturing engineer enjoys maximum ensurance that all burrs are removed when he knows that ECD and TEM have been part of the manufacturing cycle.

ELECTROLYTIC ACTION

The following statement is repeated for emphasis. During the ECD process, there is no contact between the workpiece and the "tool." Thus, there is no metallurgical change as a result of the electrochemical process. In grinding or polishing to remove burrs, the workpiece is exposed to mechanical and thermal stresses. In the electronic industry, many piece parts must maintain a flatness of 0.001 to 0.002". Sometimes these parts are warped out of specifications by a deburring step. Sometimes the distortion takes hours or even days to manifest itself. Deburring by ECD precludes this problem.

Faraday's law of electrolysis dictates how metal is removed by ECD. The amount removed is proportional to the product of time and current. Recently, the process has been simplified by the use of rotating electrodes that create turbulent flow of the electrolyte, which accelerates the deburring process. An additional advantage is that it allows the use of standard tools more often, rather than requiring custom tools for each job. Compare Figures 42.1 and 42.2.

When using a rotary electrode, the workpiece is positioned on a holding device, which is connected to the positive lead from a dc source. The cathodically connected tool can be a cylindrically shaped, brass, rotating tool that is slightly larger than the part to be finished. Some vertically machined holes should be in the tool to help improve the flow of electrolyte. The tool is often made from brass pipe.

In operation, the tool is fed downward toward the workpiece, which is immersed in the electrolyte. The rotating electrode creates turbulence of electrolyte around areas to be deburred. The spindle is reversed frequently to increase the turbulence. After a cycle of about 30 to 45 seconds, the spindle retracts and the part is removed (Figures 42.3 and 42.4).

ECD and TEM are used in nearly all of the same industries. The decision of which process to use depends on the requirements of the part and the capabilities of the

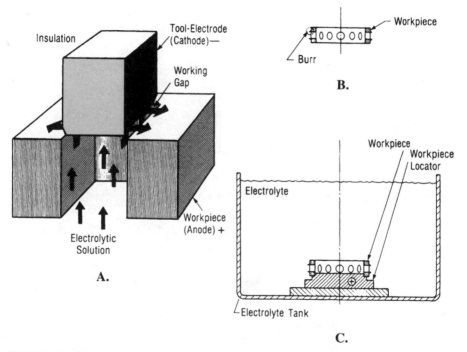

FIGURE 42.1 Steps in electrochemical deburring: A. Deburring of a workpiece by electrolytic means relies upon "deplating" the anodically connected workpiece, using a cathodically connected tool, both immersed in an electrolyte, such as salt water. B. A workpiece with a burr. C. A workpiece mounted on an anode connection in a tank of electrolyte. *(Bosch Corp., Surftran Div., Madison Heights, MI)*

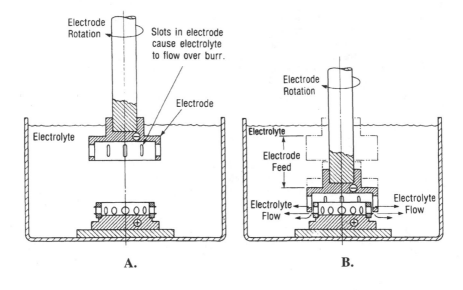

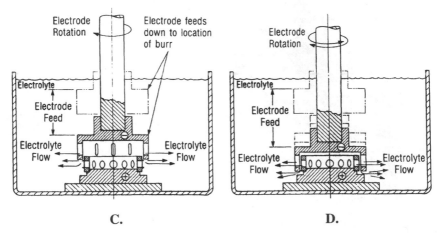

FIGURE 42.2 Depleting steps using a rotating electrode: A. The cylindrical brass tool, which has slots to cause turbulence, is connected to a negative lead from a direct-current source. B. The cathode tool is lowered while rotating, advancing toward the workpiece. C. Turbulence created by the rotating tool flushes the area being deburred with fresh electrolyte as the tool advances. D. The tool continues downward as the burr is removed. The electrode rotation is reversed frequently to create more turbulence. *(Bosch Corp., Surftran Div., Madison Heights, MI)*

processes. TEM is used for general deburring, always deburring the entire part. ECD is used for localized deburring, but it will round off edges as much as necessary; it can be used for polishing to some degree.

Now where does this leave Abrasive Flow Machining (AFM is another chapter in this book)? AFM deburrs, it smooths edges, and it certainly is used to polish dies. Apparently, there are areas where each process excels. Yet, it is possible to achieve your goal by using more than one of these processes. Therefore, you should evaluate each process just as you would before purchasing any machine tool because that's what these "deburring machines" are, machine tools. You've got to determine how many hours of work you could obtain from each.

There are two good sources of information on these processes and you should take advantage of both. First, there is the machine tool manufacturers. Visit them and bring them samples to work on. And second, there are those who have purchased these machines and normally would happily share information.

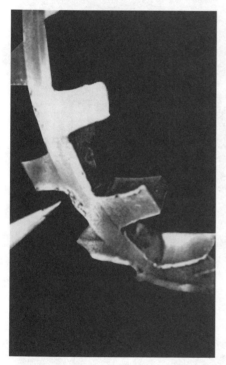

FIGURE 42.3 The work of a rotating electrode: The brass bearing cage electrochemically deburred in 10 seconds on the inside diameter, 20 seconds on the outside diameter, at 8 volts, 400 amperes, in 15 pct sodium nitate. The rotating electrode tool spins at 1200 rpm, removing burrs up to 0.07 inch. *(Bosch Corp., Surftran Div., Madison Heights, MI)*

The burrs generated by the turning and facing operations on this forged, six segment, steel pole piece half are electrolytically removed in an Electrogenics dual-station, 500 Ampere, 400 gallon system at a rate of 120 pieces per hour. Each station of the system accommodates two parts and all six segments of both parts are deburred simultaneously. The operator unloads and reloads two parts on one station while the other station is cycling. Deburring cycle time is approximately 30 seconds. The system features two-point temperature control, automatic liquid level control, and an exhaust fan for venting moist air to the outside.

Before   After

FIGURE 42.4 A sample of electrochemical deburring. *(Anocut Corp., Electrogenics Div., Elk Grove Village, IL)*

CHAPTER 43

THERMAL ENERGY DEBURRING

INTRODUCTION

Over the years, industry has made great strides using the latest in modern machining methods, which have increased productivity and improved the quality of manufactured parts. However, most of the attention has been focused on primary machining methods, but finishing and deburring parts has been largely ignored. Today, 10% (sometimes much more) of total manufacturing costs are spent on manual deburring.

When machining metal components, it is sometimes necessary to cross-drill interconnecting bores. The tail pieces of rockets and other military hardware have many such holes. And hydraulic systems have many components with such holes. Hydraulic valve bodies are a good example. There are many drilled passages to direct fluid flow. Inevitably, when these bores intersect, burrs are created. These burrs must be removed. Otherwise, they might break off and cause severe damage to the system. A good example of cross-drilled holes is Figure 43.1.

In some cases, it is almost impossible to reach and remove the burrs because they are in such hard-to-reach places. A method that has gradually gained respect since its introduction in the early 1970s is the Thermal Energy Method (TEM) of deburring. The Surftran deburring system by The Robert Bosch Corp. of Madison Heights, MI, developed a process as an alternative to the tedious job of manual deburring.

Another component-receiving processing by TEM are the many configurations of aluminum heatsinks. General Electric in Salem, VA, had a severe problem deburring 14 types of heatsinks. Ten operators worked full time in the deburring department. Operator turnover (the work was extremely tedious) led to quality problems, which were compounded by the different interpretation of what was required. The average cycle time was between 12 and 19 minutes. Annual production was 19,000 units. A TEM was installed for a cost of $190,000 and the return on investment took 18 months.

FIGURE 43.1 Cross-drilled holes. *(Bosch Corp., Surftran Div., Madison Heights, MI)*

This is the fastest deburring process in existence. The actual process time (not counting loading and unloading) is about 20 milliseconds. TEM does this by exposing components to an environment that is hostile to burrs, but is harmless to the component. The burrs are hit by a 6000°F blast of heat, which literally burns them away, leaving everything else, including threads, intact.

All designers should consider deburring because it so often presents a problem. Some manufacturing personnel insist that their products are burr free. Burrs in those parts might be tolerable, but they are generally present in some place. Very often, groups of operators are put to work hand deburring components that won't function properly with burrs present. Sometimes they perform that task because the burrs are displeasing aesthetically.

When the engineer determines that the components must be burr-free, there are alternatives. First, consider the machining methods that might preclude the necessity of deburring. Then, there are three excellent machine-deburring methods (all covered in this book).

THE PROCESS

The Thermal Energy Machine is built around three vertical posts that separate and also secure two massive triangular plates in a parallel orientation, one above the other. The pressure vessel to be processed is clamped between two sealing plates. The upper seal is positioned beneath the upper triangular plate, and the lower seal is mounted on a hydraulic cylinder, which rises to secure the pressure vessel against the seals.

Two pressure vessels are incorporated in this particular model (more are available in other models) and these are mounted on opposite ends of a rotary indexer. One station is for loading and unloading parts while the other is in the firing position. The inactive vessel can be loaded and unloaded by the operator while the other cylinder is in process.

The manufactured parts with burrs are placed in a thick-walled chamber, which is closed and sealed with a toggle mechanism, exerting a force of 250 tons (Figure 43.2). A mixture of oxygen and natural gas is admitted under pressure. The ratio of oxygen to gas is 2½ to 1. The extra oxygen is required for the second fire. The gas mixture quickly fills each nook and cranny of all parts in the chamber, even blind and intersecting holes. The combustible mixture is ignited by a 30,000-volt spark, which creates a 6000°F heat wave.

Actually, two fires must occur for the process to function. In two or three milliseconds, the gas fuel is burned out and the primary fire is extinguished. The three essential ingredients to maintain any fire are heat, fuel, and oxygen, and we lost our initial fuel. However, because most burrs exhibit a high surface area to mass relationship, the burrs were unable to transfer their heat to the parts fast enough to prevent them (the burrs) from bursting into flame, the second fire.

The only way for the heat to travel is through the root of the burr into the main body of the part itself. Because the heat is traveling through a relatively narrow cross section, its flow is restricted. Thus, the burrs themselves became a source of fuel as they burn in the excess oxygen. The second fire is now underway. The burrs will continue to vaporize until the heat finally reaches the main body and dissipates into it. This extinguishes the second fire. At this point, all burrs, chips, and contaminants have been vaporized.

You are probably wondering what happened to the burrs. Well, in the process of vaporizing, they became oxides of the metal being processed. The process turns aluminum burrs into aluminum oxide and steel burrs into iron oxide. The component's surface will

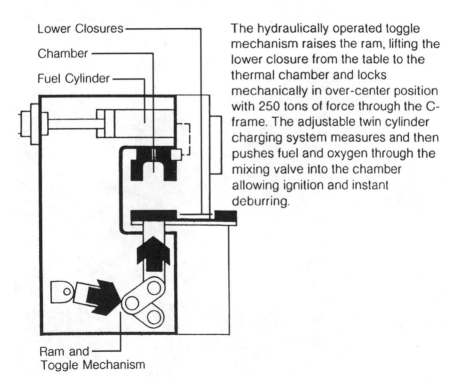

Lower Closures

Chamber

Fuel Cylinder

The hydraulically operated toggle mechanism raises the ram, lifting the lower closure from the table to the thermal chamber and locks mechanically in over-center position with 250 tons of force through the C-frame. The adjustable twin cylinder charging system measures and then pushes fuel and oxygen through the mixing valve into the chamber allowing ignition and instant deburring.

Ram and
Toggle Mechanism

FIGURE 43.2   A schematic of the thermal energy machine. *(Bosch Corp., Surftran Div., Madison Heights, MI)*

appear discolored, but the surface has not been oxidized. The process turns aluminum burrs into aluminum oxide and steel burrs into iron oxide. What can be observed is a film that can be washed off in a suitable cleaner. If the parts are to be heat treated and descaled, bright dipped, anodized, black oxided, or otherwise plated, then a post-cleaning step can most likely be avoided because these treatments would normally remove oxides anyway.

If post-cleaning should be necessary for any reason, cleaning equipment is available from the same company that manufactured the TEM machine. In fact, whenever you discuss deburring with any machine builder, you should also discuss cleaning—especially if you expect it to be required. The complete finished part is your goal. If cleaning is significant, it should be part of the manufacturing strategy. This is especially true when the cost of cleaning equipment equals the cost of the deburring machine (Figure 43.3).

The list of industries using TEM is continually growing. The most recent technical advance was the ability to process plastic, as well as metal, parts. TEM is really another manufacturing tool to be used when the situation warrants. The design engineer should consider the deburring problem before completing the design. This problem should not be left for the manufacturing engineer to solve. A few years ago, I watched a crew of six operators hand deburring rocket tails for the shoulder-fired weapon, used by the infantry. The job took an average of 20 minutes each and it left unwanted scratches. Thirty-five percent of the part cost was deburring. Now, those components are being machine deburred.

A most unique aspect of TEM is that the deburring media is gas. Normally, media refers to an abrasive of some type. Although nothing is being rubbed across the workpieces in a TEM operation, a fixture is sometimes required. That would be for very thin parts, as a support against the shock wave or, in some cases, as a heatsink. Small parts can be batch processed (Figure 43.4).

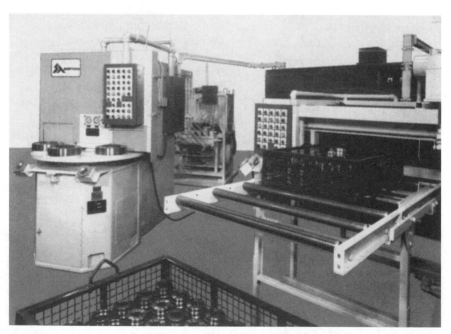

FIGURE 43.3 Deburring and cleaning machines in tandem. *(Bosch Corp., Surftran Div., Madison Heights, MI)*

FIGURE 43.4 Screw machine products are ideal for TEM. *(Bosch Corp., Surftran Div., Madison Heights, MI)*

The TEM process will not change any dimensions, surface finish, or physical properties of the parts, providing that fixtures are used when called for. That is because the parts are exposed to great heat only for milliseconds; except for the burrs, nothing feels a temperature higher than a few hundred degrees Fahrenheit. Threads are not affected by the heat because their wide roots transfer the heat quickly.

LIMITATIONS OF TEM

Despite all of the advantages of TEM, there are some drawbacks. If the parts go into the deburring chamber with oil on them, the oil will burn into the workpieces and leave a carbon smut on them that is difficult to remove. In addition, the oil might prevent gas from washing the surface properly. Before entering the chamber, all parts must be cleaned free of oil.

Blind and tapped holes must be free of compacted chips. Unless the gas can surround all chips, it is unlikely that they will vaporize completely. The final problem has to do with consistent radii. Getting consistent radii on aluminum and stainless parts is difficult to achieve. When consistent radii are important, there is another machine deburring method, Abrasive Flow Machining (another chapter in this book), whose specialty is providing consistent radii.

TEM is effective on most engineering materials, although it is more suitable on some than on others. As previously mentioned, the burr must absorb heat, reaching a high-enough temperature to oxidize the burr. This critical step is sometimes difficult to achieve if the material has a high heat transfer coefficient. Nevertheless, some companies have sufficient knowledge and experience to process many of these metals—even copper and aluminum.

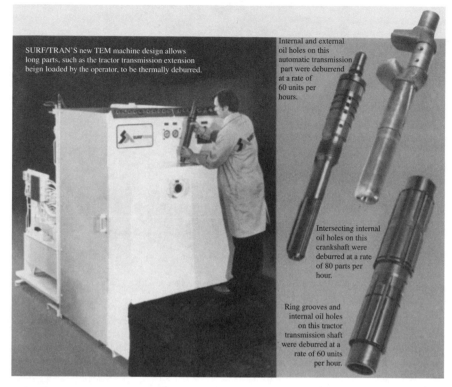

SURF/TRAN'S new TEM machine design allows long parts, such as the tractor transmission extension beign loaded by the operator, to be thermally deburred.

Internal and external oil holes on this automatic transmission part were deburrend at a rate of 60 units per hours.

Intersecting internal oil holes on this crankshaft were deburred at a rate of 80 parts per hour.

Ring grooves and internal oil holes on this tractor transmission shaft were deburred at a rate of 60 units per hour.

FIGURE 43.5   A Surftran TEM machine. *(Bosch Corp., Surftran Div., Madison Heights, MI)*

Another significant limitation is size. The newest machines cannot process anything larger than 10" diameter by 30" long. However, if a larger-capacity machine is required, it could be built for a higher cost (Figure 43.5).

Post-deburring cleaning is another consideration. Not all burrs are the same size. Most are close enough in size not to present a problem. Occasionally, a very large burr is not completely oxidized. That burr will take the form of a small molten particle, like weld splatter. That partially oxidized burr must be removed mechanically.

APPLICATIONS FOR THERMAL DEBURRING

The following are some benefits that accrue to those using the TEM:

1. Fixed manufacturing costs with no variances.
2. Eliminates costly and time-consuming manual deburring.
3. Ensurance that all burrs are consistently removed.
4. Increase quality levels and reliability.

Specific industries in particular have benefited from TEM. Screw-machine products are generally very small with holes, often cross-drilled. Whatever the material (brass, aluminum, steel, stainless steel, or plastic), screw-machine parts can readily be deburred by TEM.

The diecasting industry produces many small parts in both zinc and aluminum, which require deburring. Just consider the cross-holes in a diecast carburetor. TEM was created for parts like these.

In the fluid power industry, both pneumatics and hydraulics, it is crucial to remove all burrs. A loose burr can wreak havoc in fluid power. In these "Swiss cheese" manifolds, complete burr removal is guaranteed by TEM. TEM also deburrs spools, cartridges, valve bodies and other hydraulic components (Figure 43.6). Cast-iron parts are particularly successful for TEM, not only for deburring, but because core sand is blasted out, leaving clean surfaces.

Model P-90 Deburring System Features

Index Table:
5-Station Index Table maximizes production rates

Programmable Controller:
all machine functions are monitored and controlled by a programmable controller and can be easily adjusted.

4.75-inch diameter chamber:
capable of 400 PSI charge pressure.

Sound Enclosure:
the firing chamber area is housed in a sound enclosure to reduce noise.

Exhaust Fan:
the enclosure is continuously exhausted to remove the products of combustion.

Unitized Hydraulic System:
valves and controls are manifold mounted onto the hydraulic power unit.

I.D. windows on these aluminum automatic transmission components (3.5" × 1") were deburred at 720 parts per hour.

All loose burrs on these valve spools (2" × .75") were removed at 1440 parts per hour.

Intersecting holes on these steel hydraulic valve bodies (4" × 1.5") were deburred at 360 parts per hour.

Surftran offers a full range of TEM units from 4.75" diameter chambers to 12" diameter units. A long parts machine is available to deburr shafts up to 30" long. Most thermally deburred parts must be cleaned after TEM. Surftran also offers manual and automatic cleaning systems.

FIGURE 43.6 A small TEM machine and its work output. *(Bosch Corp., Surftran Div., Madison Heights, MI)*

Consider car-door lock plugs. Before TEM, these plugs were manually deburred by the thousands every week at high cost. They can now be batch deburred by TEM at a machining rate of about 400 per minute.

In actual production, most parts are deburred without fixtures. When parts must remain scratch free, or if diecast parts that aren't heavily ribbed are being processed, those parts should be held or supported. There are no rules about what should be fixtured. Experience is the best teacher.

There are many places in the aircraft industry where TEM comes in handy. Sometimes edges are broken manually, then are followed by a TEM treatment. In fuel pump housings, hidden trash can be eliminated by TEM. TEM also does an excellent job on jet engine vanes.

Recently, TEM has acquired a reputation for doing a good job decoring sand castings. Apparently, TEM is able to melt binders out of foundry sand, which allows the sand to be easily poured out of intricate castings.

EXPENSE

The cost of TEM equipment is high. Small machines are priced at $100,000 and up. You can easily spend $250,000 on TEM equipment and surface-cleaning machines. Maintenance and cycle cost is in the neighborhood of 10 cents per cycle, so the actual operation of TEM equipment is very inexpensive. In fact, anytime that the machine can be kept in meaningful operation, it will save money. One company in Southern California used to employ dozens of people deburring, filing, and cleaning components for the aeronautical industry. At my last visit, they had automated the operation with specialized deburring machines and only a few personnel were visible in that department.

Approximately 1000 TEMs are in operation now. No matter where your company is located, it is not too far from some other company anxious to do jobbing with TEM equipment. Jobbing is also done by the company that designed and introduced TEM to industry:

Surftran
Robert Bosch Corp.
30250 Stephenson Hwy.
Madison Heights, MI 48071

CHAPTER 44

ABRASIVE FLOW MACHINING AND ABRASIVE FLOW DEBURRING

INTRODUCTION

Abrasive flow machining (AFM), proven as a reliable and accurate process for deburring, radiusing, and polishing, embraces a wide range of feasible applications from critical aerospace and medical components to high production volumes of parts. It was originally invented in the late 1960s and has grown slowly, but steadily, in sophistication and popularity.

AFM can reach even the most inaccessible areas, processing multiple holes, slots, or edges in one operation. Advances in media formulation and tool design have established the abrasive flow process as a way of satisfying tough manufacturing requirements economically and productively.

Abrasive flow machining (AFM) or deburring (AFD) finishes surfaces and edges by forcing a flowable abrasive media through or across the workpiece. Abrasion occurs only where the flow of media is restricted; otherwise, the abrasive has no effect. The process works on many surfaces or selected passages simultaneously, finding in seconds, even those hard-to-reach burrs on cross-holes and interior areas. A sample machine is shown in Figure 44.1.

The AFM system consists of three major elements:

1. The machine to hold the workpiece and tooling in position and force the abrasive media through or across it.

2. The media carries the force applied by the machine to the workpiece edges or surfaces. The subject media covers a wide spectrum of viscosities, rheologies, abrasive types,

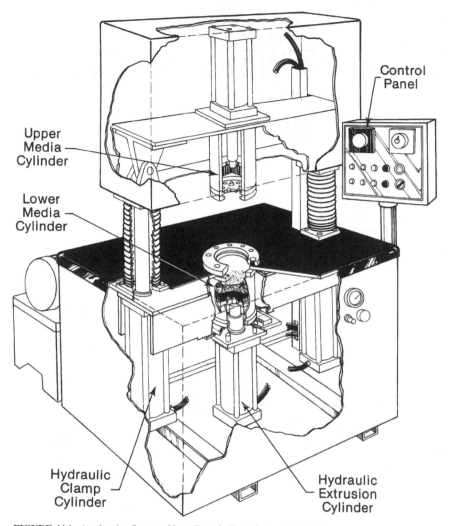

Control Panel

Upper Media Cylinder

Lower Media Cylinder

Hydraulic Clamp Cylinder

Hydraulic Extrusion Cylinder

FIGURE 44.1   An abrasive flow machine. *(Extrude Hone Corp., Irwin, PA)*

sizes, and concentrations to provide a variety of stock removal, surface improvements, or edge-finishing effects.

3. Fixtures to hold the workpiece in position and contain, direct, or restrict the media flow so that the areas of the workpiece where abrasion is desired form the greatest restriction in the media flow path.

Several parts can be abraded at one time, yielding rates of production of hundreds per hour. A variety of finishes can be produced at the same time by altering process parameters. In production applications, tooling is designed to be loaded and changed quickly. In current industrial operations, the last remaining high-cost area, is part finishing or deburring. Unless one of the new machining methods for deburring is utilized, finishing remains a labor-intensive, uncontrollable area.

Proper finishing of surfaces and edges affects more than simply the feel or appearance of a part. Performance can be dramatically improved by the correct finish. AFD is another competitor to Thermal Energy Deburring (TED) and Electrochemical Deburring (ECD) in the area of product finishing by machine. With today's focus on total automation with machine tools, AFD, TED, and ECD, all offer flexibility as an integral part of the complete manufacturing cycle.

Abrasive flow machining is used in many applications involving deburring, polishing, and edge radiusing, sometimes doing all three on the same piece. Advances in both tool design and media formulation have established AFD as a means of satisfying difficult manufacturing requirements. See Figure 44.2 for samples of work.

PROCESS FUNDAMENTALS

In AFD, two opposed cylinders extrude an abrasive media back and forth through passages formed by the tooling and the workpiece. Because the abrasive is active only where passage is restricted, the tooling is very important. One throughput can deburr one section of the part, polish another section, and establish a definite, repeatable edge radius on a third. Lapping can be done if you allow the media to gently hone a surface.

The process will handle soft aluminum, stainless steel, tough nickel alloys, or even hard materials (such as carbides and ceramics). AFM will achieve a wide range of predictable results. If the process objective is uniform polishing, the media should maintain a uniform flow rate as it passes through the passages. If the objective is deburring or edge radiusing, the flow should be increased where you want the activity to occur.

The flow pattern is a direct result of tool configuration, the media and the machine setting. The process can be applied to a wide range of part and passage sizes from gears as small as 0.060" in diameter and orifices as small as 0.008" to splined die passages several inches across.

The abrasive flow machine, available in several sizes, contains two vertically opposed media cylinders, which close hydraulically to hold a fixture or just a part between them. By repeatedly extruding media from one cylinder to the other and back again, an abrasive action is produced wherever the media enters a restrictive passage as it travels through or across the workpiece.

The machine also controls extrusion pressure. Pressures are adjustable from 100 to 3200 psi, with flow rates exceeding 100 gallons per minute. The volume of flow depends on the displacement of each cylinder stroke and the total number of strokes used to complete the job. Both of these variables are preset at the machine.

Control systems can be added to monitor and control-process parameters, such as media temperature, viscosity, flow speed, media feed, media cooling, and load-unload stations. Such automated systems can process thousands of parts per shift. AFM systems designed for production applications often include part cleaning and load/unload stations, media refeed devices, and cooling units.

TOOLING

In designing tooling, the areas where abrasion is desired are first identified. The tooling is then constructed to hold the workpiece in position and direct the flow of media to the appropriate

Where quality and performance are paramount and the cost of failure catastrophic, Extrude Hone helps manufacturers expand the limits of material strength and performance - dependably and economically.

Extrude Honing has proven to be a fast, accurate and reliable method of finishing edges and surfaces of critical components for aerospace, nuclear, medical and other applications where component performance requirements are especially demanding.

Extrude Hone eliminates hand operations and produces results more uniform, more reliable, and more precise than those achievable by hand or any other method.

Above left · Extrude Honing was initially developed to perform critical deburring of aircraft valve bodies and spools. As a result, a new quality standard was achieved, not only providing absolute burr-free internal edges (routinely passing 20x microscopic inspection) but also producing carefully controlled edge radii.

Left · This aircraft turbine fuel contol valve body has numerous internal flow intersections requiring critical deburring as shown in the photo micrograph (near left).

Above · A turbine engine fuel spray nozzle before and after Extrude Honing. To assure uniform edge radii and minimal dimensional change large hanging burrs are often removed mechanically prior to Extrude Honing.

Above (top) · Window edges of critical bearing retainers are deburred and radiused and the surface of the broached "pockets" are polished in minutes, with unequalled quality assurance. In fact, inspection can reduce from 100% to merely one or two random samples from each lot.

Above · Injector Nozzle—Miniature rocket engine component used to position space satellites. Extrude Honing is used to balance flow through each of the tiny EDM'd holes to assure uniform distribution of hydrozene fuel into catalytic chamber of engine.

Above · Critical splines are Extrude Honed to remove burrs and sharp edges, as well as heat treat scale, nitride "white layer" or other undesirable surface conditions. Stock removal is so uniform that slightly undersized internal splines can be "lapped" by Extrude Honing to restore proper size after excessive shrinkage in heat treating.

FIGURE 44.2 Examples of abrasive flow machining: deburring. *(Extrude Hone Corp., Irwin, PA)*

areas. In this latter function, it can also restrict flow where abrasion is desired or completely block the flow though areas where no change is desired.

Many AFD applications require only simple tooling. Dies generally require no tooling; the die passages themselves provide the restrictions for the flow of abrasives. For a piece with straight-through passages, such as a tube, an internal spline, or an extrusion die, the tooling merely holds the part in place between the cylinders and allows the through passages to restrict the media flow.

To process external surfaces, for instance a spur gear, it is held within a tube, so the flow is restricted to the space between the inside of the (tube) fixture and the outside of the part. Blind cavities, recessed areas, and counterbores can be processed using "restrictors" or "mandrels" to fit inside the components and restrict the flow at the desired areas.

Any number of parallel restrictions can be processed simultaneously. In some cases, two or more can be abraded as long as the cross-sectional areas of the in-line passages are equal. Very large parts can be processed in sections.

Many selected areas on a workpiece can be abraded at the same time. Several holes, slots, or edges can be deburred, radiused, and/or polished in one operation. A number of parts can be processed simultaneously. Depending on part size and machine size, dozens of parts could be processed in one fixture load, yielding production rates of hundreds of parts per hour. Jobs can be changed in a short time, ranging from minutes to a few hours. See examples of extrude honing in Figure 44.3.

Special attention is given to low-cost tooling. Wear inserts are used in areas of heavy wear. Design should consider ease of loading/unloading, as well as ease of media removal from parts and tooling after processing. Many machines have a rotary indexing table and two sets of cylinders so that the time required for the process cycle of one fixture load of parts can be productively used to unload and reload the second fixture.

MEDIA

The machining action produced in AFD is similar to a filing, grinding, or lapping operation, where the extruding "slug" of media becomes a self-forming file, grinding stone, or lap, as it extrudes through the passages restricting its flow. The maximum force that can be applied to the "slug" is the cross-sectional area of the restricted passage multiplied by the extrusion pressure.

Slug flow rates affect the uniformity of stock removal and edge radius size. Higher viscosities with nearly solid media are used to uniformly abrade the walls of large passages. Lower viscosities are preferred to radius edges and to process small passages. Slow rates passing through restricting passages are best for consistency when close tolerances are required.

Actually, some of this force is expended in deforming the semi-solid media into proper shape and getting it through the tooling to the part. Additional force is lost as the media stream moves within itself.

The AFD media is composed of a semi-solid carrier and a concentration of abrasive grains. Specific results can be achieved by varying the abrasive grain size, type, and concentration, and also the viscosity of the carrier. Higher-viscosity media are appropriate for smoothing the walls of large passages.

When abrasives enter a restrictive passage, the viscosity of the media temporarily rises, holding the abrasive grains rigidly in place. When, in this rigid condition, the media abrades the passage, then softens to its original state, producing little or no abrasion.

Below – Investment cast compressor wheels are polished to achieve maximum efficiency. Final finish is less than 20% of original. All but gross surface defects are removed — those remaining are highly visible for manual spot finishing.

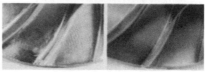

Above · Milled surfaces of an impeller are Extrude Honed to improve surface finish while retaining shape within the profile envelope eliminating hours of hand finishing and maximizing process control.

Far left · Air foil surfaces are polished from original cast finish of 60-80 microinches to below 16 microinches, even in difficult to reach areas between blades.

Near Left · Elimination of boundary layer turbulence by smoothing complex internal cast passages increases cooling air flow through hollow investment cast air foils.

Above · Removal of "recast layers" from thermally machined (EDM, laser, electron beam) air cooling passages can improve the thermal and mechanical fatigue strength of highly stressed components dramatically. The simultaneous edge radiusing produced by Extrude Honing provides an additional fatigue strength improvement. In production, hundreds of holes or slots, as small as .015 inch, can be processed in single 5 to 15 minute operation.

Above left · Scanning electron photomicrographs of electron beam drilled holes, before and after Extrude Honing .

Above center · Precise, polished edge radii are produced on the edges of fir tree slots of turbine discs by a number of manufacturers of aircraft turbine engines using Extrude Hone equipment. Uniform radii on both sides of each fir tree are generated in a single controlled operation.

Above right · Air cooling holes in a turbine disc are polished and heavily radiused to improve fatigue strength of the disc. Over 100 holes, in a disc more than 30 inches in diameter are processed in one operation.

FIGURE 44.3 Examples of abrasive flow machining: honing. *(Extrude Hone Corp., Irwin, PA)*

Boron carbide, aluminum oxide, and diamond can be used as abrasive media, but most jobs use silicon carbide. The particle size ranges from 0.0002 to 0.060", or in mesh size from 1000 mesh to 8. As you would expect, the larger grains remove stock at a faster rate, but the smaller-sized grains provide finer finishes and access to small holes.

The effective life of the media depends on several factors, starting with the initial batch of media: the abrasive size and type, the flow speed, and the part configuration. During the AFD process, the abrasive grains break and become less effective, and the abraded material mixes with and dilutes the media. Surprisingly, large amounts of cut material (as much as 10%) can be tolerated. However, this dilution doesn't happen quickly. A typical machine load of media can be used for weeks, processing thousands of parts before replacement. In some cases, media can be reconditioned periodically to adjust viscosity or to add new, sharp abrasive, thereby extending media life.

Often, a specially designed air-cleaning station is built for fast and reliable removal and recovery of media from finished parts and fixtures. Air nozzles or vacuum can remove media from internal passages and removal can be completed by means of a solvent wash. Sometimes a solvent wash station is also built within the work area to remove any final traces of media. This can be an ultrasonic or vapor-degreasing system.

Unless the parts are prefilled with media, every fixture load of parts goes into the machine empty and comes out filled with media. This must be recovered and fed back into the machine—either manually or with the assistance of several devices marketed for the job.

The depth of cut made by the abrasive grains at the surface depends on the extrusion pressure applied, the size of the abrasive grains, and the stiffness of the media. The operator presets the machine speed and extrusion pressure, then determines the number of cuts required for the job.

High production cycles generate heat within the media. The temperature should not be allowed to go above 120°F to ensure operator comfort and satisfactory media life. The heat affects media viscosity and high operating temperatures are harmful to both machine seals and consistency of operation. Heat exchangers are available to remove the excess heat.

PROCESS APPLICATIONS

Precision, flexibility, and consistency are available through AFD. A large number of applications are available for AFD in aerospace, automotive, production, and all kinds of die finishing. Diverse applications, including surgical implants and centrifugal pumps, have materialized.

The process was originally devised for critical deburring of hydraulic valve spools and bodies. It performed so well deburring edges that additional tasks were found for the process. Any parts with internal crossed holes can be quickly and consistently handled.

Most extrusion dies (new or used) must be polished prior to extruding to achieve proper flow characteristics. Most other dies and molds have a polishing step also. Until the acceptance of AFM as a viable alternative, time-consuming and expensive hand polishing was the only way to do that job. In addition to saving up to 90% of the cost, a more uniform surface is produced automatically, with far better control than that of manual polishing.

Hand polishing sometimes smears removed metal, inviting future problems. AFM completely discards this removed metal. In fact, a separate media has been developed strictly for die polishing. Currently, dies are being polished in 10 to 30 minutes. Few require more than an hour. Compare this time with the hours consumed by hand polishing.

EXAMPLES

The Ford Motor Company has been using a totally automated abrasive flow machine for deburring the external involute gear teeth on starter motor armature shafts at rates up to

1200 parts per hour. Pratt & Whitney is using this system for finishing turbine blades, fuel spray nozzles, fuel control bodies, and bearing components.

The recast layers left in electrical discharge machining operations, are now being removed by AFM. The V-22 Osprey airplane, built by Boeing, uses this process to deburr and polish small oil passageways on the transmission gears and spray nozzles. Motorcycle speed records were broken by a unit that had its engine cylinders polished by AFM.

Fuel injector bodies for automobiles have been finished by this process for more than 10 years. In fact, several car manufacturers are using AFM to process parts that they don't want to disclose to the competition.

The following variables of the process must be understood to obtain effective performance. If extrusion pressure is held constant and:

1. passage area increases, flow speed will increase.
2. passage length shortens, flow speed will increase.
3. media viscosity lowers, flow speed will increase.
4. and in case extrusion pressure is increased, flow speed will increase.

If flow speed is held constant and:

1. passage area increases, extrusion pressure decreases.
2. passage depth shortens, extrusion pressure decreases.
3. media viscosity lowers, extrusion pressure decreases.
4. flow speed decreases, extrusion pressure decreases.

Abrasive flow machining is now an established process. The Extrude Hone Corp. of Irwin, PA, has had a great deal of experience with it. It could be of tremendous assistance to any company interested in either purchasing AFD equipment or using the jobbing services of Extrude Hone to investigate the process and learn what it could do for them.

P · A · R · T · 9

MATERIALS

CHAPTER 45
CERAMICS

BACKGROUND

When humans first mastered the use of fire 10,000 years ago, they learned how to make low-temperature earthenware in open firing pits. That unsophisticated production was the start of ceramic development. Until 100 years ago, ceramics were tableware, clay pipe and brick, and also roofing tiles.

It was not until about 1820 that silica refractories were first made. About 50 years later, when mass production of iron and steel began in earnest, new refractory ceramics were developed to replace the conventional fireclays. This change was absolutely necessary for the production of iron and steel to reach today's plateau.

Scientific dictionaries report that a ceramic is any class of inorganic, nonmetallic products that are subjected to a temperature of more than 1000°F during manufacture or use. They include metallic oxides, carbides, nitrides, borides, or any combination or compounds of these.

During this late 19th-century period, K.J. Bayer discovered a method of producing aluminum hydroxide by hydrolytic deposition. Since then, there have been no major alterations in Bayer's method used to refine aluminum. Alumina is produced when this aluminum hydroxide is baked in a rotary kiln. The alumina produced this way has a purity of about 99.6%. Purities as high as 99.9% or better can be obtained by using an improvement in the process. See a flowchart of the Bayer process in Figure 45.1. For applications requiring a high degree of electrical insulation or for extremely high strength, the higher-purity material is required.

Most ceramic fabrication is a custom business. Parts are made to customer drawings; the customer carries inventory, not the manufacturer. The largest fabricators use their own proprietary ceramic powder and the manufacturing intelligence that they have gathered over many years of experimentation. They will have highly skilled craftsmen make 1, 2, or even 10 parts for you to try out. When the customer declares that the part is correct, the ceramic fabricator will automate the part, translating the work of his craftsmen into machine language.

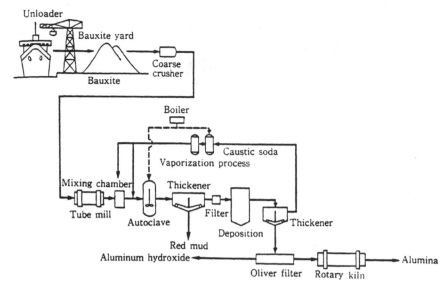

FIGURE 45.1 A flowchart of the Bayer process.

CLASSIFICATION OF CERAMICS

One possible classification of ceramics is by chemical composition. A number of oxides are segregated into binary compounds, ternary, and quaternary. Then, there are carbides, nitrides, and oxynitrides. It is also possible to classify by minerals, of which there are many.

A common classification is by molding technique. Ceramics can be molded by hand, by mechanical die pressing, by isostatic pressing, by slip casting, vibratory casting, injection molding, and a number of less-popular methods.

CERAMIC MANUFACTURING PROCESSES

The dimensional precision of a ceramic part depends on many factors: The particle size and distribution, the green density after compacting, the firing cycle and sintered density, and the inherent shrinkage of the composition. The method that enables ceramics to meet these exacting requirements must also be economical and suitable for large-scale production. Steel die pressing comes closest to fulfilling these requirements. Accordingly, it has been developed to a high degree of automation.

DRY PRESSING

When ceramic powder has to be crushed and mixed, a ball mill is the usual mixing device. However, you must be careful in selecting the correct ball mill. Because high purity of the powder ceramic is essential, the mill lining and the ball itself should be ceramic, and, if possible, the same type as that being mixed.

One of the most cost-effective ways to form ceramics is by dry forming. It permits a near net shape to be made in high volume. In this process, dry granular ceramic powder is compacted in a metal die. The moisture content of the powder varies between 0 and 4%. The free-flowing granules are crushed and densely packed to form a coherent compact. This technique varies slightly as the type of ceramic powder varies. In any case, the process requires hot isostatic pressing for maximum strength and close dimensional tolerances.

The rheology of ceramics is a study in itself. Ceramic forming depends on the purity of the material, the powder size (must be under 20 microns), the size of the part, and the tooling used. This is an area where experience is most significant. The die designer must be familiar with the shrinkage characteristics of the ceramic powder. He cannot afford to make the wrong guess because dies are too expensive.

He has a small leeway to adjust shrinkage by varying the pressure or green density of the compact. However, this is not recommended because the quality of the part might suffer. If the designer uses a higher pressure, he might end up with severe die wear. If he uses lower pressure, the low density might cause warpage.

Several problems are peculiar to this process. Cracks are the biggest of these. A crack is an indication of a malfunction, and it must be dealt with immediately. Die wear is the first element to check. Cracks can also be caused by entrapped air, bent die pins, excessive friction between the die and the powder, and improper ejection from the die. A compact will expand about 0.5% immediately after ejection. This also can cause cracks in the compact.

Several methods are used to locate cracks. Unfired, green compacts can be dipped into any low-surface tension liquid, such as alcohol or kerosene. Any crack will show because it will absorb the test liquid faster than the rest of the compact. Fired pieces can be dipped into a fluorescent dye and observed under black light.

Die wear is an important consideration when pressing ceramics. Refractory oxides and aluminas require carbide dies. Otherwise, rapid wear will occur. Sometimes, a thin carbide die interior is sweat fitted to a supporting steel exterior. The following design hints should be observed when designing both parts and dies:

1. Holes and countersinks should not be located too close to edges to avoid cracking thin walls.

2. Always strive to have uniform cross-sections. This will result in uniform density.

3. Opposing punches, like top and bottom, must have sufficient space between them to avoid damage to the punches. A few design hints are sketched in Figure 45.2.

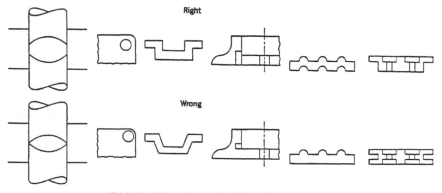

"Right" and "wrong" ceramic designs for automatic dry pressing.

FIGURE 45.2 Design hints. *(Wesgo, Belmont, CA)*

Unless the die is made of carbides, many ceramics will cause die wear because they are abrasive. There are so many precautions to be aware of, that the best step is to use the services of an experienced die designer.

ISOSTATIC PRESSING

The simplest manufacturing method is die pressing. However, that also has a simple problem. It is extremely difficult to compress the ceramic powder uniformly. This unequal pressure causes strains in the molded parts, which affect quality. In addition, any dry pressing creates friction at the walls. To avoid such problems, isostatic pressing has become very popular. Applying pressure from all sides, instead of unidirectionally, is a much more desirable method. Two other names for this process are *rubber pressing* and *hydrostatic molding*. Figure 45.3 is a schematic representation of a laboratory hydrostatic molding.

A preliminary molding at low pressure is carried out by creating a weak shape. This shape can be created using a simple metal mold. Mandrels can be used to create hollows. The molded form is then placed inside a thin rubber bag, which is subjected to a high hydraulic pressure. This is a very uniform pressure and the finished parts are more uniform in physical properties. The design of the rubber molds to produce ceramic parts requires substantial empirical experience. Different parts of the mold can be constructed of various rubbers, each for a specific purpose.

For large parts that don't lend themselves to dry pressing, cold isostatic pressing is often the means to use. Sometimes dry powder is poured into a rubber mold, which is then inserted into a liquid media; then, high pressure is applied through the liquid. Compression occurs equally in all directions. This process eliminates normal wall friction (see Figure 45.3 again).

HOT ISOSTATIC PRESSING (HIP)

Hot isostatic pressing densifies the ceramic parts, and closes internal voids, "healing" them by diffusion bonding. To HIP ceramics successfully, a few things must be known:

1. The quality of the starting powder.
2. The powder-handling and green-forming techniques.
3. A knowledge of the powder characteristics and their effect on processing parameters, microstructure, and properties are all critical to successful application of HIP in the development and manufacture of advanced ceramics.
4. There is a need for the ceramic powder size to be as small as obtainable because this is dictated by densification kinetics. It should not exceed 5 microns in size and 0.5 microns would be desirable to achieve such good properties as mechanical strength. Do not confuse this ceramic powder size with that used in powder metallurgy, which is 10× as large.

Slip Casting

The ceramic material, in the form of fine powder, is mixed with about 30% of water, by volume. The mixture, called a *slip*, is poured into a mold made of Plaster of Paris. The

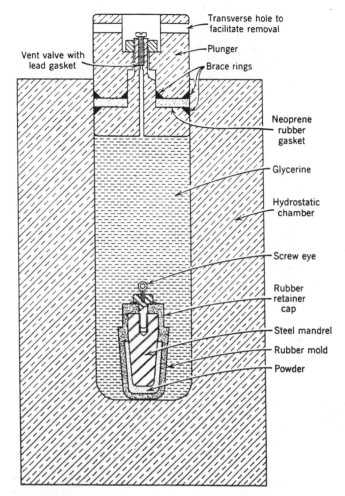

Transverse hole to facilitate removal

Vent valve with lead gasket

Plunger

Brace rings

Neoprene rubber gasket

Glycerine

Hydrostatic chamber

Screw eye

Rubber retainer cap

Steel mandrel

Rubber mold

Powder

FIGURE 45.3 Rubber pressing, hydrostatic molding, or isostatic pressing. *(Wesgo, Belmont, CA)*

Plaster of Paris mold absorbs water slowly from the slurry. If the liquid in the slip is a solvent instead of water, the mold is made of a different material (possibly filter paper) to absorb the solvent. Compared to other tooling, plaster molds are inexpensive and are ideal for short runs.

After a predetermined time period, the excess slip is poured from the mold and the part is removed. Control of the casting thickness is determined by the amount of time permitted to pass from the pouring of the slurry into the mold until the pouring off of the excess slurry. The detail of the casting is accurate and delicate. One disadvantage of the process is the long drying time it requires.

The slip casting method is unique to ceramics and has been used for many years. The method is not good for producing thick walls, but complex forms can be cast with high

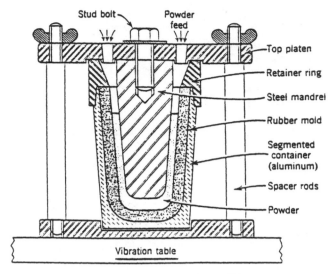

FIGURE 45.4   Vibratory casting. *(Wesgo, Belmont, CA)*

fidelity. It is a good method when a few parts for testing must be made and when the quantity required is low.

Vibratory Casting

Much interest is being shown by industry in the use of vibratory forces for the compaction of ceramic powder. This is particularly true in forming irregular shapes. The principle of this method is the agitation of many particles when there is only slight restraint, so those particles will seek the configuration that offers the closest packing. If, for example, the vibrated material consisted of a mixture of fine and coarse particles, the fine particles will tend to fill the voids between the coarse ones. If all the particles are approximately the same size, but of different shape, the vibratory action will orient the various shapes into receptive voids between other irregular particles.

In this process, ceramic powder is mixed with no more than 10% of water and poured into a mold. The mold is secured to a vibration table and is vibrated. This packs the powder very tight and the ceramic form would have a lower shrink rate than slip casting. Because the water content is so much less, and, in fact, a Plaster of Paris mold is unnecessary. The detail of parts made by this technique is also accurate and delicate (Figure 45.4).

EXTRUSIONS

A slurry that has its water content squeezed out in some kind of filter press becomes a moist cake. A die or nozzle to produce the ceramic shape desired is attached to an extrusion machine. The extrusion process is used to form long objects with a simple, uniform cross section. Normally, the exterior of the shape is not machined, although it certainly could be after it dries. For most products, the long object is simply sliced like a sausage and used as cut off.

THE FIRING PROCESS

After the parts have been pressed by whatever process available, they are in the "green" state, which means that they are weak and easily susceptible to breakage. In this state, they can be machined by conventional means, such as milling, turning, or drilling. Because the vendor knows how much the parts will shrink in sintering (the percentage of binder in the powder mix), she can machine very close to final size, leaving just a little for the finish.

Next, the parts are fired and hardened into a new state. This process is called *sintering* and is similar to the sintering process used in powder metallurgy. The sintering occurs in kilns of which there are two types: batch kilns and continuous kilns. In batch kilns, the parts to be fired are loaded onto ceramic shelves, placed in the kilns, fired for a set period, cooled, and removed. The batch kiln is flexible in regard to firing conditions and is relatively inexpensive.

The continuous kiln is generally used for high-volume production. It has three zones for preheating, firing, and cooling. The parts are loaded onto carts or ceramic carriers, which enter the kiln at one end, one at a time, at predetermined time intervals. The cart advances on a conveyor through the three zones until the parts are sintered.

REFRACTORIES

Many factors must be considered when selecting furnace refractory materials, including ease of installation, insulating characteristics, maintenance costs, and operating conditions. The furnace load can vary in one day from 300 to 3000 lbs. The temperature can vary from 1200 to 2400°F. This constant thermal shock creates extensive refractory damage.

Some refractories retain heat so long that it might require a 16-hour cool-down period before repairs can be made. To alleviate problems such as these, a ceramic fiber lining was tried and was found to be successful. The bottom line was cost and efficiency, and the ceramic liners are now being used extensively.

The previous type of castable refractory lining had two disadvantages. First, it was very susceptible to thermal shock. Second, each repair was costly because of the long downtime (16 hours) waiting for the temperature of the furnace to drop low enough for a specially dressed repairman to enter. A ceramic fiber lining is the new design. It is manufactured by the Carborundum Company Fibers Division. The 10" thick 12" × 12" blocks are fastened as a module, directly to the furnace's steel castings with power-actuated steel pins. A 3000-degree polycrystalline ceramic-fiber blanket is placed between the modules to control shrinkage and reduce maintenance.

The savings from this furnace redesign is tremendous. In a firing cycle, 38% less fuel is consumed. The cooling period is faster and so is the start up period. The average time to start the furnace and bring it to idling temperature is now 6 hours, instead of the former 12.

MATERIALS

Ceramic parts are made of main constituents and auxiliary constituents, which help the sintering of the main. The two constituents must be thoroughly mixed for best results. Mechanical pulverization followed by classification is the most widespread method of preparing ceramic powder. Two significant determinants of the sintering qualities of parts are powder size and grain distribution.

The widespread method for producing the fine powder used in ceramic manufacturing is to turn lumps of raw materials into powder by mechanical pulverization, then classification. Ball mills are commonly used to reach micrometer-sized grains of powder.

To further increase the performance of ceramic products, ultrafine powders must be made purer. Materials with these qualities are being developed by using several modifications to the basic process.

The Wesgo Company of Belmont, CA, produces a high-purity alumina, which is a proven tough performer in corrosive, heat, and wear environments. It is an excellent electrical insulator and is displacing other materials in rugged applications. Alumina is as hard as sapphire and outperforms some of the hardest known materials in situations calling for extreme hardness.

On the whole, ceramics have low ductility and high compressive strength. Theoretically, ceramics could also enjoy high tensile strength, but they do not because small cracks, pores, and other defects act as stress concentrators and their effect cannot be reduced through ductility and plastic flow. Ceramics are sensitive to even small flaws. As a result, failure typically occurs at tensile stresses between 3 and 30 ksi. Ceramic properties are shown in charts that are available from supply houses.

On the positive side, high tensile strengths could be obtained by eliminating these flaws and defects. The hardness, strength at elevated temperatures, wear resistance, light weight (specific gravity of 2.3 to 3.85), dimensional stability, corrosion resistance, and chemical inertness are all attractive properties. On the negative side is reliability. Failure still occurs as a result of brittle fracture with little warning. Accurate machining is difficult, so products must be fabricated through the use of net-shape processing.

Structural ceramics include silicon nitride, aluminum oxide, and ceramic composites. Silicon carbides and silicon nitrides offer excellent strength with moderate toughness. They work well in high-stress, high-temperature applications. Silicon is stronger than steel, extremely hard, and as light as aluminum. It has good resistance to wear and thermal shock, it is an electrical insulator and retains tensile and compressive strength up to 2550°F. A feature that provides excellent dimensional stability is its coefficient of thermal expansion, which is $\frac{1}{3}$ that of steel and $\frac{1}{10}$ that of plastic.

An area of considerable interest and research is a ceramic engine block. If perfected, it would permit higher operating temperatures, an increase in efficiency, and would allow the elimination of radiators.

Ceramic technology has advanced cutting tools significantly. Cobalt-bonded tungsten carbide quickly became popular as an alternative to high-speed tool steel. Then came the coated carbides, utilizing vapor-deposited coatings of titanium carbide, titanium nitride, and aluminum oxide to reduce wear and friction, and enable faster cutting rates.

One of the toughest ceramics known to man is silicon nitride. It is still considered an advanced engineering material. This is already meeting industry's demand for performance at levels other conventional metals and ceramics cannot meet. It has low density (less than $\frac{1}{2}$ that of stainless steel). Above 1000°C, it has a bending strength above 100,000 psi. Besides excellent corrosion resistance, it has low thermal expansion, low thermal conductivity, a high thermal shock resistance, and high fracture toughness.

The compressive strength of alumina exceeds that of glass, porcelain, or quartz. At elevated temperatures, up to 2000°F, alumina is dimensionally stable under heavy loads, but metals would warp and flow plastically. Alumina has a fairly high thermal conductivity, compared to other ceramics, being roughly equivalent to Fe-Ni or Fe-Cr alloys. This material is inert to oxidation and is virtually unaffected by water, solvents, and salt solutions.

Most companies that manufacture ceramic powders have different categories of powder purity. Wesgo, for instance, has four standard, vacuum-tight compositions, ranging from 94 to 99.5% alumina to cover a wide range of applications. As purity increases, such

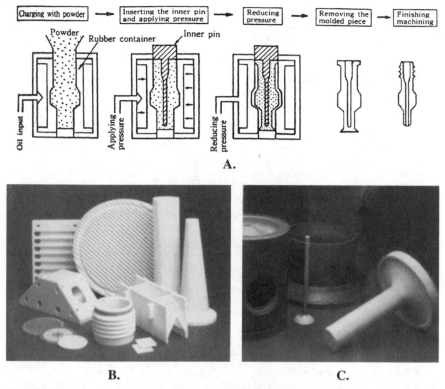

FIGURE 45.5 Samples of ceramic production: A. Making spark plugs. B. Alumina components. C. Sintered silicon nitride components. *(Wesgo, Belmont, CA)*

properties as chemical resistance, thermal conductivity, dielectric strength, and resistivity increase. But higher purity also means higher cost.

Electronic equipment and electrical appliance manufacturers use a large number of ceramic insulators. Some of these are of simple design, such as spacers and bushings. Others are more complex (such as sockets for vacuum tubes, switches, and bases for trimmer capacitors). All have certain requirements in common: they must be dimensionally accurate so that they can be used in automation, and must have uniformity of dielectric and mechanical properties to ensure the high quality of the components of which they are a part. Ceramics are used as insulators, varistors, capacitors, high electrical-conductivity ferrites (an oxide magnetic material), magnetics, chemical properties, microwave characteristics, thermal properties, optical properties, and for biological applications (Figure 45.5).

DESIGN

Successful applications of ceramic products include pistons, turbine blades, nozzles, gears, and bearings. More recently, other uses for ceramics include parts for missile radomes, turbine shrouds, electronic substrates, and pump and valve parts (especially for the chemical industry). Many parts using these ceramics are still in development; consequently, they are being made in small quantities. Yet, there are plenty of parts (such as seals in small, domestic oil burners and water-distribution pumps), which have been providing good service for the past 20 years.

Ceramics are used for electronic resistors and heating elements for furnaces and space heaters. Ceramics with semiconducting properties are used for thermistors and rectifiers. Dielectric, piezoelectric, and ferroelectric behavior can be utilized in a number of applications. Barium titanate, for example, is used in capacitors and transducers.

Cermets are combinations of metals and ceramics (usually oxides, carbides, or nitrides) bonded together in the same way that powder-metallurgy parts are produced. Cermets combine the high refractory characteristics of ceramics with the toughness and thermal shock resistance of metals. Ceramic cutting tools have become very popular. Cobalt-bonded tungsten carbide quickly became an alternative to high-speed tool steel. Then came the carbides coated with vapor-deposited coatings of titanium carbide, titanium nitride, and aluminum oxide to reduce wear and friction. Now additional ceramic cutting-tool combinations permit cutting speeds as much as 10 times faster.

They are used as crucibles, jet engine nozzles, and in other applications that require strength and toughness at elevated temperatures.

Wesgo has made a tailpipe and combustion chamber for a large combustion engine constructed from silicon nitride (Figure 45.6). Until at least 1990, these 70-pound pieces were probably the world's largest components made from this material.

Design—Brazing

Joints where one or both parts are ceramic can be made using epoxy adhesives, mechanical fasteners, or low-melting glass frits. The method selected must survive the specific conditions of service that the assembly sees. It might be thermal cycling, high temperatures for an extended period, vibrations, corrosive environments, or high stresses. The major problem with using ceramics is their brittleness or low tensile strength.

Consequently, machined surface defects (such as microfractures caused by grinding) must be removed by lapping or honing. In some cases, alumina can be refired to heal the cracks. If refiring is done, the surfaces must be used without further machining.

The best ceramic brazing alloys are spherical in shape, low in gas content, and free of impurities. They are available in the following forms: ribbon, sheets, rings, washers, special preforms, and pastes. They are used in vacuum, hydrogen, or inert gas furnaces. There is a brazing alloy to meet all of industry's requirements for high strength, hermeticity, and reliability in ceramic-to-ceramic or ceramic-to-metal bonds without prior metallization. Depending on the contents of the brazing material, the alloy system will permit brazing at temperatures of 1380, 1545, or 2050°F (Figure 45.7).

Any process must be commercially successful before it can attain popularity. In the near future, if not right now, it will be necessary for advanced ceramics to permit successful and economic brazing of ceramics to steel parts. In the matter of the ceramic engine, joints must be made between ceramics and steel, and ceramics and iron. In the electronic and vacuum-tube industries, these joints have already been made for many years.

Currently, the conventional practice of joining metal to ceramics involves a two-step molybdenum metalizing process requiring both nickel plating and brazing. This is expensive and difficult to control. Wesgo has developed a one-step process using a silver-copper brazing alloy that contains 2% titanium. Lower melting-point brazing alloys have also

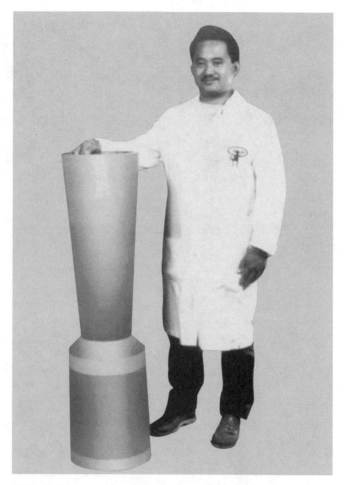

FIGURE 45.6 Large silicon nitride components: A tailpipe (top) and combustion chamber shell for a pulse combustion engine. *(Wesgo, Belmont, CA)*

been developed, which permit successive joints to be brazed near each other when the subsequent brazes are made with lower-melting alloys.

In these situations, several criteria must be met. The filler metal must have suitable ductility, permit controlled flow to a preselected position, and allow sufficient wetting of both parts. For best results, (blushing) minimum filler flow on both metal and ceramic surfaces is required.

Thermal expansion of both metal and ceramic parts should be uniform from start to finish. Both parts should be heated similarly to about 120°F below the solidus temperature of the filler metal and held there until all parts, including the brazing fixture, reach uniform temperature. Then, the heat should be increased to about 120° above the

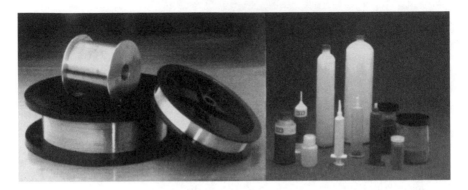

Ceramic-to-metal design guide

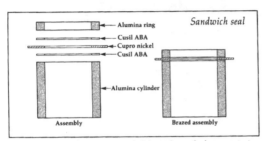

Assembly — Brazed assembly

Sandwich seal

- Alumina ring
- Cusil ABA
- Cupro nickel
- Cusil ABA
- Alumina cylinder

The sandwich seal increases joint reliability and uses a backup ceramic ring, with two purposes: it eliminates the bending movement in the metal and distributes the shear stress equally between the two ceramic ring faces.

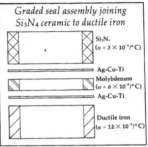

Graded seal assembly joining Si_3N_4 ceramic to ductile iron

- Si_3N_4 ($\alpha = 3 \times 10^{-6}/°C$)
- Ag-Cu-Ti
- Molybdenum ($\alpha = 6 \times 10^{-6}/°C$)
- Ag-Cu-Ti
- Ductile iron ($\alpha = 12 \times 10^{-6}/°C$)

Silicon nitride, molybdenum, and 410 stainless-steel rings are simultaneously joined by step brazing to form a graded seal. This technique can be used for joining ceramics to metals with large differences in thermal expansion.

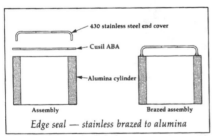

- 430 stainless steel end cover
- Cusil ABA
- Alumina cylinder

Assembly — Brazed assembly

Edge seal — stainless brazed to alumina

The metal part is attached edgewise to permit concentric distortion, with stress distributed across the total ceramic face by forming a full filler-metal fillet.

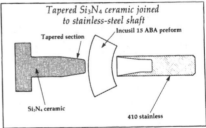

Tapered Si_3N_4 ceramic joined to stainless-steel shaft

- Tapered section
- Incusil 15 ABA preform
- Si_3N_4 ceramic
- 410 stainless

A tapered-joint brazed assembly consists of a high-expansion metal member on the outside, and a mated tapered ceramic on the inside. The tapered joint allows the alignment of both ceramic and metal shaft axes, and the tapered metal applies gradual circumferential compressive loading (starting from the knife edge) along the ceramic shaft.

FIGURE 45.7 Examples of ceramic brazing using Cusil ABA. *(Wesgo, Belmont, CA)*

liquidus temperature of the filler metal and maintained for up to 10 minutes before cooling.

A microprocessor-controlled vacuum furnace will minimize thermal stresses on the ceramic. The brazing alloy can sometimes be selected to allow simultaneous brazing and solution heat treatment of the assembly.

The major problem of mixed-material brazing is their difference in thermal expansion. If the difference is small and the parts also small, a ductile filler is used. If the parts are larger than 5", a combination of proper design, low brazing temperature, and ductile filler is necessary.

If the thermal expansion difference is large, brazing should be done with a graded seal. A two-step joining technique can be used. A stainless steel-molybdenum braze first, followed by a lower temperature braze to a ceramic ring with a lower-temperature brazing alloy. This minimizes joint stresses on cooling to room temperature (see Figure 45.8 for samples of brazing).

The brazing material available now can easily meet the requirements of aircraft manufacturers and builders of high-power vacuum tubes. The filler material is oxide free, and has low vapor pressure. Advanced brazing techniques can provide sophisticated, complex subassemblies. This information on brazing merely skims the surface. The material manufacturers should be contacted directly for esoteric information (Figure 45.9).

In this rocker-arm assembly for diesel engines, a partially-stabilized zirconia pad is brazed to ductile nodular iron with the filler alloy of Ag-Cu-In-Ti.

Silicon nitride to stainless brazed joint

A vacuum tube envelope in the first stage of subassembly. An Al_2O_3 cylinder is joined by 430 stainless steel with Ag-Cu-In-Ti. Above, an assembled vacuum tube and its disassembled components are shown. In the background are several vacuum tubes installed in a surge arrestor.

FIGURE 45.8 Some ceramic-brazed components. *(Wesgo, Belmont, CA)*

Cusil-ABA,® now a fully qualified production material, is a true alloy, not a blend or composite system. It wets and bonds to nearly all metallic and nonmetallic surfaces, such as oxide, nitride and carbide, and produces strong joints, dependent upon adequate treatment of ceramic seal faces. In addition to Cusil-ABA,® other alloys are being developed to include copper, silver-copper-indium and gold-copper, allowing a brazing temperature range from 750° to 1125°C.

Properties of Cusil-ABA®

Nominal Composition	63wt%Ag 1.75%Ti Balance Cu		
Physical Properties	Liquidus Temperature	1500°F	815°C
	Solidus Temperature	1435°F	780°C
	Density	5.2 troz/in^3	9.8 Mg/m^3
	Thermal Conductivity*	104 BTU/ft/hr/°F	180 W/m^1/°K
	Coefficient of Linear Thermal Expansion (R.T** to 932°F or 500°C)	10.3x10^{-6}/°F	18.5x10^{-6}/°C
Recommended Brazing Temperatures		1526-1562°F	830-850°C
Furnace Atmospheres	Vacuum, (10^{-5} Torr with leak rate not to exceed 5 microns/hour) Inert Gas (Argon or Helium)		
Electrical Properties	Electrical Resistivity	26 ohm cm/ft	44x10^{-9} ohm-m
	Electrical Conductivity	.038 mho ft/cm	23x10^6 mho/m
Mechanical Properties	Young's Modulus	12x10^6 psi	83 GPa
	Poisson's Ratio*	0.36	
	Yield Strength (0.2% Offset)	39,300 psi	271 MPa
	Ultimate Tensile Strength	50,200 psi	346 MPa
	Elongation (2 inch gauge section)	20%	
	Hardness	110 KHN	1100 MPa

*Calculated
**Room Temperature

FIGURE 45.9 The composition and properties of the Cusil-ABA brazing compound. *(Wesgo, Belmont, CA)*

MACHINING CERAMICS

After many years of trying various machining methods, procedures evolved that do an excellent job. Vendors know just how much the composition will shrink because of processing pressure and the sintering. This allows them to do most of the machining while the part is in the green condition. The milling, drilling, and turning can be done fairly quickly, although heavy machining forces should be avoided. After sintering, when the part has reached full strength, honing, grinding, and lapping are commonly done.

Lapping is a method of producing simple geometric forms, such as flats or cylindrical surfaces. This process is for extremely high precision. Often, hairline cracks appear on ceramic parts. Many are invisible, except when viewed under black light, after the surface has been dipped into fluorescent dye. Lapping can eliminate those scratches.

Parts without tight tolerances just get cleaned up for delivery. Parts with tight tolerances are now ground to the finish dimensions. If they are not severely stressed in use, they will be delivered. However, if they function under high stress, they will next be lapped or honed to get rid of microcracks before delivery. If the part is to be brazed and requires a good finish, it must be microcrack free or the brazing process will enlarge the microcracks and spoil the assembly.

The main factors influencing the lapping characteristics are the type of lap, the type and size of the abrasive grains, the type of lapping fluid, the lapping pressure, and the lapping speed. The critical point is to select appropriate values for the type of ceramic to be lapped.

When the grinding process is to be used, consider several things. In addition to the type of grinding wheel, you must decide on the speed of the wheel and the feed. When grinding ceramics, setting the wheel speed higher will produce better cutting and longer wheel life. With any ceramic product, the machining has a large effect on the strength of the part. In fact, when considering the machining process, contemplate the effect that it will have on productivity.

Cutting with a diamond wheel has certain advantages in terms of surface quality, accuracy, and expense. The finished surface will be very smooth. A diamond wheel would be especially useful for grinding large-diameter aspheric surface reflectors with high precision.

Then, there are a number of more exotic methods of stock removal. Electrolytic grinding is a fast way to remove stock. Ion beam machining is used on piezoelectric elements, quartz oscillators, and glass lenses. Research is being performed on the use of electrical-discharge machining of certain types of ceramics.

Lasers have been used to machine grooves on ceramics. Lasers also do a good job drilling holes and holding diameters and center distances to close tolerances. Moreover, the cost is trivial compared to drilling with a diamond drill. Because this is a noncontact machining method, it is not subject to the quality changes caused by heating, tool wear, or pressure. Thus, stable machining is possible on thin, delicate parts. Depending on laser wavelength, it is possible to machine through a transparent material. Unlike electron-beam work, lasers do not require a vacuum in which to operate.

Ultrasonic machines do an excellent job drilling holes on glass, ferrites, and quartz (see chapter on ultrasonics). Because the ultrasonic tool does not rotate, it can be used for boring or carving specially shaped holes. Cut-off methods are numerous. Cutting can be performed by ordinary grinding wheels, diamond wheels, abrasive waterjet machines, and ultrasonics.

TOLERANCES

At Wesgo, small microminiature parts and cylinders that are 20" in diameter and 5 feet long can be pressed out of ceramic powder. There are companies which can already handle even larger parts. The tolerances are a function of size. Normal as-fired tolerances are ±1% or ±0.005" on all dimensions, whichever is greater. Depending on the size and shape of the part, standard grind and lap tolerances achievable are:

Grind tolerances are ±0.0002" on all dimensions.
                    Finish 16 microinches.
                    Flatness 0.001 in/in.
Parallelism   To 0.000020".
Flat lapping   0.000012". +3 microinch is attainable.

CASES

With gas turbines for automobiles, the inlet temperature will be about 1350°C to achieve the same ratio of fuel consumption as diesel engines. Currently, only silicon carbide and silicon nitride can withstand such temperatures and both are highly promising materials. Properties required for safety are a flexural strength of about 50 kg/mm2 at 1350°C and about 70 kg/mm2 at 1000°C for turbine rotors and stators. Creep deformation of not more than 0.5% after 2000 hours under 12 kg/mm2 at 1350°C is also required. In the near future, the urgent task is to establish a manufacturing process that does not harm these qualities of the ceramic material.

Practical applications of ceramics in automobiles are just taking their first steps. No one has enough experience to know what ceramics to use where. As experience accumulates in the future, applications will broaden steadily, establishing ceramics as an industrial material.

For heat engines, the higher the cycle temperature and cycle pressure, the higher the thermal efficiency rate. Thus, the development of materials that can withstand high temperatures is desirable. The turbine inlet temperature of almost 1400°C is achieved by air cooling. Metal materials still have a melting-point limitation—even with the progress in these high-temperature alloys. It is predicted that ceramic materials with far higher melting points than metals will be practical before the year 2000.

At present, there are applied research projects on gas turbines, such as Ford's 200 HP gas turbine for automobiles, Cumming's adiabatic engine, the AGT (advanced gas turbine) 100 and 101, as well as CATE (ceramic applications in turbine engines) projects being carried out by the U.S. Department of Energy and NASA, and applied research and development in Germany and Sweden.

A mold suitable for forming the complex shape of a spark-plug insulator is used for that purpose. Spray-dried powder is filled into the heavy-wall, rubber cavity, shaped like the final article, and a central mandrel provides the central core. When the mold is filled, the filling device is replaced by a die closure and fluid is pumped around the rubber mold to a pressure of 4000 to 5000 psi. After the pressure is relieved, the part is removed and production continues. 1000 to 1500 pieces per hour are possible.

Among automobile ceramics for reciprocal engines, the following developments are expected in the future: precombustion chambers, piston heads, cylinder liners, cylinder heads, and rotors for superchargers. Partially stabilized zirconia has not only a strength of 100 kg/mm2, but also far less heat conductivity than other ceramics.

RESEARCH AND DEVELOPMENT

For years, ceramic parts could not be held to close dimensions. There were problems with powder purity and with inconsistencies of the binders. In fact, except for flatness and parallelism, which have been held very close for years, most dimensions could not be machined accurately. However, in recent years, fabrication and machining of ceramics have improved considerably. This was mainly because qualities, not available in other materials, became so significant that research and development finally solved the ceramic production problems.

In 1981, Japan's Ministry of International Trade and Industry, commissioned private industry to improve the level of ceramic production technology. The actual targets for this research were producing small-sized powders and improving its purity, molding, sintering, and machining. Companies in the USA and Western Europe recognized the importance of this effort and joined the search for answers.

Currently, great interest is being shown by industry to vibratory compacting of both ceramic and metallic powders. Vibration agitates the powders so that they seek the closest packing possible.

Silicon nitride and silicon carbide are both undergoing tests to see if they can perform satisfactorily under load at 2450°F. There are a number of places in heat engines where the automotive and aeronautic industries would like to use these ceramics.

At the University of California, researchers are trying to develop a superior method of manufacturing ceramics. They hope to handle ceramics such as pliable plastics, instead of solid, powder material. The researchers are now producing silicon-nitride coatings in 20 minutes, instead of the usual several hours, by using polysilazane precursors. This is performed at temperatures of 450 to 800°C, instead of the conventional 1700°C.

Although several American companies are performing excellent research on ceramics, several are engaged in powder-injection molding, which are currently producing small ceramic parts (as large as a golf ball) and holding tolerances of ±0.001". These parts require no machining.

What is happening now rivals what is going on in medicine. The advanced ceramic market is really many broad fragmented markets. Growth is expected in many ways. The market will come in anything from high-speed machinery for making cigarettes to sandblast nozzles, to parts for dispensing blood, to centrifuges and power-plant equipment. Many parts are involved, which is how we see the growth of ceramics for the next few years.

CHAPTER 46
ADVANCED COMPOSITES

INTRODUCTION

In 1959, while preparing a boron sample for oxidation studies by chemical vapor deposition, Claude Talley noticed that the resulting wire was quite strong and stiff. Subsequent measurements of tensile strength and modulus were the basis for the first publication of significance on this material. The persistence of Mr. Talley at Texaco and the sponsorship of the U.S. Air Force Materials Laboratory, led to the development of a continuous boron fiber that General Schriver, in 1964 stated, was "the greatest breakthrough in materials in 3000 years."

In the early 1960s, the glass fiber was the only continuous fiber of consequence available. This fiber had low modulus and was usable only in organic matrix composites. Much research was conducted on single crystal whiskers, which individually offered extremely high strength and modulus. However, problems with handling the material, fabricating it into components, low yield, low reproducibility, and extremely high cost, forecast a dark future for this type of reinforcement. Hence, the timing of the development of boron fibers was so important. This was the real basis for U.S. Air Force General Schriver's statement, which was probably slightly inflated and overenthusiastic. But this discovery was to have far-reaching consequences.

The term *composite* usually signifies a combination of two or more elements, which form a bonded, quasi-homogeneous structure that produces synergistic, mechanical and physical property advantages over that of the base constituents. Composites are created by combining two or more reinforcing elements and a compatible resin matrix. The purpose is to obtain specific characteristics. The components do not dissolve into each other, like sugar in water, but they do act synergistically.

A composite can include many common metals combined with resin and fiber. Advanced composites (AC) designates certain composite materials with considerably superior properties to those originally called *advanced composites*. Currently, industry defines AC as containing a fiber-to-resin ratio greater than 50% fiber, with the fibers having a modulus of elasticity greater than 16,000,000 psi.

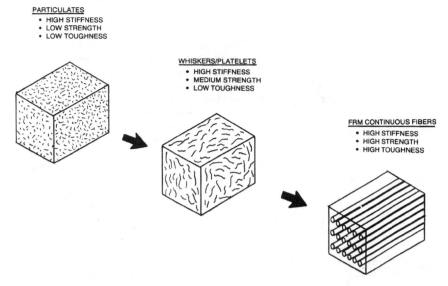

PARTICULATES
- HIGH STIFFNESS
- LOW STRENGTH
- LOW TOUGHNESS

WHISKERS/PLATELETS
- HIGH STIFFNESS
- MEDIUM STRENGTH
- LOW TOUGHNESS

FRM CONTINUOUS FIBERS
- HIGH STIFFNESS
- HIGH STRENGTH
- HIGH TOUGHNESS

FIGURE 46.1 Metal matrix materials. *(Textron Specialty Materials, Lowell, MA)*

As shown in Figure 46.1, there are three types of composites. A particulate-based material, formed by the addition of small granular fillers into a binder, increases stiffness but not strength. A whisker/flake filler does increase strength somewhat because of its higher aspect ratio, which has a greater ability to transfer load. A continuous fiber system, because of fiber continuity, provides the strength of the high-performance fiber, as well as increased stiffness.

The major significance of the comparison of three composite types is the clear distinction between continuous and discontinuous systems. Although the mechanical properties of continuous fiber composites are superior to those of the discontinuous systems, there are many stiffness-controlled applications that benefit from discontinuous reinforced material. These applications generally involve complex geometries, where it is difficult to position continuous fibers during the fabrication process.

It is only with the continuous types that both stiffness and strength are fully translated into the composite. In operation, it is very difficult to achieve the theoretical potential of the aspect ratio. Hence, when calculating the physicals of a contemplated composite, use a reasonable factor of safety. The knowledge of fabrication skill comes only with experience, and success depends quite a bit on the workmanship of the technician.

Advanced composites has come to mean a resin matrix material reinforced with high-strength, high-modulus fibers of carbon, aramid, or boron, and it is generally fabricated in layers. The four basic areas of composite technology are:

1. Organic resin-matrix composites
2. Metal-matrix composites
3. Carbon-carbon composites
4. Ceramic-matrix composites

Organic matrix composites are the most common and least-expensive types.

THE MATRIX

In a composite, the matrix serves two significant functions. It holds the fibers in place, and it deforms under an applied force and distributes the stress to the high-modulus, fibrous constituent. For maximum efficiency, the matrix material for a structural fiber composite must have a larger elongation at fracture than the fibers.

The matrix must change shape as required to transmit the force to the fibers and place the majority of the load on those fibers. The matrix also influences corrosion, and chemical, thermal, and electric resistance. The two main classes of polymer-resin matrices are: thermoset and thermoplastic. The principal thermosets are epoxy, phenolic, bismaleimide, and polyimide. There are many types of thermoplastic matrices. The matrix material of both classes must be carefully matched for compatibility with the fiber material and for application requirements.

Most structural-composite parts are produced with thermosetting resin-matrix materials. In metal-matrix composites, the most frequently used matrix is aluminum, although alloys of titanium, magnesium, and copper are being developed.

FIBER SCIENCE

The term *fiber science* applies to a material tailoring discipline that includes type of fiber, the percentage of fiber, and the oriented placement of the fiber during the production process. Fibers can run longitudinally (warp) or they can run transversely (waft). A new weaving technology allows fibers to run on a bias.

Since the advent of high-strength, high-modulus, low-density boron fiber, the role of fibers produced by chemical vapor deposition has become well established. Boron-aluminum was used for tubular truss members, which reinforced the Space Shuttle structure. However, there are drawbacks to this composite. There is a rapid reaction of boron fiber with molten aluminum and a slow degradation of the mechanical properties of diffusion-bonded boron-aluminum at temperatures greater than 900°F. These problems led to the development of the silicon-carbide fiber.

SILICON-CARBIDE (SIC) FIBER
PRODUCTION PROCESS

Continuous SiC fibers are produced in a tubular glass reactor by chemical vapor deposition (CVD), as shown in Figure 46.2. The process occurs in two steps on a carbon monofilament substrate, which is resistively heated. In step one, pyrolytic graphite (PG) about 0.000040" thick is deposited to smooth the substrate and improve electrical conductivity. In step two, the coated substrate is exposed to silane and hydrogen gases. The former decomposes to form beta silicon carbide continuously on the substrate.

The mechanical and physical properties of the SiC filament are:

1. Tensile strength = 500,000 psi
2. Tensile modulus = 60,000,000
3. Density = 0.11 lb./in.^{3}
4. Diameter = 0.0056"

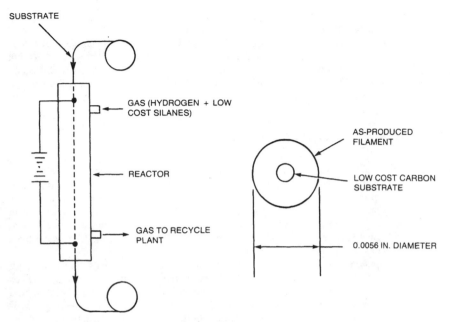

FIGURE 46.2 Fabrication of silicon carbide fiber. *(Textron Specialty Materials, Lowell, MA)*

PROCESSING CONSIDERATIONS

As in any vapor-deposition process, temperature control is of utmost significance in producing CVD SiC fiber. The Textron Specialty Materials Company process calls for a peak deposition temperature of about 2370°F. Above this temperature, rapid deposition and subsequent grain growth occurs, resulting in a weakening of tensile strength. Temperatures significantly below the optimum cause high internal stresses in the fiber, which produce a degradation of metal matrix-composite properties when machining transverse to the fiber.

Substrate quality is also an important consideration in SiC fiber characteristics. The carbon monofilament substrate, which is melt-spun from coal tar pitch, has a smooth surface with occasional anomalies. If severe enough, these anomalies can cause strength-limiting flaws in the fiber.

COMPOSITE PROCESSING

The ability to readily produce acceptable SiC fiber-reinforced metals is directly attributed to the ability of the SiC fiber to bond to the respective metals and to resist degradation of strength while being subjected to high-temperature processing. In the past, both boron and Borsic fibers have been evaluated in various aluminum alloys and, unless low-temperature, high-pressure diffusion-bonding procedures were adopted, severe loss of fiber strength has been observed. In titanium, unless fabrication times are highly curtailed, fiber~matrix interactions produced brittle compounds with drastically reduced strength.

The Textron Specialty Materials SiC is named *SCS*. In contrast to the previous paragraph, the SCS grades of fibers have surfaces that readily bond to the respective metals without destructive reactions occurring. The result is that the aluminum composites can be used in high-temperature processes, such as investment casting and low-pressure, hot molding. Also for titanium composites, the SCS-6 filament has the ability to withstand long exposure at diffusion-bonding temperatures without fiber degradation. As a result, complex shapes can be fabricated.

COMPOSITE PREFORMS AND FABRICS

Green tape is an old system that consists of a single layer of fibers, spaced side by side across a layer, held together by a resin binder, and supported by a metal foil. This layer constitutes a prepreg that can be sequentially "laid up" into a mold in required orientations to fabricate laminates. The laminate processing cycle is then controlled to remove the resin by vacuum as volatilization occurs. Normally, to make a prepreg, the fibers are wound onto a foil-covered rotating drum and oversprayed with resin. Then, the layer is cut from the drum to provide a flat sheet of prepreg.

Plasma-sprayed aluminum tape is a more advanced prepreg, similar to green tape, but the resin binder is replaced with a plasma-sprayed matrix of aluminum. This method has two advantages. This method is faster and less expensive than the green tape because hold time to ensure volatilization and removal of resin is not required. Also, there is no possibility of contamination of any resin. But they are both used in similar fashions.

Woven fabric is perhaps the most interesting method because it is a universal preform concept that is suitable for a number of fabrication processes. In this uniweave system, the SiC monofilaments are wound straight and parallel at 100 to 140 filaments per inch and held together by a cross weave of yarn or metallic ribbon. Two types of looms can be modified to produce uniweave fabric. The Rapier-type loom can produce continuous 60" wide fabric and a shuttle-type loom can produce a 6" width. Various cross-weave materials (such as aluminum, titanium, and ceramic yarns) have been used (Figure 46.3).

PROCESSING METHODS

Investment Casting

Investment casting is a well-known method of producing complex shapes. Most industries make good use of this method because it forms almost to finished size and saves much costly machining. The aircraft industry had been suspicious of the fatigue strength of the thinwalled aluminum castings until now. This casting technique, sometimes called "The Lost Wax" process, uses a wax replicate of the intended shape inside a porous ceramic mold, which is formed around the wax. After removing the wax by steam heat or another method, this provides a cavity for the casting. The SiC fibers are installed in the mold—either by opening the mold or burying them in the wax replica. The fibers strengthen the aluminum castings sufficiently to meet the approval of the aerospace industry, which is eager to use investment castings.

HOT MOLDING

Hot molding is a term coined by Textron Specialty Materials to describe a low-pressure, hot-pressing process that is designed to produce SiC aluminum parts at significantly lower

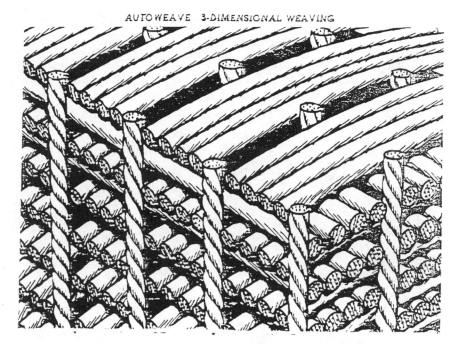

FIGURE 46.3 Above weaving is followed by a desensification in pitch. *(Textron Specialty Materials, Lowell, MA)*

cost than the typical diffusion-bonding process. The SCS-2 fibers can withstand molten aluminum for long periods. The molding temperature can now be raised high enough to ensure aluminum flow and consolidation at low pressure, thereby negating requirements for high-pressure, high-cost die-molding equipment

The best way of describing this process is to draw an analogy to the autoclave molding of graphite epoxy, where components are molded in an open-faced tool or die. The mold in the Textron process is a self-heated, ceramic tool that embodies the part profile. A plasma-sprayed aluminum preform is laid into the mold, heated to a near-molten condition, and pressure consolidated in an autoclave by a metallic vacuum bag.

DIFFUSION BONDING

Diffusion bonding of SiC/titanium is accomplished by hot-pressing preforms that are stacked together between titanium foils for consolidation. Two methods have been developed by aircraft manufacturers. One method is based on "Hot Isostatic Pressing," which uses a steel membrane to press components directly from the fiber/metal preform. The other method uses previously hot-pressed SCS/titanium laminates that are then diffusion bonded to a titanium substructure (Figure 46.4).

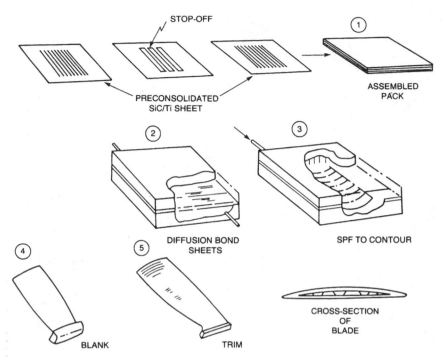

FIGURE 46.4 Superplastic forming/diffusion bonding of an SiC/titanium blade. *(Textron Specialty Materials, Lowell, MA)*

HYBRID COMPOSITES

Hybrid refers to the use of various combinations of boron, graphite, aramid, and glass filaments in a thermoset matrix. Hybrids are used to meet diverse design requirements in a cost-effective manner; better than either advanced or conventional composites.

Some of the advantages of hybrids over conventional composites are balanced strength and stiffness, optimum mechanical properties, thermal distortion stability, reduced weight, improved fatigue resistance, reduced notch sensitivity, improved impact resistance, and optimum cost (as related to performance).

The main forms of hybrid composites are:

- *Interply hybrids* These consist of plies from two or more different fibers stacked in alternate layers.

- *Intraply hybrids* These consist of two or more different fibers mixed in the same ply.

- *Interply-intraply hybrids* These consist of plies of interply and intraply hybrids, stacked in a specific sequence.

- *Selective placement* This uses a combination of fibers, as needed in any form.

- *Interply knitting* This is a form of vertical interply stitching with a polyester or aramid strand. It connects two to five plies and strengthens the composite against interlaminar shear, which occurs when the resin matrix fractures and the individual plies separate.

COMPOSITE CONSTRUCTION

Composites consist of laminates and sandwiches. Laminates are composite materials of two or more layers bonded together. Sandwiches are multiple-layer structural materials that contain a low-density core between thin faces of composite materials. Theoretically, there are as many different types of laminates as there are possible combinations of two or more materials. See Figure 46.5 for the process of making a "Z" stiffener.

Materials are divided into metals and nonmetals. Nonmetals are divided into organic and inorganic. Accordingly, there are six possible combinations in which laminates can be produced: metal-metal, metal-organic, metal-inorganic, organic-organic, organic-inorganic, and inorganic-inorganic. If the laminates contain more than two layers, there are many more possibilities.

Sandwiches consist of a thick, low-density core, such as a honeycomb or foamed material between thin faces of a high-strength and high-density material. In sandwich composites, a primary objective is a high strength-to-weight ratio. The core separates and stabilizes the faces against buckling under torsion or bending, and provides a rigid and efficient structure.

PROPERTIES OF COMPOSITES

Composites can be made stronger than steel, lighter than aluminum, and stiffer than titanium. This is possible through the careful selection and use of high-strength fibers, such as carbon/graphite, aramid, or boron, bound in a matrix of epoxy. In aircraft structures, a

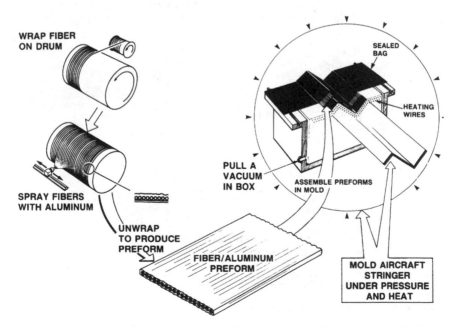

FIGURE 46.5 The hot molding of SiC/Al "zee" stiffeners. *(Textron Specialty Materials, Lowell, MA)*

graphite-epoxy composite presents about the same strength and stiffness as aluminum. However, the composite weighs 45% less than the aluminum. It has a superior fatigue resistance and a lower thermal conductivity. It is also noncorrosive and highly wear resistant.

However, corrosion could occur if graphite and aluminum are in direct contact in the presence of moisture. This reaction does not occur when graphite and titanium are in contact. The two properties most in demand when advanced composites are considered are tensile strength and Young's modulus. Aramid has a slightly higher tensile strength than carbon~graphite, but a much lower Young's modulus.

APPLICATIONS OVERVIEW

Advanced composites containing such exotic materials as carbon/graphite and aramid fibers in an organic-resin matrix are being used currently, mainly by the aircraft industry. Nevertheless, more and more is being used by other industries in applications ranging from automobiles to spacecraft and printed circuit boards, and from sports equipment to prosthetic devices. These uses are for unusual situations because the cost is still very high.

Continuous silicon-carbide titanium driveshafts are being used for airplane engine primary and auxiliary power take-off shafts (Figure 46.6). They are made by hot isostatic pressing. Textron Specialty Materials has provided a continuous production of uniform, high-quality boron filaments for use in versatile prepreg tapes, filament-wound structures, metal-matrix components, and reinforced structural elements. Manufacturing procedures are precisely controlled and constantly monitored to ensure a consistent, reproducible, standard filament in diameters of 4.0 and 5.6 mils.

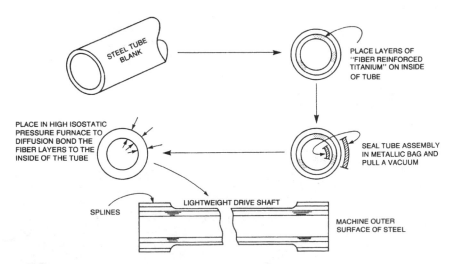

FIGURE 46.6 The process of making a drive shaft. *(Textron Specialty Materials, Lowell, MA)*

COMPONENT	MANUFACTURER	METAL EQUIVALENT	WEIGHT SAVINGS%	STATUS
F-15 Horizontal and Vertical Stabilizers Boron Epoxy 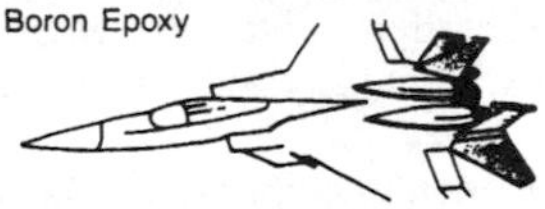	McDonnell-Douglas Mitsubishi Heavy Industries	Titanium	22% Estimated	Production
F-14 Horizontal Stabilizer Boron Epoxy 	Grumman Aerospace Corp.	Titanium	19%	Production
B-1B Reinforced Longeron Boron Epoxy 	Rockwell International	Steel-Titanium	44%	Production
Hawk Series, Sea Stallion Boron Epoxy 	Sikorsky	—	—	Production
Mirage 2000, Mirage 4000 Boron Epoxy 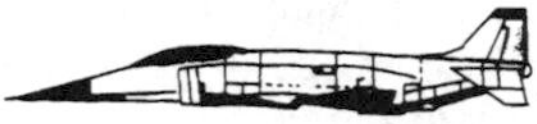	Dassault	—	—	Production
Space Shuttle Boron Aluminum Boron Expoxy 	Rockwell International	—	—	Production

FIGURE 46.7 Boron composite aerospace applications. *(Textron Specialty Materials, Lowell, MA)*

Textron Specialty Materials has developed a silicon-carbide fiber for high-performance composites in both metal ceramic- and polymer-matrix composites. Silicon carbide offers the advantages of:

1. Low cost
2. High strength
3. High heat resistance (up to 1200°C)
4. Low electrical conductivity
5. Corrosion resistance and chemical stability
6. Excellent wettability for metals
7. Compatibility with plastics

Aluminum composites, fabricated by casting and hot molding, enjoy full translation of fiber properties. Components are being made for aircraft, helicopters, missiles, bridge structures, gun barrels, and projectile munitions. Titanium composites are being made for gas turbine engine fan blades, drive shafts, aircraft landing gears, and missile structures (Figure 46.7).

DATABASE LIMITATIONS

Thirty years of development in the advanced-composite field have left us with a scarcity of data. We are very limited in knowledge of how stress is carried and transferred in complex loads. The interface bond between the fiber and matrix is not fully understood. Impact resistance is another area in which we lack knowledge.

Metal will crack or dent when struck. This doesn't happen to composites. However, when a composite is struck and there is no external, visible damage, there might be a defect on the inside. Delamination might have started internally and it might propagate until destruction occurs. Until recently, only thermosetting resins were used as resin matrices because of their high-temperature properties. However, currently several thermoplastics have become available for high-performance roles and are now part of the ongoing study.

FABRICATION

Industry has discovered that the experience acquired in years of producing fiberglass components would be somewhat useful in fabricating composites. As a result of many experiments, organic matrix composites are now made primarily by molding in autoclaves, and metal-matrix composites are formed by diffusion bonding (Hot Isostatic Pressing).

Four popular methods are used to produce continuous fiber composites with closely controlled properties: lamination, filament winding, pultrusion, and injection molding. The construction technique selected depends on the shape, size, type of part, and the quantity to be manufactured.

LAMINATION

The laminate process starts with a prepreg material, which is a partially cured composite with the fibers aligned parallel to each other. A pattern of the product's shape is cut out and the prepreg material is stacked in layers in the desired geometry. The assembled layers are then cured under pressure and heat in an autoclave. Graphite/epoxy composites are cured at a temperature of 350°F and a pressure of 100 psi. The new high-temperature composites, such as bismaleimides, are cured at 600°F.

FILAMENT WINDING

In the filament-winding process, fibers or tapes are drawn through a resin bath and wound onto a rotating mandrel. This is a slow process, but the direction can be controlled and the diameter can be varied along its length. If working with tape, it is an endless strip whose width can vary from 1" to 1 yard. With both fiber- and tape-winding processes, the finished part is next cured in an autoclave and removed from the mandrel at a later time.

In aerospace structures that are strength critical, carbon fibers are wound with epoxy-based resin systems. The other resin systems are limited to special applications. Filament winding is used to make cylindrical objects, such as missile canisters, pressure bottles, and tanks.

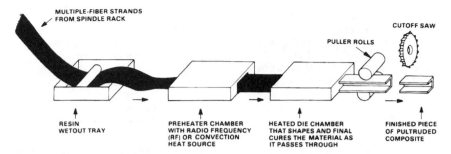

FIGURE 46.8 The pultrusion process. *(Textron Specialty Materials, Lowell, MA*

Pultrusions

In composite technology, pultrusion is the equivalent of a metal extrusion. In the process, a continuous-fiber bundle is pulled through a resin-matrix bath, then through a heated die. The process is generally limited to constant cross sections, such as tubes, channels, I beams and flat bars (Figure 46.8).

INJECTION MOLDING OR RESIN-TRANSFER MOLDING

This process fills the void between compression molding and hand manufacturing lay up. In resin-transfer molding (RTM), two-piece matched-metal molds are used. RTM uses low injection pressures, which, in turn, permits the use of low-cost tooling. The reinforcing material, either chopped or continuous strand, is draped in the cavities. The two mold halves are clamped together, then the resin is pumped into the closed mold.

COMPUTER-ASSISTED DESIGN (CAD)

Research and development programs in AC, and in fact, any mold design, can be expedited by the use of CAD. Design changes can be tried out quickly this way with minimal risk. For instance, if you show the assembled mold on the screen, it will indicate interferences and restrictions, thus precluding remachining and modifications. CAD has even been used to lay out equipment and complete floor plans for producing advanced composites. In fact, companies in the business of design and manufacture of dies and molds, should use the computer in their design effort if they want to be competitive in pricing.

MACHINING CUTTING AND JOINING

AC materials are generally unsuited for normal machining and fabrication techniques. Special methods must be used. Before doing any machining on these AC components, you must ensure that there will be no delamination, fraying, or cracking of cured composite edges. With modifications, standard machine tools can often be used. Spindle speeds and feeds should be selected, depending on the type of laminate material, its thickness, and the machine used. Whatever cutting tools are used, they must be sharp.

Uncured composite materials can be cut with shears, scissors, or carbide disc cutters. For cured composites, you might cut with reciprocating knife cutters, lasers, ultrasonics, or abrasive waterjets. Lasers work fine on cured composites, but can burn uncured material. Abrasive waterjet cutting can present moisture problems. Knife cutters can clog. The safest procedure is to question the suppliers.

If the structural composite is made with a thermosetting resin, it cannot be joined by welding as though the resin were thermoplastic. Joining would have to be by adhesive or mechanical means. Sometimes both are used.

Defects, such as cracks, in composites are sometimes easily detected. Others can be difficult to find. Voids, missing layers, delaminations, inclusions, and improper layup present problems. Most metals fail by fatigue, but composites break under load. This could lead to catastrophic failure. Because composite structures vary in point-to-point comparisons, selecting samples for destructive testing would have no value. Also, because fiber reinforcements and resins can appear the same in an X-ray, that medium is ineffective. Therefore, we are left with only a select few methods for effective quality control.

Conventional radiography will provide good resolution when the attenuation characteristics of fiber and resin are quite different, as in boron-epoxy composites. However, in aramid-epoxy or graphite-epoxy composites, where differences are small, defect determination is difficult.

Ultrasonics is useful in detecting skin and bonding problems. Liquid-coupled ultasonics is currently the most widely used inspection method. Thermography, optical holography, and eddy-current testing are also used in special cases. When processing fibers or yarns, avoid inhalation of airborne particles. Dust from this environment can cause respiratory difficulties. Dispose of dust and remnants at a suitable landfill. This material is difficult to burn.

Various grades of fibers are produced, which are based on the standard SiC deposition process described earlier, where a crystalline structure is grown onto a carbon substrate.

Substrate quality is an important consideration in the manufacture of SiC fiber. The carbon monofilament substrate, which is melt-spun from coal tar pitch, has a very smooth surface with occasional anomalies. These anomalies, if severe, can result in localized irregular deposition of PG and SiC, which could create stresses. The carbon monofilament spinning process is closely controlled to minimize production of low-strength fibers.

The PG flaw results from insufficient control of the CVD process. This flaw is caused by irregularities in the PG deposition. The two reasons for PG flaws are: An anomaly in the carbon substrate surface and mechanical damage to the PG layer prior to the SiC deposition. The surface of Textron's SiC fibers is carbon rich. This protects the fiber from surface damage. Surface flaws can be identified by an optical examination. All flaws are minimized by careful handling and close adherence to fabrication instructions. Typical mechanical properties of the Avco CVD SiC fiber are 575 ksi and an elastic modulus of 60 msi.

FIBER VARIATIONS

It is important to tailor the surface region of the SiC fibers to the matrix. There is a difference in the surface composition of three fibers. SCS-2 has a carbon-rich coating that increases in silicon content as the outer surface is approached. This fiber is commonly used to reinforce aluminum. SCS-6 is primarily used to reinforce titanium. It has an even-thicker carbon-rich coating, in which the silicon content also increases as the outer surface is approached. SCS-8 was developed to provide better mechanical properties in aluminum composites than SCS-2.

COST FACTORS

SiC is potentially less costly than boron because:

1. The carbon substrate used for SiC is less expensive than the tungsten used for boron.
2. Raw materials for SiC (chlorosilanes) are less expensive than boron bichloride, the raw material for boron.
3. Deposition rates for SiC are higher than those for boron.

In 1989, continuous silicon-carbide fiber cost $2500/lb. 30,000 feet of 5.6-mil fiber weighs 1 lb. It was estimated that the fiber could cost as little as $100/lb if full-scale production reaches 40,000 lb./year. Five contractors on projects funded by NASA have been furnished samples of silicon-carbide reinforced titanium. It is hoped that airplane manufacturers will accelerate the use of composites and bring the price down to the point where other industries will use them.

As a precaution, any fiber composite under consideration should be thoroughly investigated. They all have their place. Borsic and boron fibers have been evaluated for use in aluminum alloys. Unless complex, high-pressure, low-temperature, diffusion-bonding procedures were followed, degradation of fiber strength would result. Similarly, with titanium, unless fabrication times are shortened, fiber/matrix interactions produce brittle, intermetallic compounds that severely reduce composite strength.

The SCS grade of fiber, in contrast, has surfaces that bond readily to various metals without destructive reactions. This results in the ability to consolidate aluminum composites with investment castings and low-pressure molding. Similarly, for titanium composites, the SCS-6 filament can withstand long exposure at diffusion-bonding temperatures without fiber degradation. Accordingly, complex shapes with these composite reinforcements can be fabricated by the super plastic forming/diffusion bonding and hot isostatic-pressing process.

MAKING RELIABLE COMPOSITE JOINTS

Joining composite materials presents special problems. The usual method of joining uses mechanical fasteners, adhesives, or both. It depends on the composite and the application. Where failure could be catastrophic, as in aircraft, composites are generally joined by a combination of adhesives and mechanical fasteners. In other applications, adhesives are most common.

Many engineers, designers, and users distrust adhesives, which should be the first choice for joining composites. Actually, where large forces are encountered, adhesives spread out the stress, whereas fasteners tend to concentrate the stress.

Many factors should be considered when contemplating fasteners for composites.

1. When a spread of working temperature is expected, the difference in the coefficient of expansion could be significant.
2. The possibility of delamination, either by load stresses or the physical act of drilling.

3. Moisture invasion around or under the fasteners that might cause galvanic corrosion.

4. Sealing where necessary.

Potential changes in clamping forces should be determined at the design stage. Metal fasteners expand and contract with temperature changes. Consider this effect on high-strength joints. In many cases, this effect demands the use of adhesives.

You should perform as little machining as possible on composites. Drilling or milling, for instance, could cause delamination, fiber breakout, or resin erosion. Each composite has its own machining problems. Carbon-fiber materials require carbide drills and cutting tools; they are prone to delamination and fiber breakouts; and create much dust. Aramid doesn't suffer from these problems, except for delaminations. Instead, aramids can melt when drilled and the edges could fray.

Fasteners for composites should have large heads so that the loads can be distributed over as large a surface as possible. Clearance holes for fasteners should closely fit the hardware to reduce the tendency of fretting. Interference fits, however, can cause delamination. Special sleeve fasteners are available to provide interference fits while limiting the chances for damage.

Cutting fibers exposes ends to moisture absorption. This weakens the material and adds undesirable weight. Sealants could be used to prevent water absorption, provided that there is no necessity for electrical continuity. Because most composites aren't conductive anyway, this should not present a problem. In fact, nonconductivity can present a problem in cases where electrical conductivity is desired.

Aluminum fasteners should never be used to fasten carbon-fiber composites. Galvanic corrosion could be caused by a chemical reaction between the fibers and the fasteners. Coating the fasteners will prevent the reaction. Or, if preferred, aluminum fasteners can be replaced with titanium or stainless-steel fasteners.

Three types of adhesives are commonly used to bond composites: epoxies, acrylics, and urethanes. Generally, epoxies are used with epoxy-based composites because they have similar flow and expansion characteristics. When large parts are being adhesive bonded, room-temperature curing is an asset. Epoxies shouldn't be used to join flexible composites unless a flexibilizing agent is added to the epoxy.

Polyurethane adhesives make flexible joints. However, they are sensitive to moisture and require complex dispensing equipment. Acrylics are rigid and cure quickly at room temperature. However, they smell badly, have poor impact resistance at low temperatures, and are flammable.

Recommended composite surface preparation is to wipe with a solvent and abrade gently. When bonding composites to metal, the metal should be prepared the usual way. Clean it by abrading, sanding, or blasting, then apply the adhesive soon after (before the surface oxidizes or becomes contaminated).

FUTURE PLANS

Textron Specialty Materials Company is providing lightweight, strong, stiff, heat-resistant materials required for the toughest applications, from critical areas in high-performance aircraft to life-threatening environments on offshore oil rigs to the most advanced sports equipment ever made. This constitutes the leading edge. It is only possible because of continuous laboratory experimentation in boron filaments and composites, carbon fibers, carbon-carbon composites, continuous silicon-carbide metal and ceramic composites, and fire-protection materials.

In 1996, TSM operated the free world's only major CVD filament production plant. It is the sole commercial supplier of boron fibers, which are used in aerospace and sporting goods. The company is seeking new potential boron applications in nuclear shielding, missiles, and cutting tools.

The primary application for carbon fiber is as reinforcement of carbon-carbon aircraft brakes. In terms of purity, the fibers are 99.5% carbon, which is significant for special electronic uses. Currently, applications are being sought in ablation, high-temperature bushings, and structures.

The aerospace industry has continually sought to develop faster, more efficient aircraft. Engines for these aircraft demand materials that are tough, strong, lightweight, and able to withstand temperatures in excess of 2000°F. By incorporating SCS filaments into refractory, high-temperature matrices, the laboratory is developing reinforced ceramics for use in these high-temperature, oxidizing environments. In addition, the lab is working on a composite that provides many of the properties required for future aircraft engines. By combining SCS monofilaments with a ceramic matrix, they hope to develop a material that will meet the needs of that industry for some time to come.

The price of advanced composites has dropped enough for the automotive industry to become interested, and racing bicycles are now using them. Even tennis racquets and other sporting equipment now use that material. As the volume produced increases, the cost will decrease and advanced composites will become a much more common material.

INDEX

Illustrations are indicated in **boldface**.

ABOUT THE AUTHOR

James Brown has over 50 years' of experience in shop practices and manufacturing engineering. He holds a master's degree in manufacturing and has taught numerous college courses in engineering.